Design Research Foundations

Managing Editor

The goal of the series is to provide a platform for publishing state of the art research on foundational issues in design and its applications in industry and society. Suitable topics range from methodological issues in design research to philosophical reflections on the specificities of design rather than actual design work or empirical cases only. The definition of design behind the series is inclusive. In terms of disciplines, it ranges from engineering to architecture. In terms of design work, it ranges from conceptual issues in design through design experiments and prototypes to evaluative studies of design and its foundations.

Proposals should include:

- A proposal form, as can be found on this page
- A short synopsis of the work or the introduction chapter
- The proposed Table of Contents
- The CV of the lead author(s)
- If available: one sample chapter

We aim to make a first decision within 1 month of submission. In case of a positive first decision the work will be provisionally contracted: the final decision about publication will depend upon the result of the anonymous peer review of the complete manuscript. The series editors aim to have the complete work peer-reviewed within 3 months of submission.
The series discourages the submission of manuscripts that contain reprints of previous published material and/or manuscripts that are below 150 pages / 75,000 words.
For inquiries and submission of proposals authors can contact the series editors, Pieter Vermaas via: p.e.vermaas@tudelft.nl or Ilpo Koskinen via: ilpo.koskinen@polyu.edu.hk

More information about this series at https://link.springer.com/bookseries/13775

Gerhard Bruyns • Stavros Kousoulas
Editors

Design Commons

Practices, Processes and Crossovers

Springer

Editors
Gerhard Bruyns
The School of Design
The Hong Kong Polytechnic University
Hung Hom, Hong Kong SAR

Stavros Kousoulas
Faculty of Architecture
TU Delft
Delft, The Netherlands

ISSN 2366-4622 ISSN 2366-4630 (electronic)
Design Research Foundations
ISBN 978-3-030-95056-9 ISBN 978-3-030-95057-6 (eBook)
https://doi.org/10.1007/978-3-030-95057-6

This Springer imprint is published by the registered company Springer Nature Switzerland AG
The registered company address is: Gewerbestrasse 11, 6330 Cham, Switzerland

Acknowledgments

This volume would not have been possible without the generous support of the School of Design at Hong Kong Polytechnic University and the Department of Architecture, Faculty of Architecture, TU Delft.

We are indebted to the Design Research Foundations series' editors Ilpo Koskinen, (former) Pieter E. Vermaas, and Clementine Thurgood for their receptiveness and enthusiasm for our book project and their valuable comments and assistance. We are also grateful to Heleen Schröder for her copyediting input and overall preparation of this volume for publication.

Finally, we are thankful to all the authors in this book for their generous contributions and excellent cooperation. It would, quite literally, be impossible without them.

Contents

Author Biographies

Frank Bauer, as Elsa-Neumann PhD Fellow embedded at the Cluster of Excellence "Matters of Activity," inquires into ontological logics of digital modeling and fabrication workflows, with special interest into their operative, instrumental, and symbolic basis. He graduated in architecture from UdK Berlin (MA 2017), with stays at UIC and IIT Chicago, as well as in social science, art, and cultural history studies from ALU Freiburg, FU Berlin, and JGU Mainz (MA 2012), respectively. After positions with the Fondation Vasarely in Aix-en-Provence and KWY Lisbon, he co-founded Büro Vogel Bauer, a planning agency for contemporary fine arts production. In the Department of Digital and Experimental Design at UdK Berlin, he is currently teaching design studios within the BArch and interdisciplinary MA program in design and computation (TU/UdK Berlin).

Gerhard Bruyns is an architect and urbanist. He is an associate professor in the School of Design at The Hong Kong Polytechnic University. His research deals with aspects of spatial morphology (morpho.org) and its impact on both the formal expression of the city and societal conditions that are compressed into an urban landscape of Asia. He has published on design strategies, spatial commoning, geopolitical issues linked to spatial practices, and urban morphology.

Contingent is a multi-disciplinary research and design collective based in Vancouver and Tokyo. Current collaborators include Roy Cloutier, Nicole Sylvia, and Lőrinc Vass. Roy Cloutier and Nicole Sylvia are adjunct professors in the School of Architecture and Landscape Architecture at the University of British Columbia, and architectural designers at Patkau Architects in Vancouver. Lőrinc Vass is an architectural designer and a doctoral candidate in the Department of Architecture, Tokyo Institute of Technology.

Daniel Elkin is a researcher, designer, and maker with a decade of experience. He serves as Assistant Professor of Environment and Interior Design in The Hong Kong Polytechnic University School of Design. Educated at Cranbrook Academy of Art (MArch, 2015) and the University of Cincinnati, Elkin's research focuses on spatial agency, collaborative governance, and housing, particularly in non-normative development scenarios. His recent work in stilt house communities is published in international journals and conferences. His spatial activism work with student collaborators was published in *Cubic Journal*'s Design Social edition. His analysis of emergent spatializations was published in *Architectural Research Quarterly*. He served as first editor of the upcoming *Cubic Journal* issue on design making.

Nicholas Frayne is an emerging scholar working at the frontiers of architectural research and practice. His global background and longstanding interest in international development ground his work in an urgent need to address sustainable thinking in its cultural, economic, and political dimensions. Having won awards for his design and research work, he brings experience at architecture firms in New York, Boston, and Toronto to his understanding of architecture as a force for societal transformation, critique, and ethical action. With an epistemological philosophy rooted in experiential learning, his research is focused on sociology, historiography, and art theory in order to approach architectural praxis from new angles.

Nikolaus Gartner is an architect who graduated from TU Wien (Technical University of Vienna). In October 2019, he received the Fred Sinowatz Science Award for his master's degree thesis "Cutting Reed: Strategy for the Architectural Handling of Lake Neusiedl Reed Belt." Since 2019, he has been a member of ICOMOS Austria, analyzing the impact of architecture on the cultural landscape and world heritage site Fertő/Neusiedlersee. He was a tutor in the Department of Design and Design Theory, Institute of Architecture, TU Wien, until 2016, and is currently practicing as an architect in Vienna. His fields of interest include vernacular architecture, DIY culture applied to architecture, and local building techniques and knowledge. Throughout his work, he has been developing architectural strategies dealing with high-quality low-budget designs and sharable languages to cooperatively approach a building process. He is currently curator of the exhibition "Schilf schneiden," promoted by the Austrian association ArchitekturRaum Burgenland, to be opened in Spring 2021.

Dora Karadima is a PhD candidate in the Architectural Department at the National Technical University of Athens. Her research interests sprawl around design theory, Lacanian psychoanalysis, and the possible theorization between them. Her interest lies in understanding the inner core of design processes and practice with a particular interest in the position of the designer within them. Her work is based mostly in dismantling design theory and the mode it operates, using Lacanian theory in synergy to this quest. At the core of this work is the notion of desire and the Anthropocene, which is now, more than ever, in need of re-establishing a discourse on values with the necessary reflection within.

Stavros Kousoulas is Assistant Professor of Architecture Theory in the Faculty of Architecture of TU Delft. He has studied architecture at the National Technical University of Athens and at TU Delft. He received his doctoral title cum laude from

IUAV Venice participating in the Villard d'Honnecourt International Research Doctorate. He has published and lectured in Europe and abroad. He is a member of the editorial board of *Footprint Delft Architecture Theory Journal* since 2014.

Chi-Yuen Leung is a teaching fellow in the Department of Applied Social Sciences at Hong Kong Polytechnic University. He teaches social work subjects related to community development and macro-social work practice. He is also interested in studying social work education, practicing of community work, hawker markets, and urban poverty and social marginalization. He is currently engaged in sustainable development research in rural areas and older urban areas, including Tai O village and Sham Shui Po, Hong Kong.

Piero Medici is an architect, currently researcher and lecturer, in the Faculty of Architecture and the Built Environment, TU Delft, and in the Master of Architecture program at Fontys University of Applied Science, Tilburg. He is a founding partner of the architectural and urban practice CoPE, and winner of the international competition Europan 14. Piero holds a degree in environmental sciences (BSc in Venice) and architecture (BSc in Venice, MSc in TU Delft), and a PhD (international PhD in IUAV, TU Delft, ENSA Paris Belleville). His doctoral thesis focused on European sustainable housing and neighborhoods during the 1970s. He has 15 years of experience working as a researcher, lecturer, architect, and environmental scientist at various academic institutions and practices, including UCL Bartlett, London; KABK Royal Academy, The Hague; Ca' Foscari, Venice; Grimshaw Architects, London; and Superuse Studios, Rotterdam. Piero is author of several publications, and his research focuses on architectural approaches concerning sustainability, circular economy, degrowth, and the commons.

Gert van der Merwe graduated from the University of Pretoria (MProf Arch) in 2014. Following this, he focused on design-build, catalytic social programs, and low-cost housing. He has taught at the University of Pretoria in undergraduate and postgraduate courses (2016–2019). His research is focused on anti-gentrification and resistance tactics, and the rights of creatives and alternative social structures. He believes in embedded practice and views his praxis as action research and activism.

Katharina Moebus is a feminist designer, organizer, and researcher based in Berlin who works at the intersection of socio-politically engaged design, radical pedagogy, and do-it-together (DIT) making. She is co-initiator and co-organizer of the neighborhood laboratory Common(s)Lab, co-founder and chairwoman of the transdisciplinary research collective Agents of Alternatives (AoA), and active member of the cultural project space >top. After her BA in product and communication design at the Free University of Bolzano (2007), she completed an MA in applied art and design at Aalto University ARTS Helsinki (2011) and currently pursues a practice-based PhD at Sheffield University School of Architecture, exploring transformative economies and (spaces of) commoning practices in the urban context.

Dorina Pllumbi is an architect and a PhD candidate in the Department of Architecture, TU Delft, Netherland. Her research interest resides in studying practices of commoning as material and spatial engagement in realities of political transition. Her particular focus is to understand the role that commoning practices have played in Albania during the transitional period from a totalitarian state socialist regime to a recently consolidating neoliberal one. Currently, Dorina is a visiting scholar at Parsons School of Design in New York City.

Liana Psarologaki is an architect, installation artist, and educator with interest in meta-philosophy of space, post-humanities, and new pedagogies. She trained at the National Technical University of Athens and holds one of the first interdisciplinary PhDs in the creative arts completed at the Centre for Spatial Analysis and Intervention, UCA Canterbury. She is Associate Professor of Architecture at the University of Suffolk and the research lead for the built-environment strand at the Suffolk Sustainability Institute (SSI). In 2018 she received a senior fellowship from the Higher Education Academy in recognition of her leadership in architectural education, and in 2020, she was elected Chair of Education for the Royal Institute of British Architects (RIBA) East Region.

Maria Reitano is an architect and PhD candidate in evaluation and urban planning in the Department of Architecture (DiARC), University of Naples Federico II. Her fields of interest include urban regeneration through cooperative decision-making processes, shared and plural values of urban ecosystems, and urban commoning. She studies peri-urban territories and their analysis according to a complex system-thinking perspective. In her research, she is developing methodological approaches addressed towards the definition of multi-stakeholder spatial decision support systems (MS-SDSS) and investigating new opportunities deriving from the application of ICT and crowd-sourcing as digital social innovation tools to be implemented for the informatization of territories and the activation of sharing knowledge and co-learning processes. She is currently collaborating within the European research project HERA Joint Research Program, "Public Spaces: Culture and Integration in Europe" (PUSH).

Lasse Sehested Skafte, after studies in architecture at the Royal Danish Academy of Fine Arts in Copenhagen, graduated from the Department of Digital and Experimental Design at Berlin University of the Arts in 2017. He held positions with Raumlabor in Berlin, Herzog & de Meuron in Basel, and a docent position with the Eames Foundation in Los Angeles. There, he also worked with the Center for Land Use Interpretation on perceptual studies of the American Anthropocene landscape. For his work in theory and design, he received several scholarships, among them from the Margot and Paul Baumgarten Foundation and the Danish Arts Foundation. His collaborative practice with the Popticum collective was recently exhibited at the Maia Biennale for Contemporary Art. As registered architect MAA, Lasse is an affiliated advisor for Realdanias *Underværker* program and currently works and practices in the field of art production in Berlin.

Xiao Lu Wang is an associate in the Centre for Social Innovation at the Cambridge Judge Business School and was a former research assistant professor in the Department of Applied Social Sciences at Hong Kong Polytechnic University. Dr. Wang is an interdisciplinary researcher: she had PhD and postdoc training in behavioral economics, organizational studies, and political economy. Her research focuses on the organization of social innovation, particularly on governance and cross-sector partnership. She has examined the organization of community action and social innovation for disaster relief, social services, commoning of community space, and social enterprise ecosystem. Dr. Wang is passionate about extending her research into innovations that achieve sustainability goals. Besides research, she also actively works with nonprofit organizations in Hong Kong.

Markus Wernli's design praxis explores the intricate relationality of human and nature through the development of more regenerative, ecologically entangled ways of living and designing. His ongoing research draws connections between food systems and social, cultural, and local ecosystems to forge better relationships between what we breathe, eat, expel, wear, and grow. Much of his research might be considered participatory citizen science or citizen-design interventions that can be gathered under the umbrella of participatory research through design. He specializes in contextually applied and critical research-through-design, bringing focus to the social and ecological impact of body-technology pairings and human-biosphere interactions. Markus is a research assistant professor with the School of Design at Hong Kong Polytechnic University and has held appointments in the College of Asia and the Pacific at the Australian National University in Canberra, Zokei University of Art and Design in Kyoto, and the Multimedia Studies Program at San Francisco State University.

Chun Zheng is a landscape architect, urban designer, and currently a PhD researcher in the School of Design at Carnegie Mellon University. Her doctoral research concerning practices of commoning and urban agriculture in the USA and China is supported by the China Scholarship Council. Chun is also a Permaculture Institute of North America (PINA) certified permaculture designer. Most recently, she co-curated the digital exhibition *A dialogue must take place, precisely because we don't speak the same language* in the context of the Tbilisi Architecture Biennial and the research fellowship program MAPP (Methods, Affects & Practical Pedagogies) in the School of Commons at Zurich University of the Arts.

Stamatis Zografos is an architect and academic. He is a senior teaching fellow in architectural history and theory at UCL Bartlett School of Architecture and a visiting lecturer in critical and historical studies at the Royal College of Art. He is also the founder of Incandescent Square, an interdisciplinary platform for research and design with interests spanning from architecture and urbanism to critical heritage and curating. He is the author of *Architecture and Fire: A Psychoanalytic Approach to Conservation*, which was published in 2019 by UCL Press.

Chapter 1
An Introduction to Design Commons

Gerhard Bruyns and Stavros Kousoulas

Abstract The reasons for a dedicated edition on "design and commoning" are two-fold. First, the recent surge of renewed interest in the social conditions of design remains atheoretical. A deeper theoretical and philosophical foundation will help problematize the link between commoning and design, and in doing so define the operative theories, concepts and frameworks that influence design thinking across a series of design contexts and conditions. And secondly, design has become more ubiquitous, expanding both its domain of influence and conditions of praxis. With this expansion, design touches a variety of contested areas. Designers are continuously challenged by conflicts and edge conditions, having to mitigate between both scales of conflict and the vested interests of individuals. In the global climate of population increase and the prevalent reduction of financial resources the question and theorization of shared capacities will remain part and parcel of future of design thinking. The four thematic clusters contained here exploit the theoretical and philosophical themes related to the large commoning "problematique," providing designers better grounding in the networked context of the twenty-first century. The explicit theorization of design and the commons will explore the implicit relations through each of the collected contributions to show how this philosophical construct can be explicated in the context of network collectives and transdisciplinary approaches that currently inform design practices.

Keywords Design commons · Social · Culture · Ecology · Transdisciplinary

---

G. Bruyns
School of Design, The Hong Kong Polytechnic University, Hung Hom, Hong Kong SAR
e-mail: gerhard.bruyns@polyu.edu.hk

S. Kousoulas (✉)
Faculty of Architecture, TU Delft, Delft, The Netherlands
e-mail: S.Kousoulas@tudelft.nl

G. Bruyns, S. Kousoulas (eds.), *Design Commons*, Design Research
Foundations, https://doi.org/10.1007/978-3-030-95057-6_1

1 (De)sign Expressions

Examining the etymology of the word "design," one comes to a startling conclusion: design, simultaneously, refers to a multitude of diverse conditions. As philosopher Vilém Flusser explains, design as a noun can refer to a "purpose," a "plan," a "goal" or a "form"; at the same time, as a verb, to design means to "concoct," to "draft," to "sketch" or to "shape" (Flusser 1995, 50). In any case, and that is significant to us, design is derived from the Latin word *signum*, which literally means a sign. Therefore, design in its original *disegnare* can be directly understood as "expressing a sign." We will claim that it is of great importance to define design in its original relation with expression, rather than limiting it to specific practices (such as drawing, tracing, outlining, or modelling). Design, first and foremost, is the practice of expressing signs.

Necessarily, this leads us to a broader discussion: how can we understand signs and their expression? Let us examine them both in brief. One of the most common mistakes when it comes to signs, is to approach them strictly semantically or syntactically; in other words, to confine them only within the disciplinary boundaries of linguistics. On the contrary, we will posit that when it comes to design practices, signs should be placed in a different, third category of information. In this sense, signs belong to a pragmatic level: how can a sign affect the behavior of both a transmitter and a receiver? Consequently, we can understand signs as "meaning." Nonetheless, confusion arises, precisely because once again, the common tendency is to give to language alone the privilege of producing meaning. As philosopher Manuel DeLanda claims, our confusion regarding the word "meaning" comes from the fact that "meaning" has two meanings: signification and significance, one referring to semantic context, the other to importance and relevance (DeLanda 2006, 22). It is the second meaning of "meaning" that we have in mind here: How signs are communicated throughout living systems? How can one find meaning in the actions of another?

As such, signs (understood as meaning) can be conceptualized as the very feeling of crossing a threshold. Among an infinite number of actions and perceptions, some do indeed cross a limit that transform them to something that has a certain significance for us. In this manner, we can provide an initial reformulation of the term design: to express meaningful actions and perceptions. What about the term expression then? As sociologist Antoine Hennion suggests, it is again interesting to examine its etymology (Hennion 2016, 84). Initially, it comes from the Latin *expressare*: *ex*, "out," and *pressare,* "to press." Expression then literally means to press out, to squeeze, to extort: expression is a coming out (Hennion 2016, 84) A coming out of where however? Moreover, if to express is to press out, then towards where is this pressure oriented, where does it lead? What is being pressured, to afford something to come out of it? It would be misleading to conceive expression as a pressure in extensive terms. In addition, it is equally misleading to conceive it in spatial terms, where pressure stands merely for the force applied to a surface. Quite the opposite, pressure is not force on surface: pressure is force acting on force. In other words, expression, as the pressure to come out, belongs to the intensive: expression is always an act on the pressurized limit.

Consequently, we can now come up with a complete reformulation: design is the effort to discover, manipulate and cross intensive limits that can eventually lead to the production of meaningful actions and perceptions. Therein lies the focus of this volume. If discovering, manipulating and crossing limits is what design is about, then – by definition – design is at once both technological and collective. To be more precise, it is collective because it is technological and vice versa. To make this clear, we need to provide a broader and more inclusive definition of technology. Without exceeding the scope of this introduction, we will briefly indicate one direction, namely the thought of philosopher Gilbert Simondon. For Simondon, the dichotomy between culture and technology is based on a fundamental misunderstanding of technology which, at least in cultural terms, positions it as a foreign reality (Simondon 2017, 134). For that reason, Simondon proposes the term "technical culture," suggesting a way of thinking which surpasses that conflict. The point of departure for a way of thinking that no longer considers technology and culture apart, is a shift of focus from the usage and utility of technical objects. Aiming to provoke an awareness of the modes of existence of technical objects, one should focus on the genesis of the objects themselves (Simondon 2017, xi).

Simondon does so by developing the concept of technicity. For Simondon, technicity is fully relational since it necessarily deals with a constant becoming. If one aims to avoid reductionism, then, Simondon advises us, one should expand the scope of study beyond the technical objects to the technicity of these objects as a mode of relation between human and world (Simondon 2017, 162). The autonomy of each technical object lies in its relational technicity, since "technical objects result from an objectification of technicity; they are produced by it, but technicity is not exhausted in objects and is not entirely contained in them" (Simondon 2017, 176). In simple terms, technicity deals with how humans relate to and transform their environment through technology and how these relations transform all of them – humans, technology and environment – in turn. In this sense, one could start examining design in its technicity.

How is it though that design technicities produce collectives? It is by turning to philosopher Bernard Stiegler that we can provide an answer. Stiegler is categorical when claiming that technology is responsible for the emergence of any collective (Stiegler 1998). This is the case because technology has the capacity to potentialize particular kinds of both memory and intentionality, what Stiegler refers to as a third, epiphylogenetic kind of retention and protention (Stiegler 1998). Simply put, technological artefacts inscribe and exteriorize the actions of a collective past while simultaneously enabling future interventions. A humble table, for example, is the expression of collective efforts that lasted thousands of years aiming at literally elevating the ground from the earth, enabling a form of sociality that would not have been possible otherwise. In addition, the (fundamentally technological) inscription of plans and ideas on a piece of paper brings people together by exteriorizing the promise of a future that is not here yet. With these two examples, we can understand why Simondon suggests that we should use the term transindividual when attempting to speak of human subjects and how they evolve: the purely personal and the wholly social constantly co-transform through technology (Simondon 2020).

Design technicities, from a table to a sketch on a piece of paper, spark transindividuality, enabling the conditions for the production of a collective. Simultaneously, our design technicities in their forming of a collective also produce novel ways of thinking, novel ways of reasoning. This is why architectural theorist Sanford Kwinter claims that design is "a highly advanced form of rationality, perhaps the highest there currently is" (Kwinter 2007, 17). To this, Kwinter adds that

> If design is the dominant form of rationality in our era, it is inseparable from the grand machinery of secular striving and making identified by Max Weber a century ago; it compounds our economic, spirito-religious and sensual life into a single yarn: it is *technique* itself. To say that it is what we are, is not necessarily to celebrate, but to cast a warning and an admonition that somewhere the control of our destiny was handed to us and we failed to answer the challenge with either sobriety, ecstasy or thought (Kwinter 2007, 17, emphasis in original).

Therefore, it is of the greatest importance to examine how our design technicities produce both the world and us, the subjects that live in it. Furthermore, it is of equal importance to elaborate on how we, the (self)designed subjects of a (hetero)designed world, come together and, transindividualy, form collectives. However, and this is one of this volume's ambitions, perhaps of greater importance is to speculate on how we could – through our design technicities – produce new ways of being and becoming collective, ways that would eventually produce both a new world and a new people. In this sense, it is imperative to examine design technicities in their relation to the commons and to practices of commoning, since both have long been considered the purposeful intentionality behind the formation of any collective body.

2 Commons and Commoning

The word "commoning" derives from the wider concept of the commons, a term that has deep philosophical and theoretical roots dating back to the ideas of Plato and Thomas Hobbes. In the contemporary sense, the mechanization of commoning as an operative concept provides the foundation of an alternative and heterogeneous socio-economic model within the public sector but beyond the dichotomy between public and private. Its echoing effect has led to the fusion of the concept with a variety of design domains as for example game design, spatial design and product design.

The commons as a concept relies on an understanding of how natural recourses –referred to as "common-pool resources" – are co-shared among a number of individuals and collectives. The very act of producing, managing, sharing and distributing these common resources is what we refer to as the act of commoning. It is a concept that transverses social, economic, technological, and scalar questions. Commoning embeds its functionality within small groups (the users of a kitchen) or in a wider domain, within the civic (in public spaces and parks). As such, it is both local (a village) and global (use of the oceans) and can materialize in a restricted (a house) or boundless (immigration) format. In other words, and in connection to our entry point regarding a renewed understanding of design, commons and their commoning refer to all the collective technicities that we deploy in order to change our environment and ourselves.

Complementing our approach, this volume will build on Elinor Ostrom's original publication *Governing the Commons* (1990), which questioned the dominant models of managing and sharing natural as well as human-made resources. With the revival of the concept in economics (from thinkers such as Ronald Coase or Albert O. Hirschman) and sociology (one can think of Rosabeth Moss Kanter), Ostrom's contribution was a sociologically oriented empirical approach that helped to explain how some institutional arrangements have helped several communities to manage their commons and maximize community welfare, in some cases, for centuries.

The commons materialize where the private interests of the individual are set against the shared interests of a collective. In a historical context, this "individual versus collective" establishes specific understandings of reciprocity among kin. Closer examination of social crises has shown the effectiveness of the commons in addressing moments of uncertainty as a social problem-solving model. Co-operation in food gathering, child rearing, and defense – in whatever formats – remained co-dependent on a broader collective action. As Ostrom states, "collective-action problems pervade international relations, face legislators when devising public budgets, permeate public bureaucracies, and are at the core of explanations of voting, interest group formation, and citizen control of governments in a democracy" (Ostrom 1998, 1). The commons, in the more contemporary sense, has reverberated into the domains of political ecologies, and as such the very nature of political-economic approaches to territory, governance and types of economies (Ostrom 1998). Therefore, despite its origins in classic political philosophy, the commons has become transdisciplinary in application. It has affected discourses around asset management, environmental ecologies, urban design, geopolitical debates on human rights, and the production of knowledge. It has relied on rational choice theory, related to game theory (Ostrom et al. 2008) and the theory of public commodities to reformulate economic positions away from dominant economies of consumption, speculation and exchange.

For thinkers such as David Bollier and Silke Helfrich (2015), the structural conditions of the commons has delivered compelling patterns of engagement at three levels. First, the processes of the commons, its co-action, co-production or co-operating – either at scales of a high-rise, in an urban village deeply embedded in rural regions, in artistic communities, research settings, or related to collectives in cyberspace – remains a universal necessity. Put succinctly, one way or another, we are all in need of being involved in the very technicities that determine how we produce and manage our shared resources. Secondly, to this effect, although the commons may be regarded as a social occurrence derived from outdated principles, it still retains a modest appreciation in hyper-industrial and modernized societies. Thirdly, the commons define an "open source" paradigm shift. In this shift, the commons represents a repositioned world view, one that influences both material, formal and conceptual conditions as a process: fab labs, hacker spaces, jamming, the sharing economy, the reformation of the civic, types of governance, the private, the public and, as such, the urban, are each reframed once placed within the domains of the commons concept.

In this direction, philosophers Michael Hardt and Antonio Negri (2009) expand on the Marxist analysis of political-economic systems, setting the commons logics against the advantages and disadvantages of diverse governance models, economic systems and social movements. In the field of architecture theory, Hardt and Negri's position has been situated within the specifics of public space, linking the commons to self-organized empowerment struggles (Sohn et al. 2015). The discussion of a variety of social movements explicates how both publicness and the urban rematerialize through the self-organization of social bodies in an attempt to expose latent possibilities within the civic and urban space in times of crisis. The work draws strongly on architect Stavros Stavrides (2016), linking urban spaces to the commons in periods of urban activism.

From another angle, the commons has created a balance between different domains of knowledge in data culture. Defining knowledge as a specific commons, the information paradigm has become decentralized in both its production and ownership. Intelligences, intellectual property and the civics' role are tested through digital information which has, in the conventional sense, always been closed-off and commodified. For Ostrom, irrespective of whether it is labelled "digital," "electronic," "information," "virtual," "communication," "intellectual," or "technological," (Ostrom 2008, 5) the information and knowledge domain speaks to the sharing of a common field where materiality, know-how and data are collective by default. In this light the common-pool resources become economic as well as legal in nature, differentiating the "rights to" from the "rights from" in terms of who has access to information and who can derive rights from each data set.

References to the legalities of common property (Bromley 1998; Ciriacy-Wantrup and Bishop 1975), transference of rights and the open access of knowledge, in whatever format, remain at the heart of the questions posed in the light of the knowledge-commons versus knowledge-economies, materialized in the various licenses to use, distribute or take part in commercial enterprises. In respect to digital media and popular culture, a range of practices from social media to game modifying communities has long helped to destabilize the traditional idea of centralized authorship. Media texts or video games are not only remixed and reconfigured, redefined and deconstructed by "small" actors, but alternative economies and new ways of doing have emerged on the side. Everything from "participatory media" and "fandom" to "piracy cultures" and "Kickstarted" design education is in one way or another linked to a broader idea of the commons.

In parallel, the commons concomitantly expose certain drawbacks. As outlined by ecologist Garrett Hardin (1968) in what is termed the "tragedy of the commons," the imbalance of supply and demand exponentially affects structural as well as long term effects. Irrespective of its application in a spatial domain, in the eradication of illnesses, or in its continued advocacy of "the public good," the complexity of balancing market-driven needs and resource availability may irreversibly transform all the conditions of both common good and how various systems are brought together.

Ostrom et al. (2008) herself mechanizes the praxis of design when postulating principles for governing sustainable recourses. Among others, design remains a necessary skill when mechanizing the commons, articulating definable boundaries, determining the proportional balance between benefits and costs, making collective choice arrangements, strategizing conflict resolution, and minimally recognizing the right to organize nested enterprises and even design pedagogies (Freire 2007).

As is obvious from this short overview of the diverse lines of thinking that the commons generates, little has been done so far in terms of exploring explicitly its relation with design. We are therefore left to question the "common" thread in this conceptual field and its specificity to the design setting. Moreover, despite a recent surge of renewed interest in the social conditions of design, most accounts remain deeply atheoretical. We will claim that a more focused theoretical and philosophical foundation will help problematize the link between commoning and design, and in doing so define the operative theories, concepts and frameworks that influence design thinking across a series of context and conditions.

In the global climate of population increase and the prevalent reduction of financial resources, the question and theorization of shared (collective and technological) capacities will remain part and parcel to the future of design thinking and doing. This volume therefore exploits the theoretical and philosophical themes related to a wider field of a commoning design technicities, providing designers better grounding in the diverse contexts of the twenty-first century. As such, the theorization of design and the commons explores the implicit link through each of the collected contributions to show how this philosophical construct can be explicated in the context of network collectives and transdisciplinary approaches that currently inform design practices.

3 Design Commons…

In this context, and from the overwhelming response to our call for contributions, this book explores four areas of interest. Our selection criteria considered each submission's thematic valance, as well as crossovers with other debates. From the range of articles included in this volume, it has become clear how active spatial practices have been, and remain to be, in questioning design and the commons. This brings to question why so few voices are evident within the established domains of product design or in the emergent domains of service design, experience design, or even policy design in questioning the commons.

3.1 …and the Social

Reconsidering *design, the commons and the social* is a natural point of departure, representing the first thematic cluster. The inclusion of the commons in the social realm exposes deep rifts in both the use and application of social models, social life and of being "social." To this effect, Dorina Pllumbi reflects on the field of power relations generated by collective and emancipatory initiatives in Tirana, Albania, against the developmental pressure of state-led coalitions, questioning perpetual and long-lasting relations through spatial practices. Plumbi highlights that "collectivities" are still present, active and resistant, despite the corrosion of the notion of the collective itself after multiple decades of both totalitarian state control and neoliberal policies. In addition, Plumbi reminds us that eliminating the traditional binaries that are associated in any discussion of the commons (i.e., public versus private, individual versus collective) without overcoming them in action, can eventually cause more damage than good. As such, by referring to the Spinozian *conatus* (the driving force of each individuation) Plumbi asks what a political body can do when faced with the need to organize itself in order to tackle a specific (design) problem.

Maria Reitano and Nikolaus Gartner deepen the discussion of resilient social systems through co-design, focusing on self-production, co-production and reproduction to situate identity and technical knowledge. Focusing on co-design as a practice of "doing together," a renewed understanding of design knowledge is presented. For Reitano and Gartner, it is a collective "know-how" that matters and not just the acquisition of a factual "know-that". This collective "know-how" emerges through production itself (be it self-, co- or re-production) and by embracing the contingency of common design practices, binds a collective together. In other words, for Reitano and Gartner, a *praxis communis* is always produced and never a given, brought forth by the intricate experiential bonds of common action and knowledge production.

The contribution by Chun Zheng directs the discussion towards the commons in the framework of resilience thinking. For Zheng, the notions of scaling-out, scaling-up and scaling-deep crystallize a commons triad that unifies regional agendas, social agency and resilience strategies. By understanding urban commons as something radically different than natural commons, Zheng highlights the importance of practices of governance that emerge as the capacity to respond to disturbances and endure over time. As such, urban commons becomes a matter of sustainment and resilience, constituting therefore a dynamic social process. It is this complex dynamism that scaling-out, scaling-up and scaling-deep examine. Zheng concludes her article by underlining that the value of urban commons (and their spaces) is not merely the value of land and buildings, but, crucially, the value of people and their collective activities. In doing so, Zheng makes clear that by focusing on the importance of the collective production of new norms and values, a renewed definition of commoning can appear: to (re)produce in common.

Finally, Daniel Elkin, Chi-Yuen Leung and Xiao Lu Wang's contribution challenges the alignment between commoning practices and architecture's disciplinary limits. Their action research work in Tai O Village, Hong Kong, elaborates the role architectural products play in collaborative governance frameworks. Therefore, they question the degree to which commoning practices affect architectural design, understood now as a decision-making process. Elkin, Leung and Wang continue Zheng's claims on urban commons by highlighting how architectural design asserts its conceptual framework, where potential alignments to commoning occur and how these affect the very foundations of architecture. They do so by placing focus on architectural production itself as well as the objects (or better said, the products) that are usually associated with architecture, successfully introducing a broadened understanding of design agency.

3.2 …and Culture

The second thematic cluster examines the link between *design, commons and culture.* Frank Bauer and Lasse Sehested Skafte, close to the concerns of Elkin, Leung and Wang, probe a novel understanding of agency and commoning in the digital age. Their perspective outlines an account whereby design is decoupled from linear and sequential processes in favor of intertwined and holistic approaches. In doing so they argue for diverse approaches to the commons that may serve ecologies, economies and foundations of design more actively. By asking how digital designers can reassess their stake between individual and collective modes of production, Bauer and Skafte suggest that a potential transformation in their relation to the commons might occur, proposing a human/non-human assemblage of diverse design agents. Their concept of persistent modelling functions as a speculative design and knowledge tool that through abstraction crosses the thresholds between digital and physical, mediating design knowledge across domains. In other words, the persistent model is a process of continuous modeling that promotes simultaneously both speculation and precise interventions.

Gert van der Merwe posits that the commons should be also understood as part of indigenous systems of spatial production, viewed as an ongoing and relational process in a geography of external power dynamics. Using South Africa as case in point, van der Merwe highlights the differences between the commons of the global North in relation to the global South, where the balance between legitimacy of community is placed alongside a "socially constructed" commons. The case studies he examines stand firmly against the categorical taxonomization of mapping that sediment and immobilize bodies in space, exemplary of the occidental oculocentric representational logics and tools that confuse the map for the territory. Those taxonomies, van der Merwe argues, cannot capture the mobility of pre-colonized Africa, claiming that our tools for approaching the commons do not easily apply in the African context, expressing therefore the need for an immanent and locally bound account of the commons.

Nicholas Frayne's contribution argues for the utilization of uncertainty as a guiding mechanism for narratives that are generated from the lived environment. Linking with the work of Van der Merwe, Frayne harnesses the notions of decolonization, identity and ambiguity to situate a commons where improvisational relationships establish situational environments that support contradiction, flux and connectivity. Frayne advocates an architecture that can make one defamiliarize, suggesting therefore new ways of living (with each other). As such, he essentially proposes that the political is not to be conflated with the personal but rather with a process of continuous estrangement. Ambiguity, for Frayne, is both a conceptual framework and a social practice that results in the emergence and formation of connections. Therefore, he provides three distinct analytical modes that deal with ambiguity from different yet encompassing perspectives: living with, composing, and encountering ambiguity work in tandem to propose a different understanding of architecture, no longer as a representational practice but rather as the enterprise of forming novel connections.

3.3 ...and Ecology

The third cluster focuses on *design, commons and ecology*. Markus Wernli's approach to the commons is through the lenses of living systems and the bio-context of human-waste. Wernli discusses the regenerative, life-giving value chains, arguing for a paradigmatic shift towards bio-economic value creation, a commons distilled from food pedagogies, human nutrients and compost-friendly infrastructures. Wernli claims that human-waste commoning permits communities to regain control over their social reproduction, highlighting the essential biopolitical relations that substantiate any collective formation. An account of human-waste commons, according to Wernli, can revere the ecological use-value of land, partner communities with their non-human counterparts, link collectives with the management of their resources and challenge established food distribution practices. Through different case studies, Wernli makes clear that the often neglected human-waste commons can bring forth a novel, affirmative account of commoning that focuses on fostering modes of collective anticipation that move beyond traditional forms of communal participation and reaction.

In comparison, Piero Medici's historic lens uses the Secondary Reuse Group (SUG) as a form of critique against contemporary aspects of circular economies in light of the commons. Herein, Medici discusses the links between waste as a resource material and the social in light of the common-pool resources. Waste or the "material commons" strikes a fine balance with immaterial commons that is dependent on crafting, negotiating and design experimentations. Following a diverse number of already established accounts on commoning, Medici underlines that common-pool resources can only turn to a commons when communities can actually use them and sustain them, broadening therefore the commons to include inherited natural resources, material humanmade resources and intangible cultural resources. Through the work of SUG, Medici points that the office's design and construction processes not only suggest a move from linear to circular economies, but also manage to bind together all the diverse common-pool resources and effectively turn them to a commons.

Liana Psarologaki and Stamatis Zografos examine ecological and pedagogical models in relation to the commons, through food, fire and affordances. Their debate extends beyond the separate roles each element plays in shaping agents within environmental assemblages, presenting the ramifications of how design contributes to the consumption of environmental and material conditions. Psarologaki and

Zografos approach cooking as humans' primary technical ability, making food, fire and the regulation of their relation fundamental in the very definition of what determines us as species. Opposing the relegation of food and fire to mere infrastructural or hazard-related concerns, the authors propose that we start discussing and practicing food and fire as actual urban commons, moving beyond a state of illiteracy when it comes to the ways that those two constitutive processes define the human. As such, they outline a novel field of urban commons, where food, fire and the design of rituals of collective feasting can both highlight the historicity of our species and remind us our duty towards it.

3.4 …and Transdisciplinarity

Finally, the fourth thematic cluster, *design, commons and transdisciplinarity* explores the implicit link of commoning and how it can be explicated in the context of transdisciplinary approaches that currently inform design practices. Dora Karadima forms crossovers between the social sciences and design theory pertaining to issues of collaboration. Karadima links together three seemingly unconnected fields – design theory, the commons and psychoanalysis – by virtue of deconstructing desire, alienation and separation. Focusing on the work of Jacques Lacan, Karadima uses his psychoanalytical account as an interpretative tool that underlines the importance of autonomy as a necessary precondition for both design and the commons to co-exist in an emancipatory potential. By expanding the discussion on value of the previous thematic clusters, Karadima eventually claims that the commons are determined by the co-production of common values, analyzing this process in terms of both design objects, design processes and design agents.

In comparison, Katarina Moebus relates commoning, care and new materialism within the framework of feminism and Marxist scholarship through situated practices, drawing conclusions about design's inherent political economy to emancipate itself from the coercion driven by market forces. As such, Moebus underlines that design is not merely a problem-solving enterprise but rather acts as the mediating infrastructure through which collectives can address matters of care: what one does to maintain, sustain and repair a common world. Therefore, for Moebus, commoning is to be defined as the design and the practice of constant care. Focusing on examples from her own practice, Moebus makes clear that bringing the commons together with design essentially entails a rethinking of how a community can reproduce itself while ensuring that its common values will not be hijacked by ever looming logics and practices of monetization.

The final contribution by the Contingent Collective (Lörine Vass, Roy Cloutier and Nicole Sylvia) brings the concerns of this volume full circle. They seek, in their argument, to trace the development of the commons in architecture and urbanism to wider cosmopolitical questions that involve agency and responsibility of design. The article postulates a reconsideration of the commons on two fronts: the evolution from a discrete locus (the commons) to a process (commoning), and secondly a shift away from the primary human decisions towards a "more-than-human" ensemble. The authors ask us to conceive the commons as the politics of connection between all the heterogeneous entities that comprise what we call the social. Therefore, commoning becomes for them the means and the reason for radically rethinking our relation to relation. By developing an account of the commons that includes both common resources, commoning practices and the commoners themselves, the authors underline the contrasting trajectories of a merely managerial understanding of the commons while problematizing their (re)production. Through the practice of drawing together matters of care and drawing together those who have nothing in common (yet), the authors claim that a relational account of the commons needs both a radical rethinking of how it enunciates its relation with the future and a how it can allow for the formation of heterogenous assemblages that can catalyze that futurity.

4 Designing a World

Apart from the editorial introduction, and the threading together of the various concepts and contributions, what other reflections and implications are evident in our initial questions – the influence and new interpretations of design and the commons - in the long term? What synopsis is possible on the specifics of design, and what we outline here as the transmission of new or other design technicities? For what purpose, can these technicities facilitate design thinking?

Our observations highlight a first problem of how design and the commons merge. From either side of the divide, reflecting on the design commons raises questions on how any design – in its domain, disciplinary or material alignment – amalgamates with the domain of the commons. From applied research to more abstract and theoretical, what defines design commons seems to constitute a challenge of linking design itself to either the 'project as commons' or the 'commons as project'. The compartmentalization of either belonging to the commons in a conceptual premise alone or a material strategy that attempts to change engagement or useability, appears in many designs as well as design criticism to be a post-materialization contemplation. The fluidity of the middle ground, of being both a material endeavor as well as an abstraction within the commoning framework, remains a fluid and open challenge for design. Our experience on this specific topic over the last years, has proven that thinking commons versus producing commons will require continuous nurturing and development, as a committed endeavor with meaningful impact, spanning years if not decades.

Following a similar line, as editors we question if the design commons can rely on a priori decisions, harnessing predetermined values and processes. The different polarities between design thinking and actual materialization bring a multitude of novel transdisciplinary challenges; in how and in what way design actions can combine methods from social sciences (socio-ethnographic), critical theory (hypothetical-abductive), data sciences (artificial intelligence, big data) and fabrication processes (algorithms, computational protocols)? In this, we foresee the emergence of both design as well as research challenges that will continue to test transdisciplinary methodologies by examining how design appropriates suitable analytic methods to advance design thinking, informed from a variety of angles and research perspectives.

In the context of the ecological, biological, and technological environmentalism, the emphasis on robotics, artificial and other forms of intelligences, continues to place diverse tensions on design capacities and potentials. To address those tensions, we foresee a need to destabilize our human-centered understanding of the world by opening and relating it to the heterogeneous technicities that produce it. Positioning technicities first can help in approaching the operative aspects of designing with as well as for the commons. Without falling prey to any form of technological determinism or reductionism, design theories need to develop accounts that can examine the influence of data in design practices, underlining design in its diverse roles: speculative in the mitigation of transversal concepts, synthetic in the modulation of material forms.

As already implied, limited interest for the commons in disciplines such as game design, communication design, graphic and product design is quite telling. In comparison, specific fields, for example interaction design, seem like obvious routes for nurturing design commons, yet remain underexplored. The continuous advocacy for a human-centered world and in this, human-centered design products, shifts design interest towards the individual rather than the (human and non-human) collective. Processes meant to generate shared norms, values and intentions, have become overly homogenized. This might be explained by a fundamental misreading of the value of the commons in the broader sense, as well as the values that commoning technicities themselves can produce. Especially at times of social instabilities, focusing on the individual rather than collective, certainly undermines both the design questions and the design practices.

Specifically, our observations confirm the ease with which the spatial disciplines have taken to the challenge of absorbing the commons in their praxis. The direct link of architecture, urban design, and interior design with the commons demonstrates socio-spatial sensibilities, as spatial designers appear more prone to negotiate socio-technical challenges, from their problematization to completion. This may be due to the nature of spatial disciplines, where the design of a single building, space, or installation, relies on a multitude of parties and processes within negotiated settings, whilst drawing from economy, structural engineering, building technology and diverse (often conflicting) stakeholders to facilitate the design. We extend this question onto other design fields, presenting a challenge that hopefully will show promise over the coming years of design research, in fields as food design, experience design, transition design and alike.

Naturally, the emergence of design fields with a renewed interest in socio-technical issues has recently shown promise in embedding a thinking of the commons as part and parcel of their domains. Disciplines such as social design and service design may provide fruitful grounds from which to further explore the fusion of commoning and design thinking. The redirection from single to multi-user, not only challenges the premise of product development, perhaps driven by single-sided objectives, but equally requires new and differentiated pedagogies in how such technicities are transferred through design knowledge. The premise of conceptualizing through the commons, coupled to how design groups themselves gather, share, communicate, or produce different settings, reframes the educational premise of the design commons completely. In our view, infusing design pedagogy with the design technicities of the commons will impact the foundations from which designers conceptualize, share and gain knowledge through research processes, outcome disseminations, fabrication models, and material prototyping that sets the tone for long-term thinking in a design community.

Not surprisingly, by questioning the valance of the commons with the spatial disciplines, our conclusions further highlight their tendency to act as a device concept. The distinctions between design disciplines that choose to work with the commons versus those overlooking it, may be linked to the problem of ownership associated with design outcomes. Products, as the singular outcomes of product design, remain for the most part in shifting ownership between (mostly) individuals. The collective sharing of a product, in its use or as a resource, delivers intricacies of proprietorship, monetary values and tenure of use. Who takes ownership of a single product (say a watch or a chair), irrespective of its societal value or aim, remains a consequence of a product-to-individual and not product-to-collective association. The same holds true for the opposite process, where design notions, terms and concepts from domains such as graphic design, fashion design or architecture itself, are often metaphorically adopted from the domain of, for the lack of a better word, 'common' culture, that absorbs, dissects, and further places them away from the original intent. Whereas, surprisingly, other digital media (such as game design) that do involve multi-user scenarios in the values, engagements, and shared ownerships, show greater affinities to the potentials that the commons hold. Irrespective the reason, we foresee the need for further investigations that examine such links and design-commons crossovers.

Finally, the co-dependencies between the design fields, individual users, wider target groups, or their technological appropriation, still places design within a dilemma of delivering novel outcomes. Novel design proposals by default do not inscribe to what we define as the commons or its shared values sets. The challenges presented in the processes of adaptive reuse, recycling, reappropriating, or transforming existing elements, either as products or as material settings, may cause additional strain on the transformation processes, from one technicity to another. Added to this, and in particular relevant to reuse and recycling, other factors, such as heritage, can crystalize competing rationalities along those of the commons, amplifying the challenges of the same problem. Differentiated technicities required for the creation of new material settings, versus reprocessing existing material conditions into diverse configurations seem to shift the design approach per concept and per material scale. The reappropriation of architecture – the enveloping scale of a building or space, versus the scales of a specific product – presents distinctive challenges to the commons that demand a complex and transdisciplinary response. An account of the design commons that wishes to embrace their heterogeneity needs to be simultaneously transversal, non-reductionist and provide an adequate degree of granularity depending on the problem that if focuses. In this, we urge for the careful consideration of the commons, not as an all-encompassing design mantra universally applied, but as a design position that requires careful consideration per setting, level of complexity and material conditions.

Consequently, what becomes apparent through the ways that the four clusters of this book complement each other is the vast and diverse disciplinary fields that a study on design commons can coalesce. In this sense, and even though only the last cluster is explicitly titled so, the whole volume is an exercise in transdisciplinarity. As such, all the heterogeneous design technicities that are examined in each chapter make clear why a transdisciplinary approach is fundamentally necessary to address the complexities of our current realities. While interdisciplinary research entails the collaboration of different domains, it does so from a point of integration, where any of the disciplines involved share methodologies and theoretical frameworks to work towards a unified – thus, integrated – form of research. On the other hand, transdisciplinary research affords the production of methodological, theoretical and conceptual innovations, novel trajectories that emerge in order to address what binds each discipline: a shared problem. In other words, transdisciplinarity does not obey the constraints of any discourse, but, on the contrary, transforms them to productive opportunities. Examining the commons and their design technicities as a transdisciplinary problem underlines its importance beyond the confines of any specific discipline, focusing not on a *world of design* but rather on the complex processes, assumptions and responsibilities of *designing a world.*

References

Bollier, David, and Silke Helfrich. 2015. *Patterns of Commoning*. Amherst: Commons Strategy Group.

Bromley, Daniel W. 1998. Determinants of Cooperation and Management of Local Common Property Resources: Discussion. *American Journal of Agricultural Economics* 80 (3): 665–668.

Ciriacy-Wantrup, Siegfried von, and Richard Bishop. 1975. Common Property as a Concept in Natural Resources Policy. *Natural Resources Journal* 15 (4): 713–727.

DeLanda, Manuel. 2006. *A New Philosophy of Society*. London: Continuum.

Flusser, Vilém. 1995. On the Word Design: An Etymological Essay. *Design Issues* 11 (3): 50–53.

Freire, Paolo. 2007. *Pedagogy of the Oppressed*. Trans. Donaldo Pereira Macedo. New York: Continuum.

Hardin, Garrett. 1968. The Tragedy of the Commons. *Science* 162: 1243–1248.

Hardt, Michael, and Antonio Negri. 2009. *Commonwealth*. Cambridge, MA: Belknap Press of Harvard University Press.

Hennion, Antoine. 2016. For a Sociology of Maquettes. In *Studio Studies: Operations, Topologies and Displacements*, ed. Ignacio Farías and Alex Wilkie, 73–88. New York: Routledge.

Kwinter, Sanford. 2007. Far From Equilibrium. Ed. Cynthia Davidson. Barcelona/New York: Actar.

Ostrom, Elinor. 1990. *Governing the Commons: The Evolution of Institutions for Collective Action*. Cambridge: University Press.

———. 1998. A Behavioral Approach to the Rational Choice Theory of Collective Action: Presidential Address, American Political Science Association, 1997. *American Political Science Review* 92 (1): 1–22.

———. 2008. The Challenge of Common-Pool Resources. *Environment (Washington, DC)* 50 (4): 8–21.

Ostrom, Elinor, Roy Gardner, and James Walker. 2008. *Rules, Games, and Common-Pool Resources*. Ann Arbor: The University of Michigan Press.

Simondon, Gilbert. [1958] 2017. *On the Mode of Existence of Technical Objects*. Trans. C. Malaspina and J. Rogove. Minneapolis: Univocal.

———. [2005] 2020. *Individuation in Light of Notions of Form and Information*. Trans. T. Adkins. Minneapolis: Univocal.

Sohn, Heidi, Stavros Kousoulas, and Gerhard Bruyns. 2015. Introduction: Commoning as Differentiated Publicness. *Footprint* 16 (1): 1–8.

Stavrides, Stavros. 2016. *Common Space: The City as Commons*. London: Zed Books.

Stiegler, Bernard. [1994] 1998. *Technics and Time 1: The Fault of Epimetheus*. Trans. R. Beardsworth and G. Collins. Stanford California: Stanford University Press.

Part I
Design, the Commons and the Social

Chapter 2
Commoning as a Material Engagement of Resistance: The Struggle to Save the Albanian National Theater

Dorina Pllumbi

Abstract This article discusses the struggles of commoning as a material engagement in Tirana, the capital of Albania, as it transitions from a totalitarian state socialist regime to a currently consolidating neoliberal one. It reflects upon the field of relations generated during the collective resistance to save the National Theater, which was brutally demolished in May 2020. The story of the theater is essential in understanding the ongoing suffocation of collective emancipatory initiatives generated spontaneously in times of developmentalist pressure of a corrupt state-power-capital coalition which does not confine itself even to the material cultural heritage. The commoning process of the resistance challenges power relations while generating new extra-institutional practices of care for the spatial and material conditions of the space and the human bodies affectively related to it. It contributes to a new understanding of *collectivity* in Albanian society, an indispensable feature to achieve democracy. The theater building plays an active role in generating collective practices, carrying a political and emotional load, that affect perpetual bonding relations. The affective capacity of the material-human formation challenges the current material culture of demolition and construction in Tirana and the language around it. New architectural practices of resistance emerge as new ways of engaging in collective endeavors in city-making.

Keywords Commoning · Material-human formation · Situatedness · Political transition · National Theater Albania · Resistance

A contemporary Western revival of interest in the notion of the commons, often with enthusiasm, presents this concept as a promising niche of societal post-capitalist configuration, an alternative to neoliberal models (Havik and Pllumbi 2020; Sohn et al. 2015). Yet the commons remain fragile spaces under threat of new enclosures swallowed by the oppression of capital (Harvey 2011; Mies 2014). In

---

D. Pllumbi (✉)
Faculty of Architecture, TU Delft, Delft, The Netherlands
e-mail: D.Pllumbi@tudelft.nl

G. Bruyns, S. Kousoulas (eds.), *Design Commons*, Design Research Foundations, https://doi.org/10.1007/978-3-030-95057-6_2

places that have fragile democracies, these practices of commoning are attacked, intimidated and even eliminated (Federici 2010; Shiva 2020).

The Albanian reality is one of those labelled as *developing*, meaning an accelerated version of progress, measured by economic growth; an everlasting pursuit of a Western abstract vision of newness, cleanness, glossiness. It is one of those realities with a very particular history in relation to collectiveness, because of the entrenched legacy of a Stalinist totalitarianism that lasted five decades. This experience corrupted the belief in the power of the collective, and yet, it may offer some lingering answers regarding the role of the state and the need to keep the collective process open. An over-control of the collective, as happened during the previous regime in Albania, would result in the elimination of possibilities and the restriction of freedom, therefore, in the destruction of its essence.

This article discusses the struggles of commoning as a material engagement in Tirana, the capital of Albania, as it transitions from a totalitarian socialist regime to a currently consolidating neoliberal one. The rupture from the previous regime shifted the orientation towards individualism, private initiatives, and privatization of public property, which has created a common belief that collective activities died with the communist experiment. I aim to point out that these practices are present and even resistant to the new neoliberal regime, paradoxically with an authoritarian face, that does not recognize them, and even attacks them.

In particular, I reflect upon relational agencies generated during the collective resistance to save the National Theater in Tirana, a complex story that is essential in understanding the struggles of collectivity in Albania. The theater became a place of collective resistance to a developmentalist project that threatened the existence of the heritage building and all its material and spatial relations, collective memories and affiliations. The theater building was demolished in May 2020, amplifying these affective relations, which eventually turned out to be the most vulnerable.

In the first part I discuss how collectivity as a notion was ideologically contaminated in Albania during five decades of totalitarianism, and how this has been a favorable ground for neoliberal privatizing policies to be implemented. I point out how, during recent events, it is possible to recognize different forms of collective resistance against these policies. In the second part, I discuss the dynamics that unfolded during the commoning resistance for the theater building in Tirana.

1 An Ideologically Contaminated Collectivity

The collective effort to save the theater building is particularly relevant in Albania, a country that experienced a harsh totalitarian regime, where the perception of collectivity is still vilified. Cooperativism is connected to the memory of the agrarian reform of 1945, when the peasant-land connection was destroyed through a process of collectivization of the land in order to control and manage it centrally, resulting in a failed model that lead to starvation and extreme poverty (Sokoli and Doluschitz 2019). In urban areas, private houses were confiscated, and other than state properties were almost fully eliminated. Cooperatives, collectivism and commonality are all contaminated words that will take a while to be re-normalized in the Albanian vocabulary.

Collectivism was presented as the main ideology that would penetrate every aspect of life, up to the point of eliminating any other belief. The image of the supreme leader had to supersede that of God.[1] The implementation of the communist ideology did not empower communities, recognize their autonomous character and freedom to self-organize and self-govern. On the contrary, the authoritarian state centrally controlled the economy, industry, culture and education of every community and individual (Woodcock 2007, 2016). The state apparatus appropriated the ideology of collectivity by capturing its relational aspect, therefore alienating the relation of human-land, human-human, human-matter, by overlaying it with a controlling and paralyzing gaze. By doing so, the connection of the individual with the collective and the belief in it was corroded.

In architectural practice it was no different; it was centrally controlled and pressurized. Seen as the tool that would enable the materialization of the state ideology in the built environment, architecture had to express the spirit of socialist realism that penetrated every cultural aspect. Working under a politically oppressive regime, the figure of the architect did not have the possibility to be politically committed, to engage in discourse beyond the ideological party line (Kolevica 1997).[2] With the persecution of well-known architects, the possibility for Albania to have politically and even theoretically committed architects who would question decisions on planning and architecture, ended. The figure of the architect was reduced to that of a technician, and the school, also influenced by schools in the Eastern Bloc, gained a strong technical character.

While design was performed within centrally controlled bureaus, construction had to meet economical optimization in times of scarcity, quite a challenge for a small country that in the last decade of state-socialism isolated itself from the entire world. The so-called *aksione* (community volunteering activities) were centrally organized *forced volunteering* days of working *to construct the homeland*, be it infrastructure, housing, or cultural buildings. Most citizens, regardless of their status, class or gender, and even young ages, were forced to perform this type of unpaid work, while the living conditions were precarious (Mëhilli 2017; fig. 26).

[1] Freedom of religious practice was banned by law and Albania was declared an Atheist country by Constitution on 1976. Churches were substituted with *Pallate te Kultures* (Cultural Palaces) as places of propaganda art. The aim was to replace religious ideology to the political state-socialist ideology. On the relations of Albania with other countries of the Eastern Bloc see Mëhilli (2017).

[2] This pressure was reinforced during the mid-70s with the persecution, and even incarceration, of dissident architects that dared to experiment with form. Their work was judged as being influenced by Western modernism in architecture, which was prohibited as it represented the Western imperialist capitalist world. Meanwhile in Albania, like in other totalitarian states, the party line was persistent in the aim to create a nationalist style that would highlight national identities. The incarceration of the architect Maks Velo in 1978 marks the most vicious aggression of the regime to architecture as a profession.

This is a still fresh and entrenched memory associated with the notion of collectivity.

Thinking with Hannah Arendt, we understand that totalitarianism eliminates the division between private and public spheres. While her concern is mainly the elimination of the public sphere, Arendt sees the public and the private as crucial to each other's existence. The private for her is the home of the public; it is the place for the individual to retreat in self-reflection, to then be able to appear in public again (Arendt 1998). The myth of the supreme leader had penetrated even into Albanian homes, where having a framed picture of the dictator, along with other family portraits, became a common practice. Arendt explains:

> Totalitarian government, like all tyrannies, certainly could not exist without destroying the public realm of life, that is, without destroying, by isolating men, their political capacities. But totalitarian domination as a form of government … destroys private life as well. It bases itself on loneliness, on the experience of not belonging to the world at all, which is among the most radical and desperate experiences of man. (Arendt 1973, 475)

Here she explains how terror paralyzes our relations and abilities to act in common, to be politically active and courageous, while at the same time damaging our need to retreat in private. She understands the public and private as existing in a dichotomous relationship. The public realm for Arendt is the common that keeps us connected. At the same time, while staying in this classic binary explanation of public-private realms (she also refers to classic examples like Rome and Athens), she does not identify a separate category for the commons. Nevertheless, when speaking about the importance of acting in public, she keeps in the background those near to us, like friends and family, and certainly those belonging to a community (Brennan 2017). It was an orchestrated collectivism that forcibly replaced this public-private dichotomy in Albania during the totalitarian regime, while destroying collectiveness by seizing it.

When addressing a contemporary revival of interest towards commons, Stavros Stavrides speaks about them as a third category of space, beyond public and private: "various levels and forms of proximity" can be recognized to act as agents for setting up the production of the commons (Stavrides 2016, 260). This approach understands commoning as having a transversal character, giving room to the individual and the collective, while operating at different scales and distances, going beyond dualisms of private/public, object/subject, individual/collective, nature/culture. The Albanian experience teaches us that an elimination of these dualisms without overcoming them can be damaging. The overlapping of a totalitarian form of *one* supreme power puts at risk the existence of commoning and the potential it has to emancipate. According to Silvia Federici, this is a continuous struggle for resistance. She argues that communalism comes as a necessity of coming together to resist forces of capitalism, but also as a reaction to the lack of the possibility to fully rely on the state itself. Therefore she identifies a need to re-enchant the world through the politics of the commons (Federici 2019). The attempt to capture, enclose and regulate from the outside of the commoning practices results in the suffocation of the common space and loss of its capacity to be both productive and to produce itself.

After five decades of centralized collectivism that over-controlled every single practice of daily life, like in other post-socialist countries, with the fall of the regime, a tendency in the complete opposite direction has been an understandable counter-reaction. Everything had to be redefined: the role of the state, the individual and the collective. The orientation towards individualism right after the fall of the regime has created a general idea of a total lack of commoning practices in city-making. There is a need to understand and delineate the difference between the state-communism experiment during the previous regime, and the extra-state commoning practices that manifested with the fall of the regime and that currently occur as a form of resistance against a new regime that is being established.

During this transitional period, from 1991 to the present, different forms of commoning can be recognized as playing a role in the city formation. Self-building, as a tradition but also inherited as an embodied capacity from the times of forced volunteering has been a commonplace after the 1990s. This time, free from the control of a supreme power, it is used, instead, for the immediate self-interest of the individual or one's close circle. Although commonly considered informal and chaotic, these construction practices point to a material claim towards an architecture of the city by the common people, a reaction against the long-lasting oppression of an anomalous totalitarian state.

The year 2006 marked the end of a laissez-faire period with the state initiating the legalization process of self-built structures and regaining control of the territory.[3] The last decade has seen a new neoliberal state consolidating its hold. Albania is now going through the installment of a new regime, a combination of first-stage capitalism in the milieu of consecutive corrupt administrations, with a weak, almost nonexistent social welfare, a growing gap of inequality, and total lack of entities to represent the interest of the powerless categories in the society. As a reaction to the pressure of the oligarchic concentration of power and capital and the pairing of the state with this power, the last decade has seen several instances of civil unrest as the society has started to wake up from its political apathy. In 2018, the new law for the higher education that would favor private universities and the commodification of the public ones triggered massive student protests that sparked hope in society at large. It gave another status to the protest as a possible instrument that can now be utilized in the public interest (Qori 2019). Different forms of activism for the preservation of cultural and environmental heritage under threat have taken place, such as reactions against public-private-partnership management of archeological sites like Butrinti, activism for the economic rights of mine and oil workers, against planned hydroelectric power plants to be built over several wild rivers in Albania, feminist activism, and so on. All these are forms of commoning that are temporal

[3] More on the legalization process in Potsiou, Chryssy A. 2010. *Informal Urban Development in Europe – Experiences from Albania and Greece*. United Nations Human Settlements Programme (UN-HABITAT), Nairobi: United Nations Printshop.

but consistent, commoning with nature and culture, a constitution of human and non-human relations.

2 The Albanian Theater

The replacing of the old theater with a brand new one would be part of a PPP (public-private-partnership) where a private developer would invest in a new theater while using two thirds of the public land to construct their private high-rise buildings (Fig. 2.1). In this case, one third of the public land would be used for a new theater. The masterplan was designed by the architectural firm BIG, commissioned by the state-private coalition. Bjarke Ingels was in Tirana to present the project and was introduced as the most successful contemporary star-architect in the world.[4] When the debate about the theater started to intensify in 2018, a group of artists and citizens started a civil resistance campaign which has unfolded in multiple forms of action. The group is named "Alliance for the Protection of the Theater"; I will refer to it as *Aleanca*. This triggered the creation of a community that started to take care of the building, practicing all sort of maintenance rituals and populating it with artistic activities. They guarded over the building day and night, even during the pandemic, until it was brutally demolished.[5]

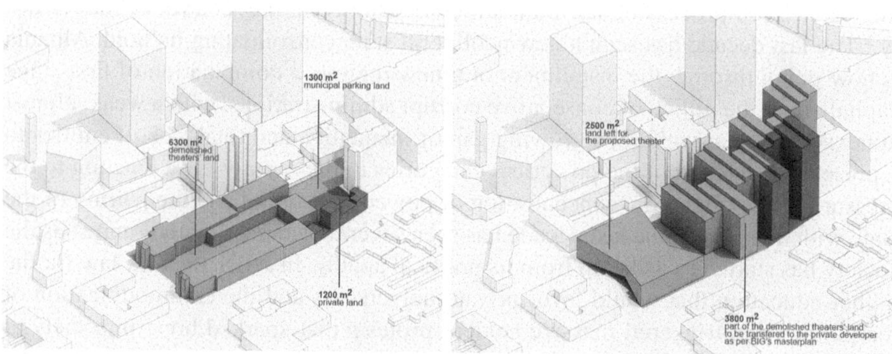

Fig. 2.1 Diagram produced by the author based on masterplan of BIG presented to the Albanian public and on the graphic interpretation that Aleanca did to this masterplan. (Source: Author)

[4] Commissioning star architects is a strategy used for various projects in Albania. From the local architects this is seen as a way to silence the public debate on these projects. For more on the engagement of BIG in the Theater story see this open response letter published initially as Pllumbi (2020a).

English version: Pllumbi, Dorina. 2020. Is Bjarke Ingels deeply naïf or a liar? Arkitekteza. http://www.arkitekteza.com/SQ/response-to-the-article-of-bjarke-ingels-published-in-politiken/. Accessed October 2020.

[5] More on the history and the chronology of the resistance can be found here. Alliance for the Restoration of Cultural Heritage (2020).

2.1 An Agential Material-Human Formation

The picture I draw here is rooted in subjectivities generated by my own relation to the place and the collective struggle to save it. Its material condition, its cultural embodiment and the ecology of the practice of resistance that has unfolded in all its vulnerability, is emblematic of the character of commoning practices in a saturated reality like that of Albania (Fig. 2.2). My own relationship to the theater building and the resistance to save it has different levels. The building itself had always had an attractive force. As an undergrad in Tirana, like every architecture student who sharpens the senses on the material reality that surrounds them, the theater building was one of those artifacts with which I had a special encounter. Later, during my years of teaching at the Faculty of Architecture of the Polytechnic University of Tirana, with students and colleagues, we had a yearly ritual of starting the second semester with a visit to several cultural buildings in Tirana, one of which was the National Theater.[6] We would analyze the functionality of the place, the foyer, the seating area, the stage, the dressing rooms, and so on. No lecture in class could compare to the level of knowledge that the theater building would provide us. It was an open, welcoming and generous book, a simple and humble one.

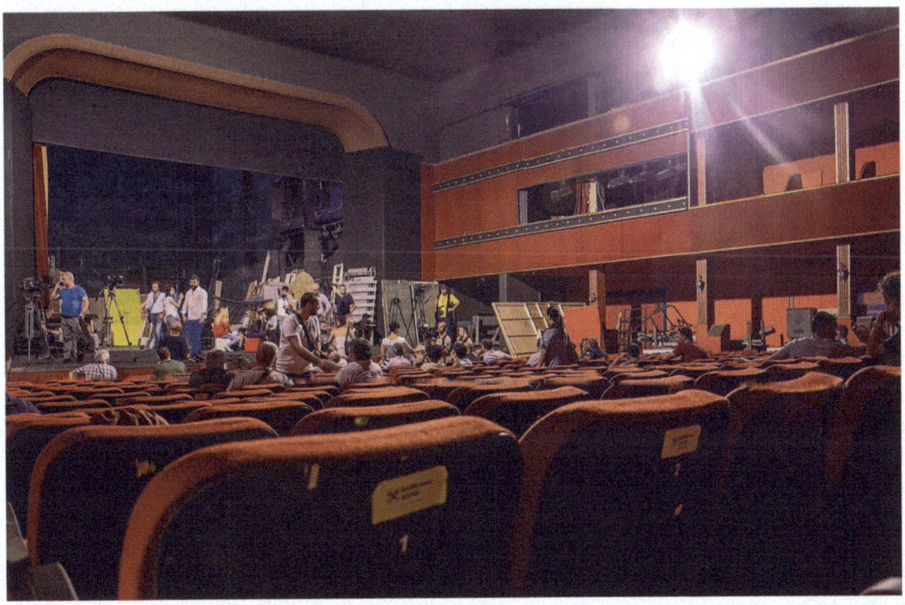

Fig. 2.2 After a clash with the police, Aleanca manages to occupy the Theater on July 24, 2019. From that moment on the building has been safeguarded around the clock by the activists for 10 months until its demolition. The whole resistance lasted 27 months. (Source: Gert Izeti)

[6] Pllumbi (2020b).

When listening to grieving voices for the loss of the theater, the details of these repeated encounters would come back to me with the same intensity. I heard people speaking about the scent of the building. In architecture school we never spoke about it. In the architecture history books no one speaks about the scent. How did we miss mentioning it? Furthermore, the red velvet of the theater hall, the stage that accommodated the freed word during the resistance, the breeze in the courtyard; how sweet and painful these sensations sound verbalized by the people close to the theater! I never thought that I would revisit these sensations in such times of shock when the building would be gone.

As one of many people living far from Albania who at the same time were present at the theater building during the resistance, never before have I experienced the unique position of being so intensively in a place, although physically far from it. The theater was one of the places where I studied the commons in Tirana, the loudest one compared to the other unnoticed places in the marginal areas of the city. Before the Covid-19 outbreak, I conducted visits and conversations with many of the citizens involved in the resistance. The intensive round-the-clock resistance was broadcast daily on social media, mainly Facebook, reaching out to every citizen, near and far, to those that could not join, for several reasons, and especially to those that did not have the courage to show up.[7] Every single event was important, every single decision was communicated. My architect colleagues, including my former classmates, were doing activism in full force. Unfortunately, not all, but lots of architects were raising their voice against the demolition of the building. Finally, there was an opportunity to understand what acting collectively means in poststate-socialist Albania, and whether it still make sense to talk about collective acting.

The discourse went beyond that of a building. We were speaking about a collective formation of matter and human bodies. This formation was not only material; it was not only human. Never before has Tirana seen human bodies attached so affectionately to an edifice. It was so hard to detach these human beings from the building that thousands of armed police bodies were required to do the job. The authorities seized the opportunity created by the global pandemic and struck in the early hours of the morning, thus avoiding the crowds.[8] It was difficult to distinguish where boundaries of the material ended and where human flesh began. What kind of *conatus* is this that pushes human bodies to cling to the roof when looking at bulldozers approaching and armed troops forcibly bringing out other bodies from the building?[9] It was something strongly relational, an affectionate entanglement with some rare agency.

---

[7] Political parties in Albania have adopted an opportunistic approach to recruit membership by providing jobs to the party members in exchange of their blindly adherence to the political line. This is a tactic used by all the political parties, and widely spread in all fields, corrupting this way the freedom of political engagement of the citizens. Professional and even academic jobs, including those from the architectural field, are not immune to this phenomenon.

[8] Van Gerven Oei (2020) and Di Liscia (2020).

[9] In his prominent book Ethics, Spinoza refers to *conatus* as the driving force that is present within humans and things. It is this conatus that makes them act within their capacities and desires.

The material-human formation was generated by a mutual love between artists and theater that has gone on for decades. When endangered, this relation provoked the immediate attachment of more bodies in solidarity, caring and loving bodies, intelligent bodies, human and more-than-human bodies (in association with technological equipment). We witnessed more than 2 years of resistance in the name of love, in the name of care, of freedom, of emancipation, of feeling alive. We met the spirits of our ancestors and the voices of our future grandchildren. We felt pieces tearing down from our souls when the theater was ripped apart.[10]

The debate on the fate of the building was characterized by all kinds of fluctuations. The politicians in power tried to denigrate it, to annihilate it. They tried to identify it as a one thing, but they knew it was more than one.[11] The collective material-human formation kept on reconfiguring itself. "*Unë jam Teatri*" – "I am the theater" – was the motto of the initial campaign that attracted numerous citizens to identify themselves with the theater, with its struggle and its power; a resistance that transversally made room for the individual and the collective. Political parties, agents, and other organizations infiltrated to make use of the moment for their own political interest.[12] Despite persistent attempts to annex it, the marvelous entanglement remained *a multi*, fairly autonomous, without converging into *a one* (Hardt and Negri 2004). It demonstrated a rising belief in the power of self-organization, while becoming an emblem of the struggles that organizing has in society as such. This made this material-human formation a multitude, with its strengths and weaknesses. Without a strong and confined identity, as politically queer as it can be, it became a place of multiple affiliations: born as fascist, appropriated by the communists, a temporary colonial elitist entity, claimed and taken over by the people, the first house of Albanology, the womb of Albanian art, a theater of the people, an open door for all coming from the street, a place of retreat, used and abused, loved

The demolition started while there were people inside and at the rooftop of the building. Activists were forcibly brought out by armed troops without identification number, something confirmed by the Minister of the Interior Affairs. Time 02:55 of the video cited below shows actors Juliana Emiri and Neritan Liçaj shouting out from the roof that this is an installation of a dictatorship.

Komani and Musai (2020).

[10] Mehdi Malkaj, well-known actor: "The theater was not demolished, it was ripped apart!"

Report TV. 2020. I përlotur aktori Mehdi Malkaj: Populli është në gjumë, hienat e shembën Teatrin [cited 2020 October 12]. Available from https://www.youtube.com/watch?v=PNjzK0SXEdk

[11] The attempt to identify the protest as pertaining to the opposition party, was with the purpose of reducing the discourse into a polarized discussion of position-opposition. The opposition parties did support the resistance, and they took extensive room for their own rhetoric, but the resistance still remained plural, multiple, compound of a varied political spectrum.

[12] Romeo Kodra, Albanian artist, expresses in his blog the frustration of participating in these protests, like the one for the theater, because of the penetration of these forces that try to absorb the movement imposing their own political agenda, attempting to give it a one identity-profile, therefore limiting the potential of these emancipatory moments to contribute for the democratization of the society.

Kodra (2020).

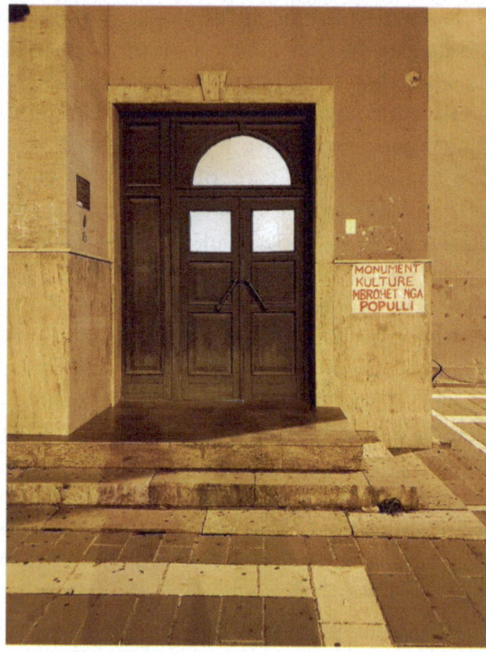

Fig. 2.3 Handwritten sign "Cultural Monument, Safeguarded by the People" that later was transformed into a big banner. Picture of the main entrance of the Theater. (Source: author, site visit on October 31, 2019)

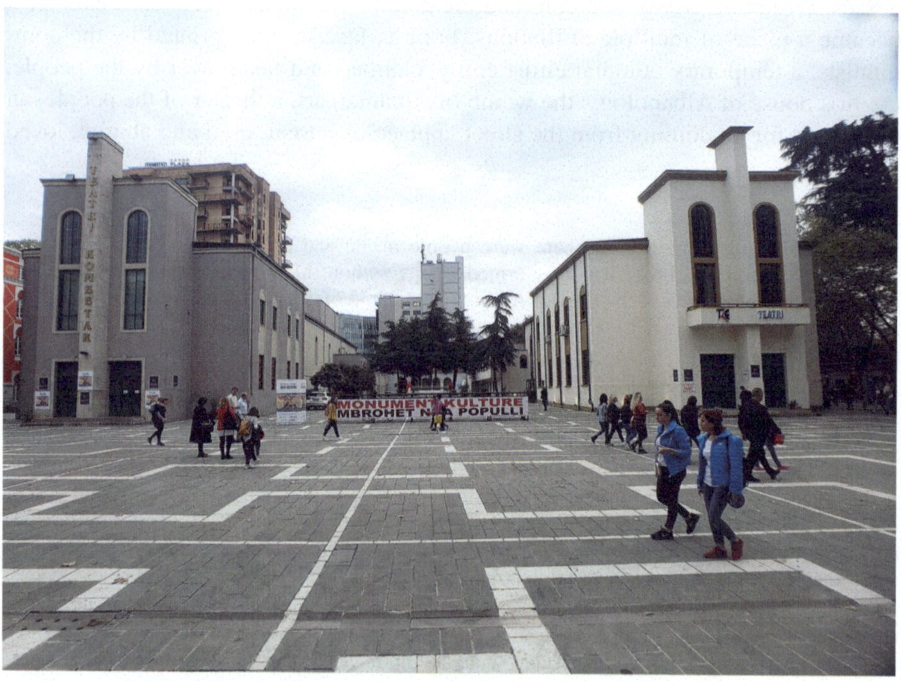

Fig. 2.4 Banner installed at the entrance of the courtyard stating "Cultural Monument, Safeguarded by the People." (Source: author, site visit on November 1, 2019)

and despised. A cultural monument, safeguarded by the people (Figs. 2.3 and 2.4).[13] That explains how horrified the politicians in power were by its agency. That explains the urgency they had to kill it, to finish with its ever-growing power as soon as possible, in such a cowardly manner (Fig. 2.5).[14] At the end, all we are left with is the Spinozist fascination of what a body (bodily formation) can do (Fig. 2.6).

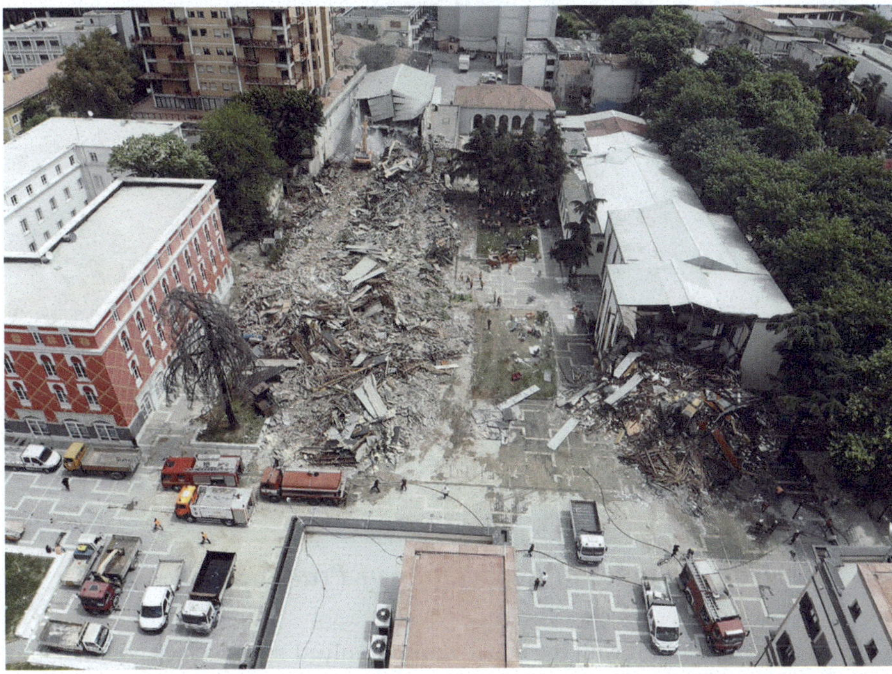

Fig. 2.5 Bird's-eye view picture taken during the early hours of May 17, 2020. The demolition had started on 4:30. (Picture taken by drone. Source: Behaudin Dobi)

[13]The theater was built by an Albanian government affiliated with the Italian government at 1938–39, right before the occupation of Albania by the Italian fascist state. After the Second World War, at the very start of the installation of a new Albanian Republic of the People (that later turned out to be a State-Socialist Regime), while still in the spirit of the Anti-Fascist War, there was an appropriation of the theater by amateur theater groups of *Çeta* (Guerrilla Albanian partisan groups), to then establish the Theater of the People. Later, in 1991, the institution was named as the National Theater.

[14]The theater building had a light structure of a combination of concrete columns, prefabricated wood roof and an infill with composite material. If they really had to, it could have been possible technically to dismantle the structure, showing this way some respect for its history. The act of bulldozing was a clear act of brutality against the building and what it stands for.

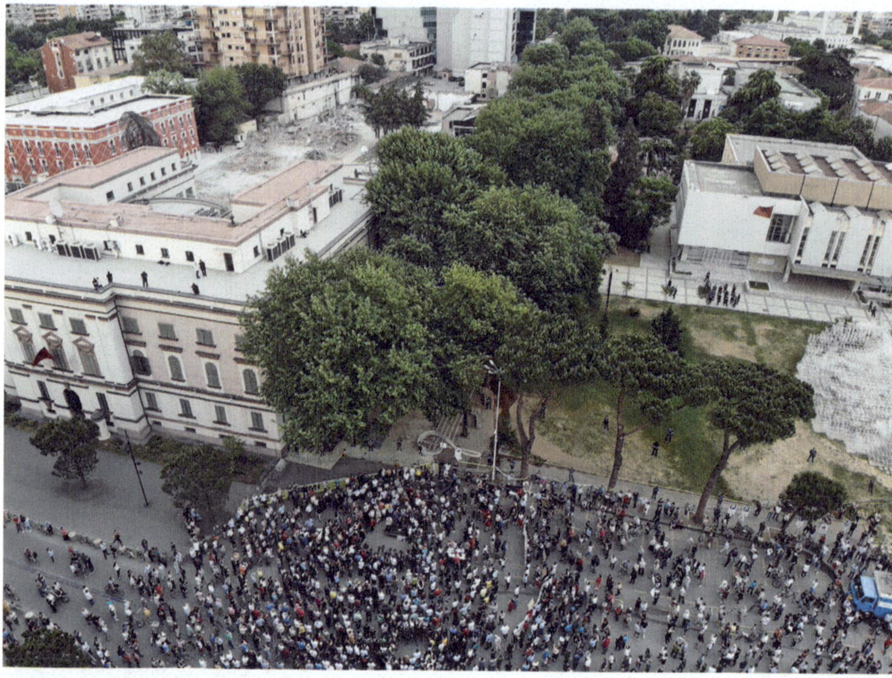

Fig. 2.6 Picture taken during the early hours of May 17, 2020, taken by drone, showing citizens gathered in protest against the demolition. Several were beaten and arrested. (Source: Behaudin Dobi)

2.2 *Survival Struggles of an Ecology of Practice*

In the process of learning with the agent-actors involved in resistance commoning, I have been speaking and sharing thoughts constantly with my colleagues involved in the process, but also others that were part of the resistance group. With that, this contribution on the discourse remains a partial perspective that does not claim to tell absolute truths.[15] Both Doriana Musai, architect and urban designer, and Kreshnik Merxhani, architect and heritage specialist, have been actively involved in the process of resistance. The adrenaline of activism has been a significant component that has affected our connections and actions. In October–November 2019 I visited the theater intensively for several days. It was a time when a Festival for Saving the Theater had already started. An open microphone was active every evening from the start of the resistance. The urban configuration of the theater site played a role in the spontaneous positioning of the microphone, which had started as a moving one. The U-shaped building had a central courtyard where the microphone was located. The

[15] I have entered in conversation with some of the key figures of the group like the artists Neritan Liçaj, Edmond Budina, Juliana Emiri, and more. They always stressed that there is no leading group, but that this is a horizontal organization.

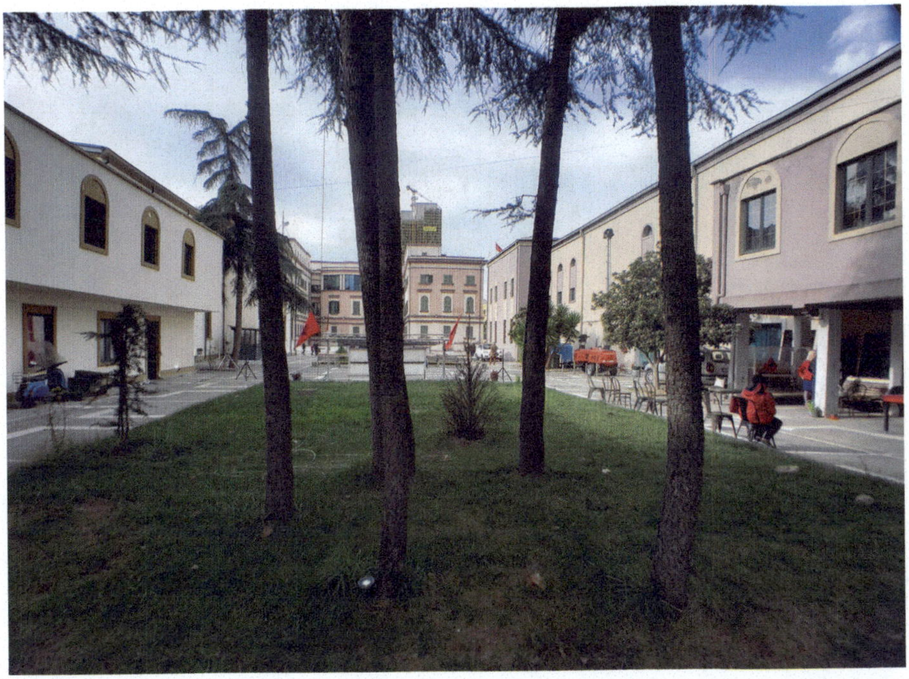

Fig. 2.7 The intimacy of the courtyard. (Source: author; site visit on November 17, 2019)

microclimate of the court was very favorable for the gatherings, with the presence of pine trees and a pool; it was a pleasant place to be in summer. While it was open to another square and pedestrian area, at the same time, it had a kind of intimacy (Fig. 2.7). Keeping the microphone in working order was a struggle in a situation when the electricity had been cut off. It was the initiative of an individual to voluntarily bring all the phonic equipment, two power generators, and maintain them to make sure that the microphone was on every single evening during the whole period, rain or shine.[16] The microphone became "the thermometer of the day, the real voice of the protest in Albania, beyond what is shown in the media, which is all filtered," says Ervin Goci, an activist and lecturer at the Department of Journalism and Communications at the University of Tirana.[17] This microphone became the place of convergence of all the protests, from the students to the inhabitants of a displaced community (at the *Unaza e re* neighborhood, known as *Astiri*), to personal individual indignations. Many had found psychological support at this microphone. It became a form of collective therapy (Fig. 2.8).

[16] Doriana Musai in conversation with the author and Ervin Goci, 8 November 2019 at the Theater site in Tirana.

[17] Ervin Goci in conversation with the author and Doriana Musai, 8 November 2019 at the Theater site in Tirana.

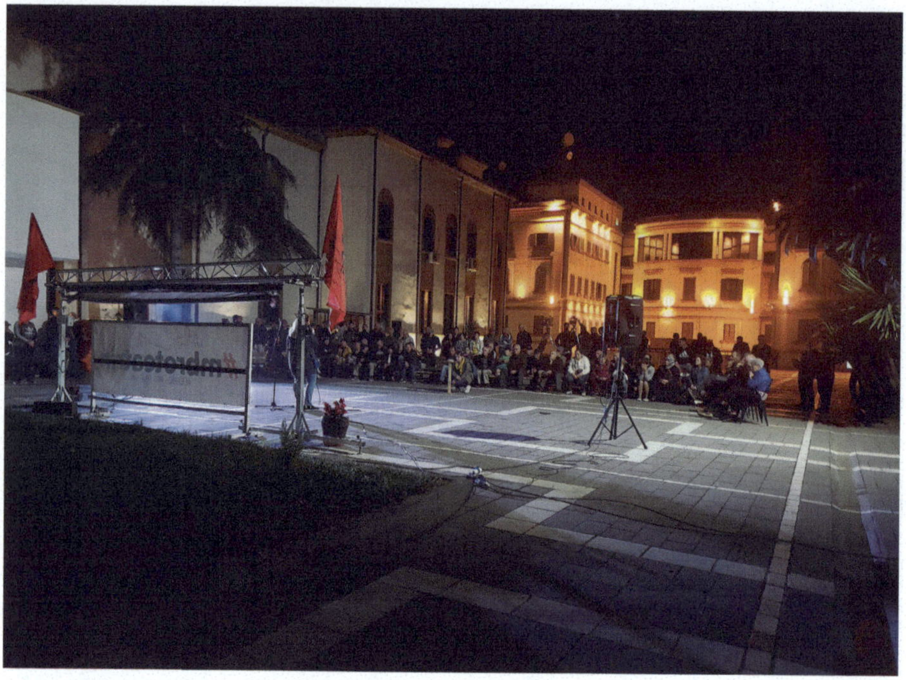

Fig. 2.8 Gathering around the microphone continued every evening throughout the 27 months of the resistance. (Source: author, site visit on October 31, 2019)

Aleanca included intellectuals and professionals of different fields, who worked intra-dimensionally, where the roles were not fixed within their professional boundaries. There were several clusters to be recognized within the group, specialized in law, journalism, architecture and engineering, sociology, and so on. Independent researchers outside of *Aleanca* like historians were also involved in shaping the public debate. The front line of artists conducted the resistance through art performances, exhibitions, and speeches. This kind of convergence from different fields is something quite exemplary for the Albanian reality. And then, there were the citizens, among whom the most prominent group was a number of retirees, who took care of the building and the activists, showing their love and resistance through cleaning and cooking. Edmond Budina, a well-known artist, speaks passionately about how physical work, cleaning and maintenance formed a moment of connection for him and the group, to perceive the place as their own, feel a sense of belonging (Fig. 2.9).[18]

The clusters of volunteers shared ways of working, creating a common situated knowledge that grew in the everyday practice of the resistance. Doriana Musai, for example, explains how the involvement within the community gave her the opportunity to experiment in other mediums beyond her field of expertise. She was able

[18] Edmond Budina in conversation with the author, 16 November 2019 at the Theater site, Tirana.

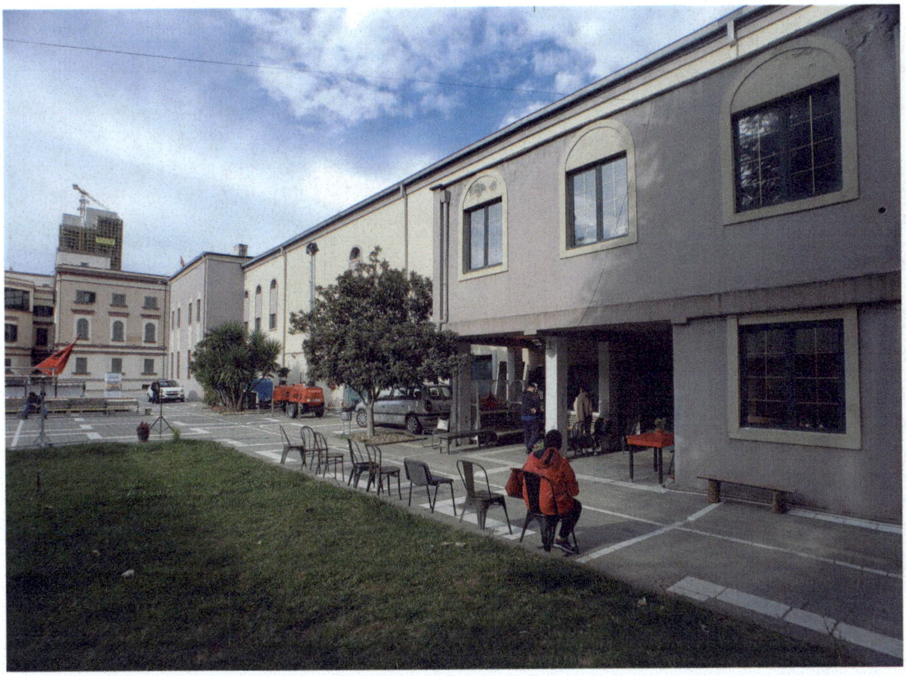

Fig. 2.9 Side entrance of the theater, a niche where the activists were stationed. (Source: author, site visit on November 17, 2019)

to explore new tools and roles, not only as an architect, but also became one of the curators of the Festival for Saving the Theater, taking care of several aspects of the artistic performances, visual exhibitions, and so on (Fig. 2.10). In collaboration with a geodetic engineer she conducted a laser scan of the building that provided a highly detailed virtual model of its physicality. Doriana has been one of the most active agents that have given voice to the material conditions of the building through photography. "The more I stay with the building, the more I discover parts of it that have particularities worth capturing," she stated.[19] She was inside the building at the moment of the attack and broadcast it live on social media, which enabled a plain transmission of the moment of shock to the wider public, which otherwise would not have been aware of the brutal level of intervention of the police forces.

The architect and heritage specialist Kreshnik Merxhani shares how a group of architects and architecture students were involved in the process. He is an activist-architect who constantly fights for heritage causes in Albania. At the theater he explains how at the very beginning they conducted a detailed survey of the building. He compared the impact that two survey techniques used – laser scan and hand drawing – had on the connection with the material conditions of the building. For

[19] Doriana Musai in conversation with the author and Ervin Goci, 8 November 2019, at the Theater site, Tirana.

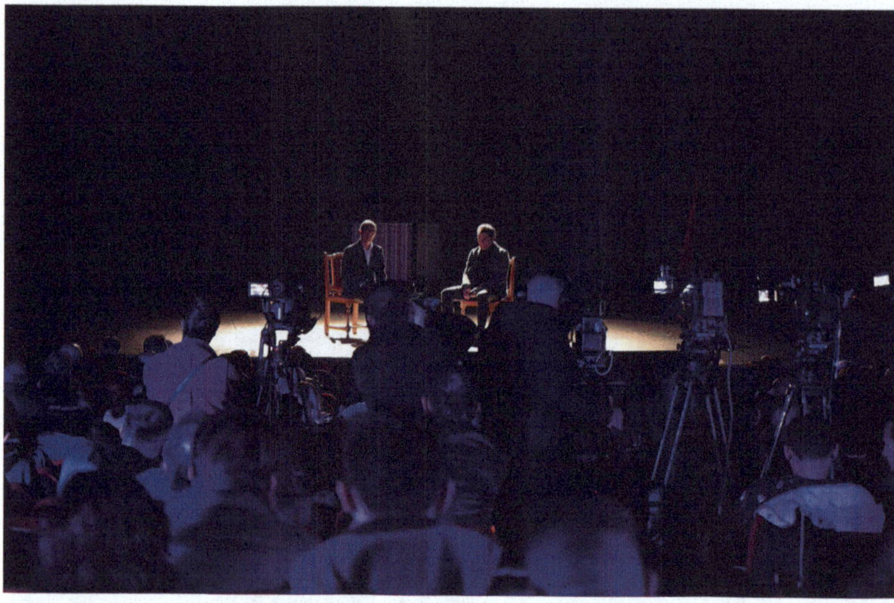

Fig. 2.10 Many performances were organized for the Festival for Saving the Theater, like plays, monologues, exhibitions, and so on. Pictured on stage are Adi Krasta and Eldon Luarasi, son of Edi and Mihal Luarasi, actor icons of the Albanian Theater. (Source: Gert Izeti)

him, the hand drawn survey was a thrilling experience and a very special moment of encounter with the building. It played a role in the resistance engagement:

> We needed to get in touch with the building. When some officials show up and want to destroy it you can ask them: 'Have you ever been inside? Because I have. Have you been at the roof? Have you seen the authentic trusses?' You have to go up there, to feel the magic of the space under the roof, just like in Gjirokaster. It is a magical feeling in itself. It is only a thin wooden skin that divides you from the world, but it has created a whole other world inside. The way the light comes in, that cut on the roof, the wood, the sound of your steps, the smell. That's architecture. That's how you experience architecture.[20]

Kreshnik recognizes that it is this atmosphere that one senses, that makes us the greatest defender of the theater, doing the outmost to save it. "How can they remove these feelings from you? How can you allow them to erase in you the sensations that this building provokes? Not another building, but this one."[21]

[20] Kreshnik Merxhani in conversation with the author, 12 November 2019, at the Theater site.

Gjirokaster is a city in the south of Albania, and its old town is a UNESCO World Heritage. The architect has the experience of working for several years with projects for the preservation of its monuments, an affective experience of care for the past, present and the future. He describes this as the most influential experience of his life, which has brewed in him the love for authentic architecture and the inner will to fight for it when in danger.

[21] Ibid.

The detailed laser scan of the building gave the community a reassurance that even if the worst were to happen and the building would be gone, they would at least have its virtual model. After the demolition, renderings coming out of this scan would become visual material for the group to articulate the demand for justice, asking that the building be rebuilt *as it was, where it was*.[22]

The architects saw an opportunity to apply for the Europa Nostra yearly program "7 Most Endangered" heritage sites, an application proved successful as the building was selected as a European heritage site at risk.[23] Kreshnik shares how the process of finding financial opportunities for the project, other than waiting for what politics decides, has demonstrated that alternative ways are possible, meaning that architects can see the profession as a mission instead of just a service to a client and politicians in power. Recently, in Albania, architects have been heavily criticized as being tools in the hands of politics and capital. Different ways of engagement in public causes would help restore belief in the profession.

Understandably, the debate for the theater has triggered an ethical discourse regarding the role of the architect in the society. Many architects have been vocal in their support for the building and the resistance community, while some others in administrative and political positions enabled the demolition through their signatures. Some well-known architects who initially were against the demolition have experienced pressure from political powers. Two of the main architectural associations, the Albanian Association of Architects and Albanian Union of Architects and Urban Planners, gave declarations against the demolition of the theater at the beginning of the discussion and, more likely, the first one was pressured to keep silent afterwards. This shows how vulnerable the position of the architect is in these situations, how limited and manipulated their professional freedom is when confronted to oppressive economic and political interests. The power struggle that can be recognized is not only in terms of material violence to the city but also limitation of the possibilities of professional engagement with minor voices, other than those in power.

The erasure of the theater, apart from protests, triggered both domestic and international reactions that occupied television broadcasts, written and social media. The prime minister attempted to soothe the shock effect caused by the brutal act of demolition. That morning, while the demolition was in progress, he published on his Facebook page renderings of the theater designed by BIG and declared in parliament that the demolition was not a big deal as they may replicate the historical building at some other site in the city, as a museum for children. He even used a purist language of rebuilding it as per its initial design by the architect Giulio Bertè,

---

[22] Will return to this discussion later in the article.

[23] EuropaNostra (2020).

Several other organizations for the heritage showed their support for the saving of the building. SIRA and Do.Co.Mo.Mo. had also appealed for the saving of the building.

DoCoMoMo (2018, 2020).

eliminating the changes made to it over time.[24] On the other side, the community that resisted for 27 months against the demolition articulated the demand to *rebuild the theater where it was and how it was.* The public debate on the future of the site is currently reduced to this binary discussion of two camps, those demanding the rebuilding of the theater, and those in power pushing the initial plan that enabled its demolition and the public land grab by private developers.[25] Unfortunately, the theater building has become a political trophy, the victim of a transitional moment of a society that aspires to democratize itself but struggles to achieve it in its integrity. At this point, the demolished theater demands justice and dignity, just like those bodies that affectionately attached to it. With the loss of the theater building, the different constituent parts of Aleanca followed different reaction paths. The building enabled the convergence of people with different affiliations. The elimination of the building exerting disproportionate force of the state lead to the un-powering of the collective formation, although several new activist figures entered the political public debate.

2.3 *Politically and Emotionally Loaded Matter*

In 1938–39 when Albania had a close connection to Italy – which later transformed to the occupation of Albania by the fascist Italian state – prefabricated pieces of the theater building were transported from Italy to be mounted in Tirana, as if they were installing pieces of Italian colonialism, planting modernism, giving a western face to the ottoman city of the time. A building of this kind served a new Albanian elite and the Italian invaders. And yet, the less mentioned feature of the history of the theater was that of the re-appropriation by "the people" during the very first years after Second World War, as Romeo Kodra puts it, beautifully "evidencing its fundamental value, … the power of the poorest to convert a symbolic building of colonization in a symbolic building of culture."[26] Edmond Budina, when asked why he is such a passionate defender of the theater, explained how this theater is a place that has the potential to emancipate the society, to be a model of autonomy and direct democracy, become the theater that people feel as theirs, that cultivates a sense of common ownership. He explains how during the resistance, they have encouraged those activist-amateurs who wanted to achieve their life's dream of performing in

---

[24] Exit.al. https://exit.al/en/2020/05/21/prime-minister-says-government-will-rebuild-albanias-national-theater-for-kids-to-visit/. Accessed October 2020.

[25] The media, which is also a reflection of the political power influences, has its contribution to reduce the discourse in these two binaries and the association of them with the two political major formations, position VS opposition.

[26] Kodra (2020).

The historical building is the material witness of important moments in the Albanian history. A dark one, but important to be witnessed, is the 1945 occurrence of special trials that condemned with death important figures that were against the installation of the totalitarian regime of the dictator Enver Hoxha.

front of a public, to experiment with acting on the stage themselves. The theater became their place of becoming and entanglement with others and the building.[27]

Beyond the physicality of the building, its aesthetics, its form, there is an array of historical practices, techniques, research, debates, exchanges, transportations, labor, and power relations that are embedded in its materiality (Thomas 2007). The recognition of these social and material practices that have made the building come to life, to then be appropriated and transformed, is substantial to avoid its fetishization. Matter is loaded with memory and at the same time has knowledge embodied in it. Crucial to the discourse is the need to read stories that matter has to tell, and to craft a new material language.

The theater building came to life in the times of Italian autarky, when there was a heated debate and experimentation on the techniques and materials to be used to glorify the image of the Italian Fascist era and at the same time to satisfy the ambition for self-sufficiency (Avilés 2009; Menghini 2013). The theater is one of a series of buildings erected at the same period in Tirana, while it simultaneously was different from its coevals because of its experimental construction technique and lack of monumentality. The Italian presence in giving a modern Western face to Tirana has been extensively discussed (Gismondi 2012). The monumental buildings were adopted during the post-war socialist regime, although they represented the opposite ideology. Both regimes, fascist and state-socialist, had monumentality and self-sufficiency as their central concern (Mëhilli 2017). The struggle to find construction techniques that would respond to these ambitions is also a common characteristic. *Populit*, an ecological and acoustic material which was used as the infill of the structural frame of the theater building, seems to have been an attraction even for the socialist technicians that were trying to adopt low-cost, high-quality techniques as part of their attempt to master fabrication and standardization:

> [*Populit*] was a high-resistance composite made of wood shavings bound with cement, which could be pressed to obtain relatively lightweight slabs providing durability and insulation. *Populit* seems to have arrived in Albania from the Italians before the Second World War, though the details are obscure. Some accounts mention the involvement of an Italian engineer by the name of Dario Pater, known to Mussolini's wife, and a fixture at Villa Torlonia, *Il Duce's* residence. The Fascist regime reveled in material innovations that showed Italian ingenuity and self-reliance, so *populit* fit into an obsession with giving autarky material form. (Mëhilli 2017, 179)

Interestingly enough, as pointed out by Kreshnik Merxhani, *celenit*, a new contemporary patented ecological material, with similar composition, is used nowadays in the newly renovated Opera and Ballet Theater in Tirana which is promoted as one of the successful projects of the current administration.[28] Here it is possible to distinguish the double standard of the language, when principally the same material is

[27] Edmond Budina in conversation with the author, 16 November 2019, at the Theater site.

[28] Merxhani, Kreshnik. 2019. "Kjo është serioze", arkitekti Merxhani: Me pupulit po restaurohet dhe Opera, është i mirë a helmues.

Dosja.al. https://dosja.al/kjo-eshte-serioze-arkitekti-merxhani-me-pupulit-po-restaurohet-dhe-opera-eshte-i-mire-a-helmues/. Accessed October 2020.

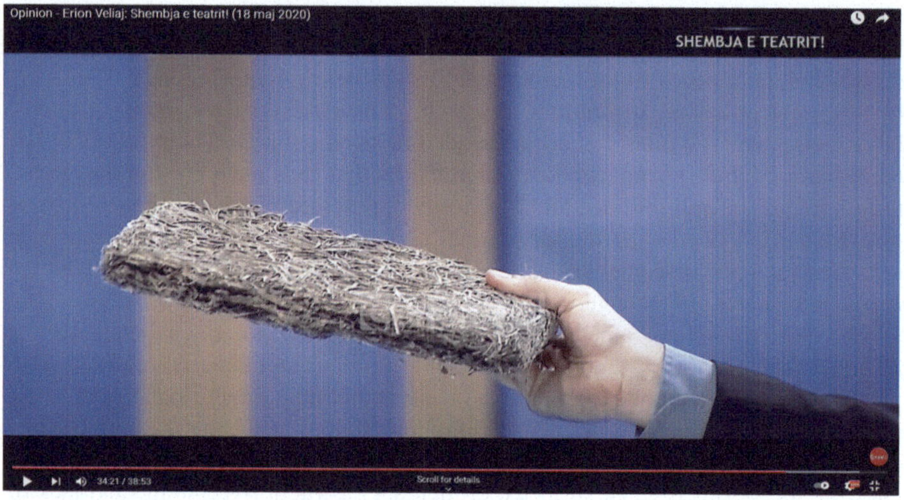

Fig. 2.11 The mayor of Tirana, Erion Veliaj, appeared in several television studious holding the material *populit* in his hands, alleging that it was *tallash* (sawdust). Here he is on the TV Show Opinion of May 18, 2020, on the national television station RTV Klan. (Source: YouTube https://www.youtube.com/watch?v=Y297-McaHHU&t=2077s)

denigrated in the case of the theater building, and praised in the case of a project brought about by the administration itself. The forefront of the rhetoric to devaluate the theater building was extensively sustained by the allegedly degraded material condition, a result of the decay intentionally left to happen due to the lack of maintenance. *Populit* was constantly denigrated, being labelled as *tallash* (sawdust), implying that it was weak, had no good bearing properties and posed a risk to the health of users, because of the presence of mold and probably asbestos (Fig. 2.11).[29] The presence of asbestos was never proven by any technical report, although the sickness of some actors was speculatively blamed on the material. Even if asbestos were found, we known that it is not a reason to demolish a building, when its removal is a commonly known practice. The intentionally offensive language used towards the material was associated with a derogatory depiction of the defenders of the building as being emotional, non-rational artists who romanticize everything old. They were accused of stopping progress and hating everything new. The more the building and its matter was attacked with this populist language, the more capacious the agency of the material-human formation became throughout its resistance.

[29] Several politicians debunked the material in videos or reactions on Facebook. Deputy major of Tirana, Arbjan Mazniku, appeared propagating the bad properties of the material, holing and refracting it when speaking, min 1:00. Video appeared on the Facebook page of the Prime Minister. https://www.facebook.com/edirama.al/videos/985013495163240

The day after the demolition the Mayor of Tirana, Erion Veliaj, continued to carry Populit in an interview in the popular TV show Opinion:

RTV Klan. 2020. Opinion – Erion Veliaj: Shembja e teatrit! (18 maj 2020). [cited 2020 October 12]. Available from https://www.youtube.com/watch?v=Y297-McaHHU&t=2077s

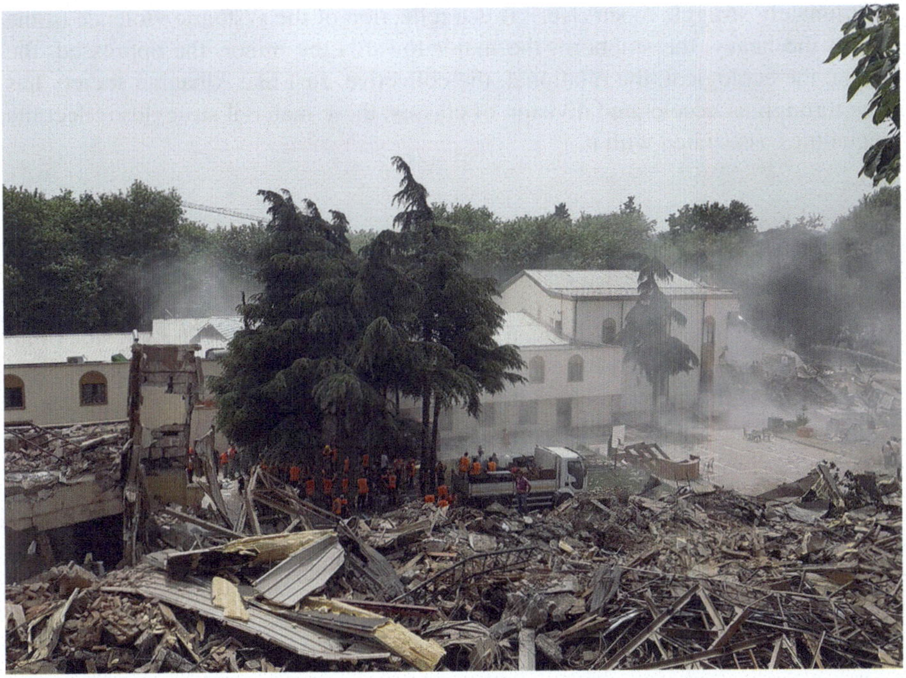

Fig. 2.12 The early morning hours of May 17, 2020. (Picture taken by drone. Source: Edmond Dingu)

The building was declared without any value, a *karakatine* (a hovel). The public was warned that this *karakatine* was in such poor condition that it would fall in on itself. Ironically, the light structure of the theater performed perfectly well during the devastating earthquake of November 2019, showing no damage whatsoever. During this event the building demonstrated that it was light but not weak. The material-human formation acted as a place of mutual humanitarian aid for those in need during the aftermath of the earthquake, when donations from far and near were collected inside the theater building and in its courtyard. This was another indicator that the official institutions had lost the trust of the people and that the theater had become an alternative institution of hope.[30]

The violence towards the theater building is the demonstration of a material culture of a so-called progress, translated in an overwriting by concrete on top of light and ecological materials like *populit* and wood (Fig. 2.12). The theater building is situated in a kind of material unrest in Tirana, when other Old Town areas

[30] Taylor, Alice. 2019. The Heroes of the Albanian Earthquake Are Civil Society. Exit.al. https://exit.al/en/2019/12/02/the-heroes-of-the-albanian-earthquake-are-civil-society/. Accessed October 2020.

unfortunately struggle to survive.[31] It is a reflection of the systemic violence of the strong, the heavy, the stubborn, the major towards the minor, the oppressed, the fragile, the ecological, the relational, the collective. Just like Albanian society has gone through an accelerated division of classes, these material struggles reflect the inequalities associated with it.

3 Conclusion

As described in this article, Albanian society has experienced the darkest side of an ideologically contaminated collectivity during the state-socialist totalitarian regime that over-controlled and paralyzed it. Contradicting the perception that, as a consequence of this totalitarian experience, there is an apathy of acting collectively in Albania, several forms of collective resistance demonstrate a perpetual endeavor of the society to regain belief in collectivity as an indispensable basis for democracy to thrive. Despite these efforts, at present, this time from another type of regime, we witness another form of attack on the initiative to engage collectively, and on the possibility of material-human activism having a say in the formation of the city. Authoritarian, politically oppressive powers continue to suffocate, control and manipulate the collective.

Several commoning practices unfolded during the long 2 years lifespan of the resistance for the saving of the theater. The nightly repeating ritual of gathering around an open microphone located in the courtyard of the theater created a place of collective therapy, a place to cherish the free speech. The "*Unë Jam Teatri*" – I Am The Theater – campaign triggered the bodily relation of a multitude of citizens that would identify with the building and its matter, meaning the attack to the theater would be the attack to their own selves. The trend of spreading a self-portrait picture with the tag "*Unë Jam Teatri*" spread beyond the activists located on the actual site; many from the diaspora and foreigner supporters joined the campaign as well. The Festival for the Saving of the Theater re-turned the place into an artistic venue, attracting more supporters and demonstrating that the building was perfectly functional and plenty of life. An entire process of cleaning and spatial organization took place before the Festival, and the maintenance of the building continued daily, consolidating the bond of the group with the building and amongst each-other. The theater became a place equal to home for the group, as they had to safeguard it around the clock. Care and solidarity for the building was paired with the collective and spontaneous humanitarian aid for the victims of the devastating earthquake of November 2019; a time when unexpected donations of all forms were collected on the theater site, demonstrating increased trust to the theater group from a broader public. It is here when the theater collective gained a new dimension, that of an

[31] An open source database initiated by Jora Kasapi to collect images of objects in dangerous or already demolished.

Preserving Tirana. https://preservingtirana.city/. Accessed October 2020.

autonomous alternative to the official institution. Young architects and engineers did a continuous visual documentation of the building and its material conditions through hand survey, laser scanning and photography. These were relational commoning practices of entering in dialogue with the building, of knowing it closer, of giving it a visual component beyond the debunking narrative used against it by the politicians in power. The recognition of the theater as a heritage site in danger by EuropaNostra and the successful finding of alternative source of financial support for its preservation was a major lesson of securing some autonomy from top-down decisions. Overall, although the theater building is not there anymore, these practices have created a common situated knowledge, which is present and active in playing a crucial role in the ongoing process of the Albanian society to re-gain belief in acting collectively.

Although in a hostile environment, the commoning practice for the saving of the National Theater building, the longest resistance of this kind in Albania, has nourished values of self-organization, of the courage to engage collectively, of care and mutual aid of matter and humans, challenging corrupted institutions and power relations. This resistance took the form of a material-human formation that played an emancipatory and educational role in producing a situated knowledge located in the daily practice of commoning, while generously keeping the process open and horizontal, without hierarchy, without reducing it to a one political identity. The process has triggered ethical questions regarding the role of the state, the individual, the collective, the intellectual, the professional and the citizen in society in general and city transformation in particular. It has exposed the role of foreign star architects who ignore and even facilitate through their design the attack on these self-organized material-human ecologies. It has provoked questions on the role of the local architect, institutional functionaries, academics, and students in the field of architecture. It triggered an ethical awareness on the practice of architecture and its social, political and cultural implications. Architectural professionals and researchers engaged in the process have been a substantial part of it, setting an example of other ways of doing architecture with others, as activists of engaging with matter and the community close to it.

Matter has carried both a political and emotional load, becoming the main agent in the process. It has been the anchor of this material-human formation of resistance. Its loss has been highly affective and has not been accepted by one part of the community, for whom the idea of reconstruction still keeps the resistance alive, although dispersed and divided. The pre-electoral atmosphere of the elections of 25 April 2021 divided the multitude into several pieces, although its agency is still present in the political and professional discourses. There is an ongoing questioning of what happened, but most importantly, a reflection of the immediate need to craft other narratives, different from the problematic language that normalizes a material culture of demolition, reconstruction, violence to the material conditions of the city, which has dominated the accelerated transition of Tirana towards a cruel so-called progress that does not spear even the collective will to actively participate in city decision-making.

References

Alliance for the Restoration of Cultural Heritage. 2020. Teatri Kombëtar (National Theater). https://www.archinternational.org/2020/08/11/teatri-kombetar-national-theater/?fbclid=IwAR0VTFCgnMTk-5zQZViUvGhPexLE-kZ_2nySqdYGlWKPiiXR27TL1FuIcEo. Accessed 12 Oct 2020.

Arendt, Hannah. 1973. *The Origins of Totalitarianism*. New York: Harcourt, Brace, Jovanovich.

———. 1998. The Public Realm: The Common. In *The Human Condition*. Chicago: The University of Chicago Press.

Avilés, Pep. 2009. Autarky and Material Contingencies in Italian Architectural Debate (1936–1954). *Footprint* 4: 21–34. https://doi.org/10.7480/footprint.3.1.697.

Brennan, Daniel. 2017. Considering the Public Private-Dichotomy: Hannah Arendt, Václav Havel and Victor Klemperer on the Importance of the Private. *Human Studies* 40 (2): 249–265. https://doi.org/10.1007/s10746-017-9424-x.

Di Liscia, Valentina. 2020. Open Letter Condemns the "Artwashing" of Albanian Prime Minister's Politics. https://hyperallergic.com/565114/open-letter-condemns-artwashing-albania/. Accessed 12 Oct 2020.

DoCoMoMo. 2018. National Theater of Albania. https://www.docomomo.com/2018/04/06/national-theater-of-albania/. Accessed 12 Oct 2020.

———. 2020. National Theater of Albania. https://www.docomomo.com/2020/05/22/national-theater-of-albania-demolished/. Accessed 12 Oct 2020.

EuropaNostra. 2020. *National Theater of Albanian*. Albania: Tirana. 7 Most Endangered. http://7mostendangered.eu/sites/national-theater-of-albania-tirana-albania/. Accessed 12 Oct 2020.

Federici, Silvia. 2010. Feminism and the Politics of the Commons. In *Uses of a WorldWind: Movement, Movements, and Contemporary Radical Currents in the United States*, ed. Craig Hughes, Stevie Peace, and Kevin Van Meter. Oakland: AK Press.

———. 2019. *Re-Enchanting the World: Feminism and the Politics of the Commons*. Oakland: PM Press.

Gismondi, Simone, ed. 2012. *Architetti e Ingegneri Italiani in Albania*. Florence: Edifir Edizioni Firenze s.r.l.

Hardt, Michael, and Antonio Negri. 2004. *Multitude: War and Democracy in the Age of Empire*. New York: The Penguin Press.

Harvey, David. 2011. The Future of the Commons. *Radical History Review* 2011 (109): 101–107. https://doi.org/10.1215/01636545-2010-017.

Havik, Klaske, and Dorina Pllumbi. 2020. Urban Commoning and Architectural Situated Knowledge: The Architects' Role in the Transformation of the NDSM Ship Wharf, Amsterdam. *Architecture and Culture*: 1–20. https://doi.org/10.1080/20507828.2020.1766305.

Kodra, Romeo. 2020. The Curtain of the Albanian National Theater Came Down. 'Starving, Skinny Ribs Wolves' Are Coming. Aksrevista. https://aksrevista.wordpress.com/2020/05/24/the-curtain-of-the-albanian-national-theater-came-down-starving-skinny-ribs-wolves-are-coming-romeo-kodra/. Accessed 12 Oct 2020.

Kolevica, Petraq. 1997. *Arkitektura dhe diktatura*. Tiranë: Marin Barleti.

Komani, L., and D. Musai. 2020. What Happened on the 17th of May in Tirana? An Outcry for Help (video). Radar Ost Digital (19–21 June): Deutsches Theater Berlin. https://www.youtube.com/watch?v=fzpSmo0hyeA&feature=emb_title&fbclid=IwAR0R15jz1oZXpTvFQ2QzqdjqohjnHigImnOqk7YVHmbwscKtgnuO3bOoB1Q. Accessed 12 Oct 2020.

Mëhilli, Elidor. 2017. *From Stalin to Mao: Albania and the Socialist World*. Ithaca and London: Cornell University Press. https://doi.org/10.1080/13507486.2020.1731141.

Menghini, Anna Bruna. 2013. Experimental Building Techniques in the 1930s: The 'Pater' System in the Ex-Circolo Skanderbeg of Tirana. In *2nd International Balkans Conference on Challenges of Civil Engineering, BCCCE, 23–25 May 2013*. Tirana: Epoka University.

Mies, Maria. 2014. No Commons Without a Community. *Community Development Journal* 49 (SUPPL.1): 106–117. https://doi.org/10.1093/cdj/bsu007.

Pllumbi, Dorina. 2020a. Er bjarke ingells dybt naiv eller bare luddoven? *Politiken*, 11 August 2020.

———. 2020b. Po tani, çfarë t'u themi studenteve? https://peizazhe.com/2020/05/16/po-tani-cfare-tu-themi-studenteve/. Accessed 12 Oct 2020.

Qori, Arlind. 2019. From Faculty to Factory. *Jacobin*. https://jacobinmag.com/2019/02/albania-student-movement-higher-education.

Shiva, Vandana. 2020. The Enclosure and Recovery of the Commons: Biodiversity, Indigenous Knowledge and Intellectual Property Rights. In *Reclaiming the Commons: Biodiversity, Traditional Knowledge, and the Rights of Mother Earth*. Santa Fe: Synergetic Press.

Sohn, Heidi, Stavros Kousoulas and Gerhard Bruyns, eds. 2015. Commoning as Differentiated Publicness: Emerging Concepts of the Urban and Other Material Realities. Footprint 16 Delft Architecture Theory Journal 9 (1).

Sokoli, Olta, and Reiner Doluschitz. 2019. Cooperative Evolvement Through Political Era/Epoch: Albanian's Case and Comparisons. *Ekonomika Poljoprivrede* 66 (1): 189–204. https://doi.org/10.5937/ekopolj1901189s.

Stavrides, Stavros. 2016. *Common Space: The City as Commons*. London: Zen Books.

Thomas, Katie Lloyd, ed. 2007. *Material Matters: Architecture and Material Practice*. New York: Routledge Taylor & Francis Group.

Van Gerven Oei, Vincent W.J. 2020. Comment: The Rotten Birth of a Covid Dictatorship. https://exit.al/en/2020/05/17/the-rotten-birth-of-a-covid-dictatorship/. Accessed 12 Oct 2020.

Woodcock, Shannon. 2007. The Absence of Albanian Jokes About Socialism or Why Some Dictatorships Are Not Funny. In *The Politics and Aesthetics of Refusal*, ed. Caroline Hamilton, Will Noonan, Michelle Kelly, and Elcine Mines, 51–67. Newcastle: Cambridge Scholars Press.

———. 2016. *Life Is War: Surviving Dictatorship in Communist Albania*. Bristol: Hammeron Press.

Chapter 3
AutoCostruzione-SelbstBau: Design as a Practical Knowledge Translation Process

Maria Reitano and Nikolaus Gartner

Abstract Two key concepts can open a theoretical investigation about the fields of design and commons: practice, understood as acting together, and knowledge, or the activity of communicating practical information. These terms are also referred to as *praxis communis* and *practical knowledge*. This chapter aims to develop a design thinking approach able to shift the concept of a design project from its traditional self-referential definition towards its interpretation as activity of collaboratively producing specific knowledge about the designed object's physicality, use and production process and to translate it to a broader community of actors involved in the production process. The adopted methodology is to be considered as an in-the-making design thinking and building process, strictly connected to the experience gained through its application to the case-study, a small building for workshops and material storage, shared by a small Italian community of friends. The approach proposes three phases: *self-production*, as common knowledge producing, *co-production*, as common knowledge translating and *re-production*, as common knowledge growing. As a result, the study presents the opportunity to start both a theoretical investigation and a design process based on the possibility of producing, communicating and growing a common knowledge by defining: a common praxis of self-building a design typology; a collaborative co-designing process; and a resilient social system able to grow both a collective sense of identity and technical knowledge.

Keywords Self- Co- Re- production · Practical knowledge translation · Design thinking-and-building · Shared system of norms · Adaptive typology · Emerging commoning processes

---

M. Reitano (✉)
Department of Architecture (DiARC), University of Naples Federico II, Naples, Italy
e-mail: maria.reitano@unina.it

N. Gartner
Independent Scholar, Vienna, Austria

G. Bruyns, S. Kousoulas (eds.), *Design Commons*, Design Research Foundations, https://doi.org/10.1007/978-3-030-95057-6_3

Within the fields of co-design, a claim is becoming more and more frequent: a "design should start with the situated actions of people doing work" (Greenbaum and Kyng 1992, 184). That is to say that a project, its strategies and principles, are to be gained and accomplished through a practice of doing together and shared with the community of users, in the form of practical knowledge. Regarding the studies on commons, three main interpretations of the concept can be summarized throughout literature: the first and fundamental is *common-pool resources* (CPR) (Ostrom 1990); a second, often referred to as *knowledge/information commons* (Hess 2008), which indicates the right of equal access to all to information, particularly in the digital era; and a third and most recent, defined as *action*, activist movement (Harvey 2012; Bollier and Helfrich 2014; Stavrides 2016), according to which the commons is to be understood as a process of social changes. The last definition is related to the term *commoning* (Linebaugh 2009), which, beyond the field of common natural stocks and resources, specifies the concept of commons through its active and procedural dimension. Emphasizing the shared terms of both fields of investigation, practice and knowledge, this chapter develops a design thinking approach that refers to the design project as the activity of producing together a specific technical knowledge.

The process of acting together establishes a *praxis communis*, a social praxis, as a system of practices and actions of a community, that, by the daily modification of its social and material structures ends up always *re*-producing them. Indeed, through collective action, communities develop, sharing an "emerging common space" that is collectively used and defined (Stavrides 2016, 149). Every common organization or form of cooperation that has the features of being physical and active allows the production of the commons in its materiality or physicality (Deleuze and Guattari 1987; DeLanda 2006; Bennett 2010), emphasizing the social productive relation between human activities and objects.

The commons is self-produced by a complex social system, is co-produced through collective action, and is continuously re-produced through the gained practical knowledge. Some of the key concepts mentioned above are differently exemplified through the activity of the Do-It-Yourself (DIY) design movement (Wolf and McQuitty 2011), which has drawn a link between design research and practical use of the product, between designer and user, who, often, being the same person, is able to address the design scope towards the users' need (Reich et al. 1996). DIY culture has also focused on both the physicality of things and objects (Binder et al. 2011) and on their capacity to provide users with social and technical infrastructures (Schuler 2013), that are shared tools, languages, techniques, structures of possibility, enabling the conditions for self-production and co-creation. Complementary to DIY is the Do-It-Together (DIT) motto, which, underlining the co-production – doing together –more than the self-production aspect, involves another fundamental term in the production process, the *Geld-Light-Commoning* (Helfrich and Bollier 2019, 74, 77), that is the independence of the commons from money and markets logic, because of the community's creative problem-solving capacities. Therefore, this "horizontal doing in common" (De Angelis 2017, 10) provides the actors with a specific object of use, designed according to the users' needs, to the time and place of the realization, and to the economic availability. Besides, by setting shared goals and strategical phases of the process, they can determine new forms of economies, based on the commons networks, on the willingness to cooperate, on mutual-aid, or, simply, on low-budget design solutions. Furthermore, the DIY and DIT fields derive many of their principles from the theorization of vernacular design (Finizola et al. 2012), the local ability of building together and producing use objects. The literature on vernacular design and architecture (Gutkind 1953; Rudofsky 1964) has, indeed, marked a theoretical tradition addressing the interpretation of the vernacular design product as a result of the relationship among a community, their ancient practical knowledge and a place, that, over time, defines reproducible and recognizable typologies. It is, then, evident how this field of investigation underlines not only the communal and aggregational spirit of the anonymous building initiatives, but also their material drivers, such as political and economic goals, as well as the advanced and sustainable technologies developed by the communities of non-professional designers. Rudofsky (1964), for example, cites "prefabrication, standardization, elastic construction, light control" (p. 5) among the most frequent vernacular building techniques and technological solutions.

The research in this chapter deals with the consideration of an architectural DIT process and the conceptualization of its achievements and constraints, asking how to co-produce a low-budget building suiting particular shared needs and local specificities. How should, then, a design thinking process integrate economic constraints, technological solutions, as well as social and spatial needs? Could a technical expertise be disclosed and made broadly accessible and intelligible? How is the materials' availability and standardization to influence the final architectural design and the building process? Aiming to define a design methodology that enhances the described terms, as well as design praxis as an act of *translation* (which, according to the Latin etymology, means to bring something to someone from another place), producing and re-producing the commons, the study presents a three-step design process based on the possibility of producing, adjusting and growing common technical knowledge about the produced object.[1]In the first part of this chapter, the case study is presented; in the second part, the methodological approach is explained as a design-thinking-while-building-together process, being organized into different phases, derived from the case-study experience; in the third part, the methodological results allow for understanding the research as a strategic step towards an interpretation of the design-commons theory, always linked to a sharing practice; in the concluding remarks, we report further purposes of the research that need to be developed and enhanced.

[1] *Traducĕre*: composed of the preposition *trans*, beyond, and the verb *ducĕre*, to bring. Source: *Treccani* dictionary, 'tradurre', http://www.treccani.it/vocabolario/tradurre/ (Cited 31 August 2020).

1 *AutoCostruzione-SelbstBau*: Case Study as Product

AutoCostruzione-SelbstBau is the prototype of a modular and flexible building model for small garden sheds, to be self-built and site-specifically modified by the users, according to their needs (Fig. 3.1). It is located in a rural area, 80 km north of Naples between two mountainous regions, both nature reserves. The opportunity to experiment with a low-budget and small built space manifested itself when a community of friends needed a place to store gardening tools and timber for the fireplace, to park their bikes and to do small carpentry jobs. According to these requirements, the project was conceived as a multi-functional space, which could both allow its mixed use and a diversified and changeable use during different seasons.

The building plot was a narrow rectangular area, located within a garden, with its longitudinal dimension parallel to a street. As a consequence, the first concept was to provide the users with a low-height building that enabled the enclosure of an open courtyard-like space, resulting from the closeness to a single-family house, and protected the open communal space from the street, inviting, at the same time, neighbours and friends to come in and share it.

The building was realized by the architects themselves, with the community's collaboration, during 2 weeks – from December 2019 to January 2020 – and in three construction phases. At first, the foundation was laid and timber strips were fixed, providing a base for the perimeter and inner walls. The second phase consisted in the realization of the timber structure, first by assembling the perimeter and inner

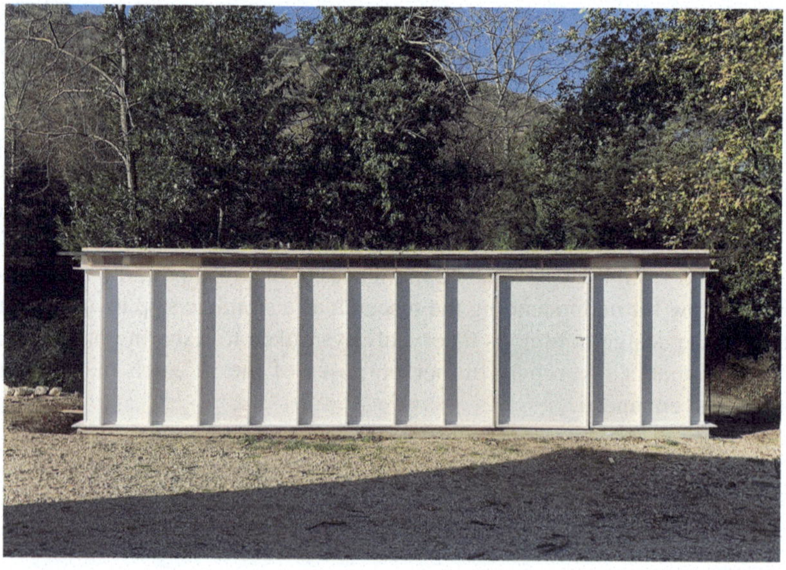

Fig. 3.1 *AutoCostruzione-SelbstBau*: prototype of the design process, photo of the longitudinal prospect. (Source: authors)

wall frames on the ground and then securing them on the base and, afterwards, by fixing the timber beams in correspondence with the structural axes. The transversal walls, defining the inner space distribution, functioned as static stiffening during the construction phases as well. The components were subjected to industrial production measures, such as the OSB panels and the construction timber, which had a maximum length of 2 meters. Finally, OSB panels whose modularity determined the adopted wheelbase measure were used for the walls and roof, which was built with a slight slope towards the extremities, allowing water drainage.[2] The roof waterproofing was ensured with a thin layer of grass, providing a heat shield during the hot season. A strip of polycarbonate panels was fixed all over the perimeter walls to allow the natural lighting, corresponding to the height of the beams. The typology shows a rigid structure, its form is a result of the repetition of a structural cross section, made of pillars and beams, and modular panels for walls and ceiling.

The spatial distribution is articulated in three different spaces: one, accessible from the longitudinal façade, constitutes a small garage for bikes and garden tools storage; the other two are connected and one constitutes the gateway to the other. These last two rooms are to be used in a much more flexible way. The inner one provides a laboratory-like space for different kinds of jobs related to gardening and carpentry, and can be used to store materials and garden furniture during winter. The other is an open room for fireplace timber storage that can also be used to keep tools and objects used daily.

2 Design Thinking In-the-Making: Methodology as Sharable Praxis

Because of the active design process adopted for the realization of the prototype, the methodological steps, which we will present separately and in sequence, are the result of a design thinking process, continuously adjusted according to the increased technical awareness gained during the on-site building work. Three main phases can summarize the design and building process. The first, *self-production*, refers to the foundation of the whole process, that aimed at programming the operational steps and organizing the building process, according to the territorial, spatial, technological, temporal, and economical variables. This step resulted in the definition of situated strategies and techniques, which were eventually adjusted in the building phase. The second phase, *co-production*, regards the actual building phase, that consisted in a DIT process and implied the progressive definition of a shared technical knowledge, based on the practice of communal building. This step allowed the testing of all technological solutions and their context-fit adjustment. The third phase, *re-production*, aims at opening up to a broader community of potential users of

[2] *Oriented strand board* (OSB) is a type of engineered wood. Source: *Wikipedia*, https://en.wikipedia.org/wiki/Oriented_strand_board (Cited 31 August 2020).

similar designs and of potential actors in design commoning processes. This last step constitutes a call for ideas and experiences to test and improve the empirical instruments developed in this research, growing common practices and sharable knowledge about design as a process of building in common.

2.1 Self-Production*: The Role of Materiality and Situated Techniques in the Programming Phase*

The first methodological step towards the definition of the design thinking process consisted in the organization and programming of the building process, considering the specific needs that the project had to fulfil. In order to recognize the different involved variables, the collective problem was identified, common goals were defined, and possible solutions and strategies were proposed to achieve the goals. Nevertheless, before finding answers, the key to identifying collective problems and goals was the setting of the right questions. Indeed, the following questions turned out to be the drivers of the design process: *To whom?* To whom (of the final users) is the design product addressed; in what ways are they going to use it; are these uses permanent or temporary? *Who?* Who among the involved people is willing to self-build; are they going to be the final users? Answering these first two questions required an analysis of the uses and activities related to the building. In particular, it is worth noticing that even though the final users coincided with a family and its community of friends, people who volunteered to help during the building phase involved a broader group, among which the architects themselves and a professional electrician. Furthermore, from the analysis of uses and needs it was not really possible to attribute specific functions to the building, which had to be a multi-functional and flexible workshop-like space for storing different kinds of things and doing carpentry jobs. The remaining questions are: *Where?* Is the project site private, public or common property; what are the specificities of the building site; what kind of technical expertise is required? *How?* What are the most appropriate technical and technological solutions to be adopted? These last two questions led to site-specific analysis and a building typological study in order to investigate similar structures and their technical characteristics.

After an analysis of the territorial context, a design and building program was developed according to six variables and factors, from which the definition of different strategic projects derived, allowing the self-build with the available resources and existing limits. This operational program is to be considered as the actual design thinking process, the strategic moment in which the involved people were asked to think about the impacts of the design through different dimensions. Indeed, the project was developed according to different spatial scales, functional questions and technological details. The first variable consisted in the consideration of the site-specific context and surroundings. Indeed, the urban analysis allowed the designers and the involved community to foresee the impact that the building would have on the shared open yard, the building site, which was used by the small community for different activities, such as parking for cars and bikes, meeting neighbours, carpentry, gardening and so on. The second variable concerned the specific needs deriving from the functional and spatial distributions. As a consequence, the architectural design defined the spatial organization according to the connections to the entrance of the yard and of the family house nearby, so that the orientation of the small shed could best strategically enable the mobility of people, bikes and cars, and fit the different activities taking place in the collective space. Moreover, a fundamental factor influencing the decisions through the different spatial scales included the specific characteristics of the materials and their supply chain at the local level. In fact, the technological design defined the modularity of the built space and its dimensions according to the availability of materials and their standardized measures. Then, the developed situated technical details reflected both the best fit to the environmental conditions, the daily light cycle, the site specifics such as slope and nearby trees and vegetation, and the limitations imposed by the use of prefabricated materials. As regards the provisioning and transportation of building materials, this was planned according to the local availability of resources and materials. This aspect of the project involved technical detail thinking to be continuously adapted to the available local materials, which sometimes happened to have slightly different measures from those designed. This phase resulted in the mapping of a network of local manufactures and sellers that was also useful to control transportation costs. Another relevant variable in the building process was time. The construction site had to be prepared to allow the realization of the building within a limited time: 2 weeks. It consisted in the setting up of temporary workstations where materials could be cut and assembled, and in setting up temporary support systems for the period when the bearing structure would be built. The organization of the construction site was fundamental for a more efficient work logistics, resulting in less time lost to bad weather and other unexpected problems. Moreover, a further strategical aspect involved the project's low budget. Indeed, in order to respond to the users' needs and the limited budget, a detailed metric calculation was developed, including building materials, construction instruments and transportation costs. A key factor in cost cutting was the choice for standardized, industrial materials with fixed measures, which determined the modular definition of the design: for example, timber slats were used as beams and pillars in their original dimensions without being cut. This prevented material waste, which would have been caused by cutting timber pieces and, as a consequence, reduced the amount of materials needed to the standardized pieces. It also defined the repeatable module and the construction sequences and accelerating the building process. Indeed, the module allowed us to first experiment on-site with the best-fit approach to assemble the materials and then to repeat the process as many times as needed to obtain the total measurements. The definition of a module constituted a fundamental step, strictly linking the design with the building phase and enabling its future repetition, modification and improvement. Then, the last fundamental variable to take into account consisted in the self-building skills and possibility of the involved people. This implied the design of specific building techniques, which could easily and safely be performed and

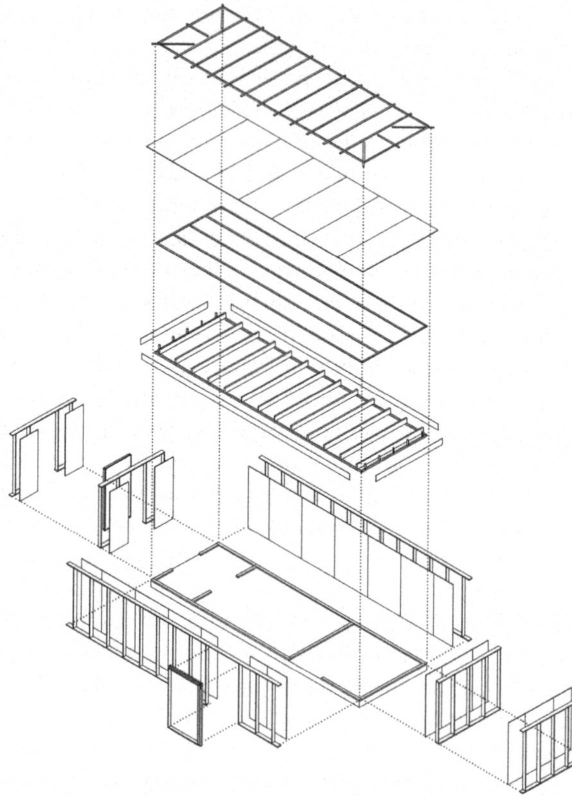

Fig. 3.2 *AutoCostruzione-SelbstBau*: prototype of the design process, split axonometry. (Source: authors)

repeated by non-professional building workers. The building structure was designed not to need the intervention of professional carpenters: standard tools such as a conventional saw and an electric drill, which can be found in every home, were used to assemble the structure. This simple structural design thinking, involving basic structural materials to be put together through DIY methods, guarantees the design to be accessible by a community of non-professionals, thus, to be translated and self-built (Fig. 3.2).

2.2 Co-Production*: The Definition of a Common Practical Knowledge*

This second step refers to the actual building phase. However, this moment not only represented the actual process towards the realization of a built product, but the assessment and test of all the technical and technological decisions. Indeed, the built structure is to be considered as the prototype of a building system to be further developed, and all the different technological details the attempts to provide maintainable and removable solutions to be later improved and adapted during the life cycle of the building. Even though the building process had been programmed and documented through executive drawings, these, eventually, constituted just a blueprint of the actual structure, which was the result of continuous adjustments in-the-making. As different unforeseen logistic and technical constraints came up, the building program had to be changed and re-elaborated according to the contingent situations. This process defined the point along the way when the deepest awareness about the technological achievements and constraints was gained by all the people involved in the construction phase, as well as by the architects themselves. Then, the shared practical knowledge developed during the first construction steps allowed the workers to avoid mistakes in the further ones, to collaborate without misunderstandings, and, as a consequence, to be more precise and efficient. Nonetheless, the practice of doing together highlighted the social relevance of the productive process. If, indeed, on the one hand, the latter aimed at the realization of a material object in space, on the other, it established new social bonds and dynamics, based on the communal sharing of a practice in space, the choral building work, as well as of the specific technical knowledge acquired through that practice.

2.3 **Re-Production***: Towards the Repetition of a Self-Building Typology*

The third proposed methodological phase constitutes the fundamental link that allows the definition of the design methodology as a *praxis communis*. It is, indeed, an invitation to test the developed building system, and contribute to the obtained practical knowledge about it. This would initiate a process of growing technical knowledge based on a repeatable and mutable practice. In particular, the design product would be re-produced by copying and modifying the prototype, while the gained technical and material knowledge about the product would be translated and continuously assessed and increased. In short, the re-production process would be repeated while enhanced, defining a complex adaptive process, hence resilient to contingent and unforeseen problems. Thus, re-producing would mean defining a typology of emerging processes, combining both the terms of the commons and design.

Regarding the case-study, the deep situated technical knowledge gained by all the involved people, as well as by the users themselves, allowed the latter to periodically do maintenance job and modify the spatial distribution of the garden shed, according to the different seasonal changes. Indeed, after almost 2 years since the construction, the building has been differently used and modified. The modular bearing structure has been implemented with a single pitch roof, to better cope with the bad weather during winter. Some perimetral panels and polycarbonate windows have been replaced. The users were also able to modify some technological details, as the external joints between the walls and the horizontal foundations, which had been damaged by winter rainfalls, and the door locks, that needed more secure systems during night. Moreover, the spaces have been differently used as storage room for materials and gardening tools, bike garage, and workshop activities. Then, the spatial distribution has been adapted, by modifying the carried structure and differently accessorizing the internal walls.

3 Design Thinking as Adaptive Process and Practice

As concerns the results of the exposed process, different aspects should be taken into consideration. On the practical level, it provided a community with a material and spatial object to be used according to its needs. However, another material outcome consisted in the codification of the case-study structural assembly process. In Fig. 3.3, reproducing a *co-designing toolkit*, the materials and instruments required are shown, together with a diagram explaining the modular structure assembly. Aiming to provide further users with a common language for the design to be reproduced and modified, this toolkit is also a means to communicate the accessibility to all of the initiative, involving standard materials and do-it-yourself carpentry tools. This toolkit thus constitutes a practical guide, showing within one picture, the materials, tools and assembly techniques required. Then, beneath the possibility to simplify the building process through a diagram, lies the precise definition of the

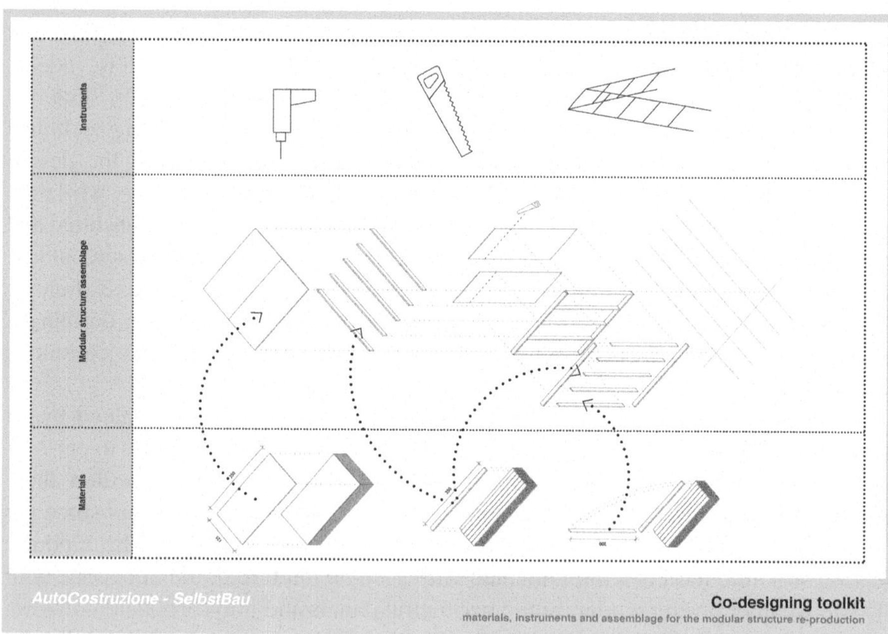

Fig. 3.3 Co-designing toolkit: materials, instruments and assembly for the modular structure reproduction. (Source: authors)

building system, from its bearing structure to its technological details. If, indeed, on the one hand the building model could be considered as an outcome as well, it is, on the other, the technical awareness gained through the practice of building that represents an important result of the experimental research and fieldwork. Finally, it seems that the overall outcome of this study should be represented by the approach of the design thinking and building phase itself, which was developed as a complex and situated process of material and contingent practices and collective work.

4 Conclusions

By investigating the theoretical and practical relationships between the commons and design, we have developed an argument deriving from key concepts in recent literature about commoning (Hardt and Negri 2009; Harvey 2011; Stavrides 2016; De Angelis 2017) and co-design (Huybrechts et al. 2017): action and practice, information and communication, materiality and physicality. The analysis of some related terms, such as production process, typology, and communication tools has turned them into useful instruments to define the methodological framework, linking commons and design thinking within the case-study experience.

The presented methodology, developed from a specific case study, presents an immanent standpoint, assuring it not to be detached from the empirical data and enabling a local and non-referential design typology to be specified in its spatial, technical and functional characteristics. Moreover, some adopted methods responding to contingent problems remain related to the particular and cannot be considered as instrumental general answers. In order to develop the proposed methodology and its theoretical position founded within a specific practice of doing in common, further research should test its validity by adopting the main phases and specifying them according to particular and commonly posed questions.

As regards the specific design matters, this contribution investigated the development of a building typology, by using standardized materials as a way to reduce costs and broaden the potential to produce common technical knowledge. Then, constraints and opportunities deriving from this choice have been considered, concerning the structure, the modularity, and the architectural design. Consequently, the modular components defined the structural language, enabling the users to progressively add and build on the structure itself, transforming its possibilities of use, without changing its statics and without the need of engineers.

Finally, the study dealt with a theoretical position within design and architectural studies that considers a non-referential role of the design product, and, instead, takes into account its building process as a practice of doing in common and a choral work. The relevance of this process is to be found in both the affective and relational networks defined by the communal practice, and the disclosure of a technical knowledge to a broad group of actors, who, by engaging in the building practice, actually, succeeded in developing that specific technical awareness, integrating all the different situated variables.

References

Bennett, Jane. 2010. *Vibrant Matter: A Political Ecology of Things*. Durham/London: Duke University Press.

Binder, Thomas, Giorgio De Michelis, Pelle Ehn, Giulio Jacucci, Per Linde, and Ina Wagner. 2011. *Design Things*. Cambridge, MA: MIT press.

Bollier, David, and Silke Helfrich, eds. 2014. *The Wealth of the Commons: A World Beyond Market and State*. Amherst: Levellers Press.

De Angelis, Massimo. 2017. *Omnia Sunt Communia: On the Commons and the Transformation to Postcapitalism*. London: Zed Books.

DeLanda, Manuel. 2006. *A New Philosophy of Society: Assemblage Theory and Social Complexity*. New York: Continuum.

Deleuze, Gilles, and Félix Guattari. 1987. *A Thousand Plateaus: Capitalism and Schizophrenia*. Trans. Brian Massumi. Minneapolis: University of Minnesota Press.

Finizola, Fátima, Solange Coutinho, and Virginia Cavalcanti. 2012. Vernacular Design: A Discussion on its Concept. In *Proceedings of the 8th Conference of the International Committee for Design History & Design Studies-ICDHS 2012*, 557–561. São Paulo. https://doi.org/10.5151/design-icdhs-107.

Greenbaum, Joan, and Morten Kyng. 1992. *Design at Work: Cooperative Design of Computer Systems*. Hillsdale: L. Erlbaum Associates Inc.

Gutkind, Erwin Anton. 1953. How Other Peoples Dwell and Build, 1–7. *Architectural Design* 23:2–4; 31–34; 59–62; 121–34; 159–62; 193–97.

Hardt, Michael, and Antonio Negri. 2009. *Commonwealth*. Cambridge, MA: Belknap Press of Harvard University Press.

Harvey, David. 2011. The Future of the Commons. *Radical History Review* 2011 (109): 101–107. https://doi.org/10.1215/01636545-2010-017.

———. 2012. *Rebel Cities: From the Right to the City to the Urban Revolution*. London: Verso Books.

Helfrich, Silke, and David Bollier. 2019. *Frei, fair und lebendig: Die Macht der Commons*. Bielefeld: Transcript Verlag.

Hess, Charlotte. 2008. Mapping the New Commons. Governing Shared Resources: Connecting Local Experience to Global Challenges. In *The 12th Biennial Conference of the International Association for the Study of the Commons*. 14–18 July 2008. Cheltenham, UK. https://doi.org/10.2139/ssrn.1356835.

Huybrechts, Liesbeth, Henric Benesch, and Jon Geib. 2017. Institutioning: Participatory Design, Co-Design and the Public Realm. *CoDesign* 13 (3): 148–159. https://doi.org/10.1080/15710882.2017.1355006.

Linebaugh, Peter. 2009. *The Carta Magna Manifesto: Liberties and Commons for All*. Oakland: University of California Press.

Ostrom, Elinor. 1990. *Governing the Commons: The Evolution of Institutions for Collective Action*. New York: Cambridge University Press.

Reich, Yoram, Suresh L. Konda, Ira A. Monarch, Sean N. Levy, and Eswaran Subrahmanian. 1996. Varieties and Issues of Participation and Design. *Design Studies* 17 (2): 165–180. https://doi.org/10.1016/0142-694X(95)00000-H.

Rudofsky, Bernard. 1964. *Architecture Without Architects: A Short Introduction to Non-Pedigreed Architecture*. Albuquerque: University of New Mexico Press.

Schuler, Douglas. 2013. Creating the World Citizen Parliament: Seven Challenges for Interaction Designers. *Interactions* 20 (3): 38–47. https://doi.org/10.1145/2451856.2451867.

Stavrides, Stavros. 2016. *Common Space: The City as Commons*. London: Zed Books Ltd.

Wolf, Marco, and Shaun McQuitty. 2011. Understanding the Do-It-Yourself Consumer: DIY Motivations and Outcomes. *AMS Review* 1 (3–4): 154–170. https://doi.org/10.1007/s13162-011-0021-2.

Chapter 4
Scaling Out, Up and Deep Understanding the Sustainment and Resilience of Urban Commons

Chun Zheng

Abstract The inquiry of how to manage common resources has always been a central topic in the discussion of the commons. However, our understanding of what can be discerned as the sustainment and resilience of urban commons – urban spaces that are managed by citizen groups – is under-evaluated and simplified if we only ask whether they last long enough or are able to be expanded in size and duplicated in multiple locations. In this chapter, I introduce the urban commons and the importance of discussing their sustainment and resilience, wherein I refer to the notion of sustainability and resilience in design and ecology spheres. A review around existing literature on governing the commons and expanding urban commons then provides the basis for the further explication of the sustainment of urban commons from three scaling perspectives: scaling out, scaling up and scaling deep. Lastly, I discuss the co-existence and correlations of these three scaling dimensions and open up the question of what the role of design is in promoting the scaling processes.

Keywords The commons · Urban commons · Scaling out · Scaling up · Scaling deep

Around the world, urban developments are still dominated by market and state-driven forces. Urban commons – spaces that are mainly defined by their use value and maintenance costs, rather than the market-driven value are struggling through its way of exploring an alternative for urban life. As increasing numbers of do-it-yourself tactics, bottom-up citizen groups, and grassroots commons initiatives are taking root, urban commons have shown us their potential to be a viable vehicle in reshaping urban lives into more desirable and resilient forms (Baibarac and Petrescu 2017). An urgent question for, among others, associated designers, urbanists, architects, and citizens, arising in tandem with the emergence of urban commons, is the sustainment and resilience of urban commons themselves.

C. Zheng (✉)
School of Design, Carnegie Mellon University, Pittsburgh, PA, USA
e-mail: czheng1@andrew.cmu.edu

G. Bruyns, S. Kousoulas (eds.), *Design Commons*, Design Research Foundations, https://doi.org/10.1007/978-3-030-95057-6_4

The inquiry of how to manage common resources has always been a central topic in the discussion of the commons, from Elinor Ostrom's common-pool resource (CPR) management principles to the most recent analysis by Farjam et al. on the patterns of commoning activities. The main focus of these studies is on the preservation of the material resources and extending the duration of the commons. Most of these case studies are long-term "material and inherited commons" such as forests, irrigation systems and land used for pasture, while urban commons are "social commons" that are (re)produced and social (Bauwens 2017). Compared to natural commons, urban commons are situated in the urban context and also weigh the social attributes of the commons differently. Thus, I find that our conceptions on what constitutes the sustainment and resilience of urban commons can't directly adapt from the ones for the natural commons and requires a more complete explication.

Our perceptions of what can be discerned as urban commons' sustainment and resilience is under-evaluated and simplified if we content ourselves with asking whether they last long enough or are able to be expanded and duplicated. Arguably, these criteria easily let urban commons fall into the trajectories of gradually enclosing the shared resources to maintain a static status and/or focusing on propagating their physical formats to maximize networked impacts. My concern is that such conditions may not be as sufficient as is generally assumed. While the temporal duration and scaling-out – the geographic or numeric disseminations – are inseparable and necessary for the discussion of managing urban commons, they alone are not sufficient, as I shall argue in this chapter.

I claim that the criteria which help to qualify the urban commons as sustained and resilient concern more complex and dynamic social impacts. I construct my proposition based on the creation and evolution of urban commons encompassing scaling-up and scaling-deep processes. These processes are partially identified and narrated but are more often under- or unnoticed by initiators, organizers, and participants of urban commons. In addition, I propose that more attention should be paid to the coexistence and correlation of the scaling processes and the role of design in these co-formation processes. The processes of such combinations are where design and designers can intervene and promote, in the pursuit of a more sustained and resilient development of urban commons.

My proposition results in an explication of the sustainment and resilience of urban commons based on three scaling dimensions. I build on the distinction between concepts of scaling out and scaling up, while providing a further explanation of scaling deep of urban commons. The chapter aims at identifying the relevant and necessary scaling conditions by which we may discern an urban commons as a sustained and resilient one. Moreover, inspecting urban commons through the scaling lens underscores the complexities and complementary nature of the processes of urban commons, opening up new avenues for research on how scaling in multiple dimensions relate to one another and illuminating the role of design in enabling and facilitating the scaling processes. The chapter is a conceptual inquiry and thus aims at expanding the theoretical understandings in managing urban commons.

I begin the chapter by briefly introducing the urban commons and the importance of discussing the sustainment and resilience of urban commons, wherein I refer to the notion of sustainability and resilience in the design and ecology spheres. I continue with an exploration around existing literature on governing the commons and expanding urban commons, which are the basis for the further explication of the sustainment of urban commons from three scaling perspectives: scaling out, scaling up and scaling deep. Drawing on Manzini's assertion that scaling out and up are essential for increasing impacts of small collaborative initiatives (Manzini 2015, 178–180), I emphasize developing the depth of urban commons. In the final section of the chapter, I discuss the co-existence and correlations of these three scaling dimensions, and lastly raise the question arising from these scaling processes regarding the role of design.

1 Sustainment and Resilience of Urban Commons

The commons originally was an economic term used in Europe to determine the land use of farming and grazing. Lasting from the thirteenth to mid-nineteenth century, it refers to common lands that were owned by the local manors within a feudal system, while the local community had only a range of use rights (Wall 2014, 8). Over the development of the commons in the past decades, a more inclusive and self-governing commons regime has thrived and is prevalent worldwide. In addition to a community and a shared common resource as two main elements of the traditional commons, contemporary commons opens up to include a third and more complicated aspect – a form of governance. Michel Bauwens (2017) outlined a synthetic definition of the commons:

> The commons has been defined as a shared resource, which is co-owned and/or co-governed by its users and/or stakeholder communities, according to its rules and norms. It's a combination of a 'thing', an activity, commoning as the maintenance and co-production of that resource, and a mode of governance.

Individuals' right to participate in the governance fundamentally shifts the discourse of the commons because the notion of "an owner" or the two poles of public–private ownership is diminished (Benkler 2006, 61).

Subsequently, urban commons come into being as a relatively new typology of the commons, following the rapid urbanization in the twenty-first century. The power asymmetry in urban spaces makes the urban commons the emerging channels for citizens to seek equal spatial rights and have their voices heard. In formats of community gardens, cooperative housing, crowdsourcing community campaigns, and so on, urban commons are avenues where a collective of citizens experiment with solutions towards issues such as food insecurity, urban ecological sustainability, housing affordability, and immigrant rights. As David Harvey (2013, 4) claimed, the right to the city, in the pressure of the market and(or) the state, is a collective rather than individual process of (re)making the city. Dialectically, despite the hopeful rhetoric, his words also imply the external and internal challenges to the sustainment and resilience of urban commons to serve as a continuous canal of citizens' right to the city.

On the one hand, urban commons arise in response to imbalanced developments and various crises. On the other, reactive urban commons often have to begin with small scales and interim spaces, which reinforces their precarious status compared to other types of commons. Enclosure, individualistic mentalities or financial limitations can all lead urban commons to the loss of their participatory and emancipatory qualities, and thus, end their life as urban commons. There have been numerous instances of urban commons facing challenges to sustainment and resilience: self-initiated housing projects amidst the Vienna Settlers' Movement in the 1920s were taken over by the state and devised into public housing; community gardens in Hong Kong struggled to generate revenues to maintain their everyday operation; let alone the common occasions when individual participants experience conflict over not getting equal shares from common resources.

We tend to affirm the contribution of urban commons to the resilience and sustainability of our cities but forget that the premise for them to continue to fulfil that function is their own sustainment and resilience. Sustainment, as defined by Sutton (2004, 4–9), is the process of assessing and improving a system's ability to preserve its function and value under continuous operation, maintenance, and unexpected change. Resilience, derived from socio-ecological theories, is a system's ability to absorb disturbance and continue to function (Walker et al. 2009, 5). The concept of "sustainment" stresses the stable and enduring management of a system in its lifetime. Resilience, on the other hand, widely applied and advanced in different fields, from ecology to social science, psychology, businesses, and planning, highlights the system's capability of recovering to either static or dynamic equilibrium (Folke et al. 2010). As an urban commons is a social, political, and spatial system, drawing on the basic definitions of sustainment and resilience, we can come to a broad definition of the sustainment and resilience of urban commons: it is its ability to respond to disturbances and endure over time. However, since these two concepts are both general and widely applicable terms, a combination of both will only be the foundation for us to further explore what exactly constitutes them.

2 Governing and Expanding Urban Commons

The struggle of maintaining the commons was brought to the public's attention by Garrett Hardin. Hardin's main concern in the essay 'The Tragedy of the Commons' was the discrepancy between the finite common resource and the exponential growth of the population. He used a pasture and herdsmen metaphor to explain that when the external social environment is stable, individuals' greed will ultimately exhaust the common resources. This tragic pattern is observable in many systems such as agriculture, pasture, and air. To end the tragedy, he argues, it is necessary to enclose and coerce, either through privatization or state control (Hardin 1968).

Twenty years later, in the book *Governing the Commons: The Evolution of Institution for Collective Action*, Ostrom summarized eight principles that are shared by successful commons regardless of their settings, which include: clearly defined boundaries; congruence between appropriation, provision rules, and local conditions; participatory decision-making and conflicting solving; and monitoring and penalizing mechanisms. The empirical case studies analyzed in her book were all enduring natural commons: meadows, forests, and water bodies. Ostrom offered a framework for the analysis of self-organizing and self-governing CPRs in the final chapter. Her focus was largely from an economic and policy-making perspective on the evaluation of benefits, costs, and the process of institutional change (Ostrom 1990, 90). Most recently, Farjam et al. (2020) conducted a study of over 3800 activities in eighteen European commons and discovered a U-shape pattern of the long-term management of the commons. It indicates that the commons, mostly, has an initial, high dynamic institutional-definition phase, followed by a stable period, and ends with a burst period in response to challenges.

Hardin pointed out that the problem of managing the commons is a "no technical solution problem" but provided assertive judgments on the response to the problem. Ostrom challenged Hardin's conclusion and shifted the question to how people collectively manage the common resources outside imperatives of both the market and the state. It is worth noticing that Ostrom's case studies were all large-scale natural commons. Although she referred to them as small-scale CPRs, they were still large systems compared to urban commons. Farjam et al. further attempted to advance Ostrom's study by expanding the research samples. All these abovementioned literature and studies have been productive in exploring the governance of the natural commons and are still prevalent in the commons discourse today. However, when it comes to urban commons, should we be satisfied with and directly adopt these theories?

The difficulty of understanding the sustainment and resilience of urban commons lies in its differences with traditional commons. Unlike natural commons which can last for decades and rely on the maintenance of natural resources, urban commons is a dynamic social process. As Linebaugh (2009) stated in the *Magna Carta Manifesto*, we need to consider "the commons" as a verb rather than as a noun:

> To speak of the commons as if it were a natural resource is misleading at best and dangerous at worst – the commons is an activity and, if anything, it expresses relationships in society that are inseparable from relations to nature. It might be better to keep the word as a verb, an activity, rather than as a noun, a substantive. (p. 279)

Many architects follow this line and study the mechanism of how groups of people collectively manage shared urban resources. They discern that an urban commons treats and establishes urban space as a medium through which institutions of commoning take shape (Stavrides 2016, 83). Stavrides uses "commoning" instead of "commons" to emphasize that rather than a static form, urban commons is a changing process. He attempts to explore the meaning and production of spaces of commoning in the context of today's urbanized world from an architectural perspective. He believes that commoning practice and spaces of commoning are keys for alternative thinking towards our cities. De Angelis (2017, 11) also pointed out that commons are not just resources held in common, or commonwealth, but social systems whose elements are commonwealth, a community of commoners, the ongoing interactions, phases of decision making and communal labor process that together are called commoning. Petrescu and Trogal (2017) questioned who the commoners in the commoning process are, what kind of relations the people can generate through commoning, and how the spatial organizations represent these relations. Through their practice of R-Urban, a citizen-run network of resilient urban infrastructures, Petrescu and Trogal try to seek answers to what new kinds of "social" architecture can (re)produce and what kinds of politics, values, and actions are needed from both the commoners and the designers (Petrescu and Trogal 2017). If we return to the basics of urban commons, their three elements are the shared resources, the community, and governance (Bollier 2014; Džokić and Neelen 2015). Fundamentally, the urban commons is gregarious. The social interactions among the commoners define the urban commons itself. Urban commons provide an arena for the interactions to happen and in turn, these interactions constantly shape the commons. As a result, the dynamic reciprocal relation developed by the urban commons and its participants is what we call "commoning."

In the light of the uniqueness of urban commons as a type of dynamic social commons, there are three questions that we should rethink when discussing the sustainment and resilience of urban commons in comparison to the natural commons. First, in the thread of the research on governing the natural commons, the temporal duration always stands in the center of all goals of the commons. However, as urban commons are emergent responses to urban issues, they are established in a fairly short time to respond to the needs so that the life span is not sufficient to define their success. Second, when small-scale formats put urban commons in uncertainty, geographic and numeric disseminations tend to become the immediate strategies for urban commons to achieve wider impacts. Nevertheless, wouldn't these strategies reinforce a capitalist model of reproducing the same product and thus commodifying the commons eventually? Thirdly, since the social dynamics are essential for urban commons to be differentiated from other commons, how can we include incommensurable social aspects into the study of urban commons? What emerges from the brief review above is that while the governance of the natural commons has been heavily researched, the rising small-scale, uncertain, and complex urban commons reveal the limitation of defining the success of the common through the lens of scaling over time and space.

3 Scaling Out, Up and Deep

Ezio Manzini's effort to understand the success of small, local, open, and connected social innovation projects suggests that scaling out and scaling up are two main strategies to achieve impacts. I argue that the sustainment and resilience of urban commons should involve a third scaling aspect, that is *scaling deep*. Therefore, to fully constitute the sustainment and resilience of urban commons should come from scaling in three dimensions – scaling out, scaling up and scaling deep. In reality, the three dimensions of scaling happen simultaneously. In this section, a clarification and an example of each scaling dimension will be given based on existing literature and case studies. The selection of these three examples is to some extent arbitrary and the result of personal interests and encounters. My intention in including these examples is not to provide standards of commoning practices but rather to illustrate the conceptual discussion of the three ways of scaling. They probably demonstrate all three scaling modes but I will focus on elaborating the most outstanding way of scaling in each case.

3.1 Scaling Out

Van den Bosch and Rotmans (2008) characterize scaling out as generating new iterations through replication and multiplication. In the scaling out process, each duplication is a new consideration of the adaptation in different contexts. In *Design, When Everybody Designs*, Manzini (2015) also points out that scaling out is not a simple replication of ideas but the design of a new and locally appropriate solution. He argues that

> when we discuss how to replicate collaborative organizations, we are in reality discussing how these ideas may spread and how different groups of people may recognize, adopt, and localize them (that is, adapt them to different contexts). (p. 180)

To summarize, scaling out of urban commons consists of three possible forms. First, with the original project at the center, urban commons can expand concentrically, showing an increasing size and impact circle. Second, urban commons scales out by geographic and numeric disseminations. Each duplication has a similar function. Third, deepening the second scaling out type, each duplication adapts to its own context and ramifies to new urban commons that respond to specific local needs.

The Knowledge and Innovation Community Garden (KICG) in Shanghai, China, initiated a Seeding campaign to facilitate community trust-building during the Covid-19 pandemic. The campaign encourages people to share plant seeds and home-grown produce in non-contact ways (Fig. 4.1). Conversations among neighbors are instigated along with the material exchanges. The Seeding campaign's purpose is not only to advocate for self-sufficiency in a crisis but also to reconnect neighbors in the time of physical separations (Yang 2020). Since the end of January 2020, the campaign quickly expanded from a small group of people based in KICG to a network of over two hundred people in Shanghai by the end of February. Gaining even wider social attention during the months of March to July, more and more activists and home gardeners across the country participated in the Seeding activities. Initiators in Beijing, Tianjin, Guangzhou, and Chengdu called for the

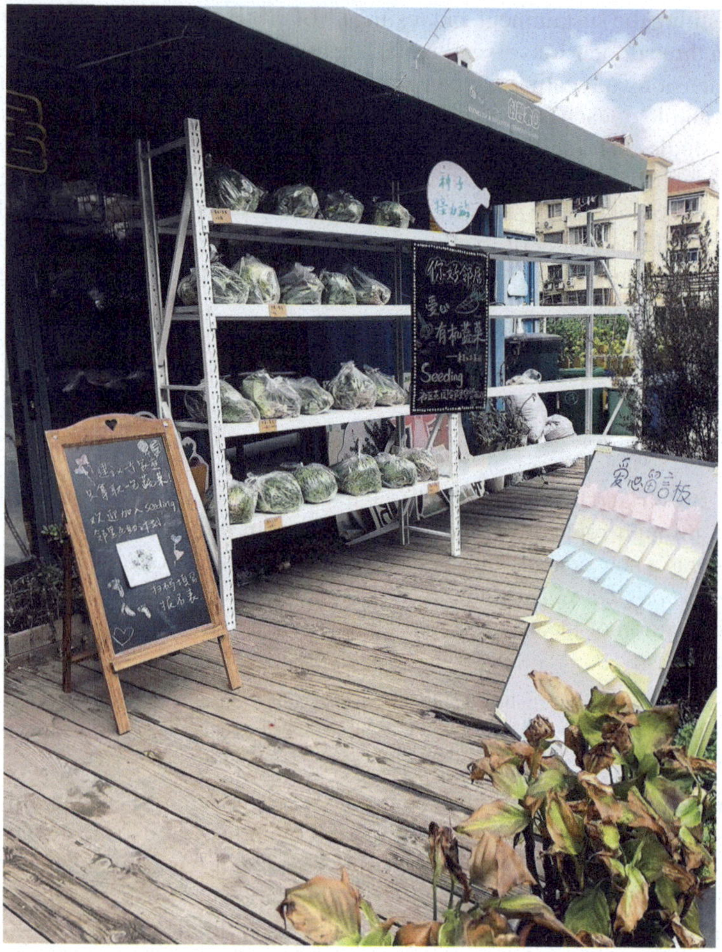

Fig. 4.1 Seeding station in the Knowledge and Innovation Community Garden, Shanghai. (Available via Clover Nature School: https://wemp.app/accounts/ac2a61b5–ebc6–4e30–9c63–53a824297b11. Accessed 10 Aug 2020)

joint collaboration of local organizations to set up satellite projects one after another[1] (Fig. 4.2). The accessible participation requirements of Seeding determined the ease of its spreading. The duplications in other cities copied the same operation model of the initial campaign in Shanghai. Participants can physically share seeds and vegetables with neighbors, share home-gardening experiences online, or attend hands-on workshops hosted by the initiator organizations. These low-threshold activities

[1] The information was acquired from posts published by the Knowledge and Innovation Community Garden's WeChat official account – Clover Nature School (https://wemp.app/accounts/ac2a61b5–ebc6–4e30–9c63–53a824297b11).

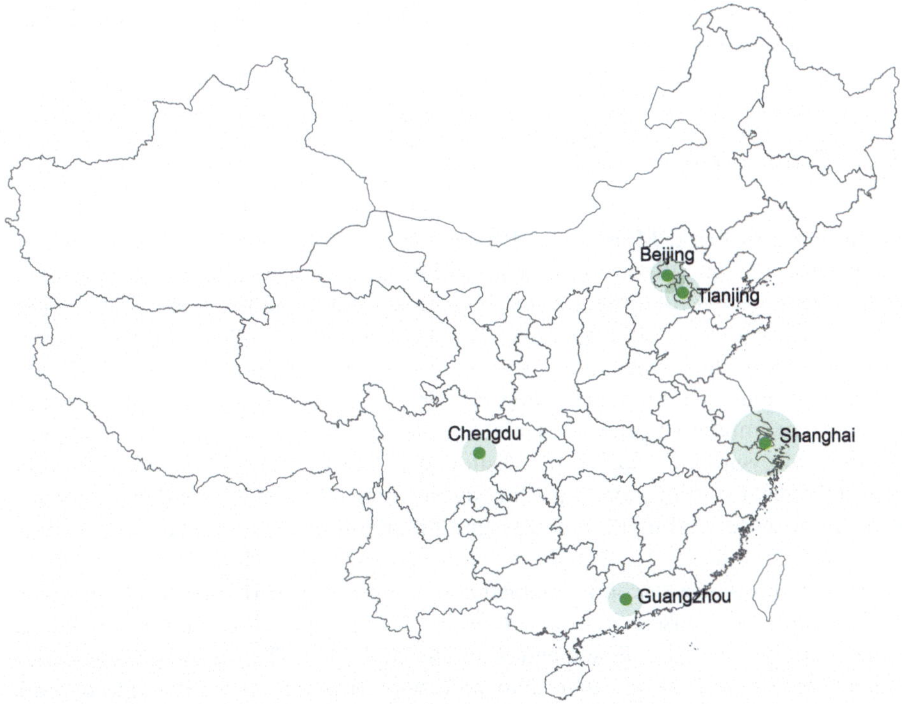

Fig. 4.2 Satellite duplications of the Seeding campaign across China. (Diagram: author, based on the WeChat posts by the Clover Nature School)

enabled the fast scaling out of the Seeding campaign as a typical example of the geographic duplication of urban commons.

Scaling out is usually the first strategy that urban commons projects will come up with when attempting their expansion. However, scaling out shows its limitation if it's used in isolation. In a study by the J.W. McConnell Family Foundation where more than a dozen innovative initiatives participated in examining the scaling strategies in their operation, one participant noticed that the number of duplications is not as important as the impact and the sustainability factor and stated that "if you cannot replicate your programme and ensure it is done with high integrity and fidelity and ensure the programme can be sustainable, then your efforts of scaling (out) are fruitless" (Moore et al. 2015, 78). During an online sharing session, the initiator of the Seeding campaign, Yuelai Liu, mentioned his worry about the fast duplication of the campaign across China. "We have had increasing media coverage when the project keeps expanding. However, it seems we are merely including more people into the group but not able to fully understand what exactly we are doing. We have reached a bottleneck period," says Liu. As stated by Nooteboom (1999, 132), a new practice "becomes more and more differentiated across contexts, causing efficiency losses, lack of standardization, economies of scale and increased complexity because of ad hoc add-ons." A singular usage of scaling out strategy can easily drive the urban commons to drain its energy of variation and turning everything into a simple pursuit of reproduction.

3.2 Scaling Up

In transition theories, Multi-Level Perspective is a three-level framework to consider when managing innovation or transition for systematic change (Moore et al. 2015). Scaling up often refers to the transition from niche to regime and/or landscape in the Multi-Level Perspective. The differences of emphasis on niche development versus the connections between niches and regimes, from which two conceptual understandings of scaling up derive. The first thread interprets scaling up as an accumulative process of niche scale practices and eventually leads to a regime and even landscape shift. Westley et al. (2014) refer to scaling up as the organizational effort to address the broader institutional or systemic roots of a problem. In this case, scaling up requires an ultimate influence on higher-level institutions through policy or legal change. Another thread sees scaling up as a translation of a niche-level practice into mainstream regime- or landscape-level practices. Scaling up in this sense is not the activity itself but a spread of "cultures, practices and structures" (Van den Bosch and Rotmans 2008, 33). The second understanding emphasizes the impact of a marginalized or unnoticed practice's rise into a dominant position. In reality, two threads of conceptualizations of scaling up both manifest themselves, but the second understanding is closer to the abovementioned "scaling out" concept. Here, my definition of scaling up builds on the first type of understanding. Scaling up for urban commons is the process of a niche-scale project to gradually build its impact and intentionally acquire cross-sector collaborations to intervene in higher-level social and political systems. In Manzini's notion (2015, 186), scaling up requires the local organization to connect and coordinate vertically with larger actors to trigger changes on various scales. Scaling up crosscuts the multiple levels of stakeholders from grassroots commoners to policymakers. The idea of hierarchy and horizontality is often conceived as a hegemonic organizational structure (Harvey 2013, 70). However, here I want to acknowledge that from a perspective of efficiency, scaling up is not a top-down hierarchical structure but a structure that allows more agency and helps niche-level actors to pool resources from higher levels. The horizontal scaling is a bottom-up empowering process.

Scaling up is recognized as a main strategy of City in the Making *(Stad in de Maak)* in Rotterdam.[2] After the 2008 financial crisis in Rotterdam, artists Ana Džokić and Marc Neelen from Stealth.unlimited were struck by the imbalance between too many vacant buildings and a lack of affordable housing. They persuaded a real estate developer to spend upfront the money that would otherwise be lost for maintaining the vacant properties for the following eight to ten years and

[2] Information acquired from City in the Making official site: https://www.stadindemaak.nl/en/.

allow them to temporarily use the buildings in exchange for maintaining them. City in the Making is concerned with developing a business model in which each building has a micro-circular economy able to afford its own upkeep. It has created a mix of live-work and common spaces in each of the buildings. The ground floors are made available for production and collective use while the upper floors are made suitable for living and/or working. Economically, each building keeps itself running from both financial and non-monetary contributions by its inhabitants. Besides, each building takes care of its own governance, which means the inhabitants make their own rules. The properties also offer some resources to the neighborhood as a contribution to the diversification of street and neighborhood life.

Starting from their first two properties in 2014, City in the Making now has an expanding network of seven buildings (Fig. 4.3). They remove the properties from the real estate market, own and manage the properties for a period of five to ten years. City in the Making made intentional endeavors to scale up because they realized that to survive the state-controlled and privatized housing market, vertical scaling is the only way out. It has been seeking diverse encounters and dissolving into larger undertakings to add to itself more significance and weight in the housing system. In 2016, City in the Making was awarded the Job Dura Award and attracted

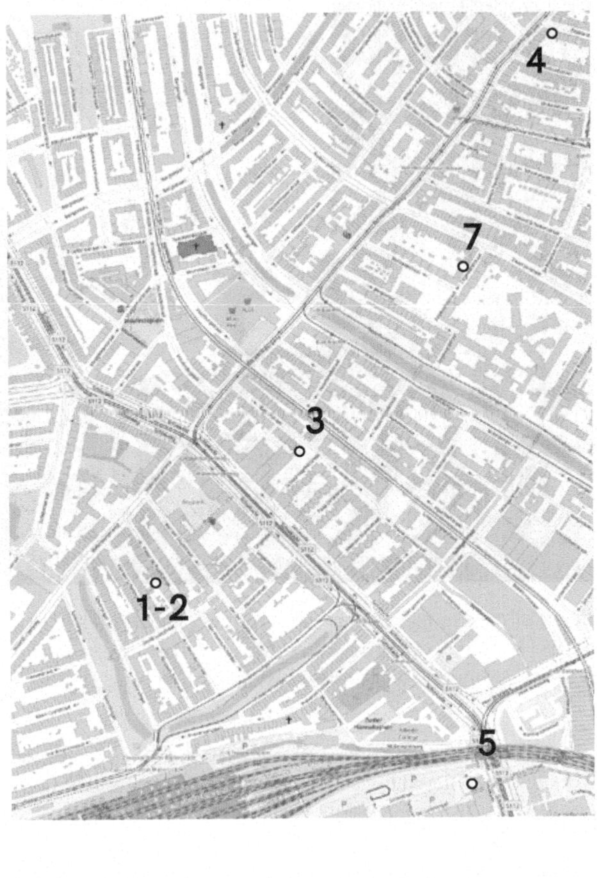

Fig. 4.3 Project locations of City in the Making in Rotterdam. (City in the Making brochure, available via Stad in de Maak: https://www.stadindemaak.nl/wp-content/uploads/2018-SidM-NL-A5-web.pdf. Accessed 10 Aug 2020)

a lot of attention (Fig. 4.4). It seemed the the adventure of building a new affordable housing model was finally acknowledged by the mainstream. However, new concerns arose whether absorption into the mainstream would undermine its commoning qualities, because all the media coverage focused on the novel mixed living and working conditions in the project instead of its recognition of values of unpaid labor, community participation and political engagement (Džokić and Neelen 2018).[3]

Scaling up is largely influenced by the political context in which an urban commons is situated and therefore can be desirable to some and impossible to others. Only when the environment is favorable can urban commons successfully scale up. Since the local and the neighborhood are usually the most comfortable scales for urban commons, larger-scale regional resources are traditionally managed by planning departments, which can be an inherited obstacle for urban commons to scale up in a foreseeable future. The boundary of scaling up to gain enough political agency and being institutionalized is ambiguous. Many urban commons are compromised in face of state appropriation and market interests. As Ostrom elaborated in her later research of large-scale forms of governance, the capacity for participation diminished rapidly when local administrations began to intervene (Harvey

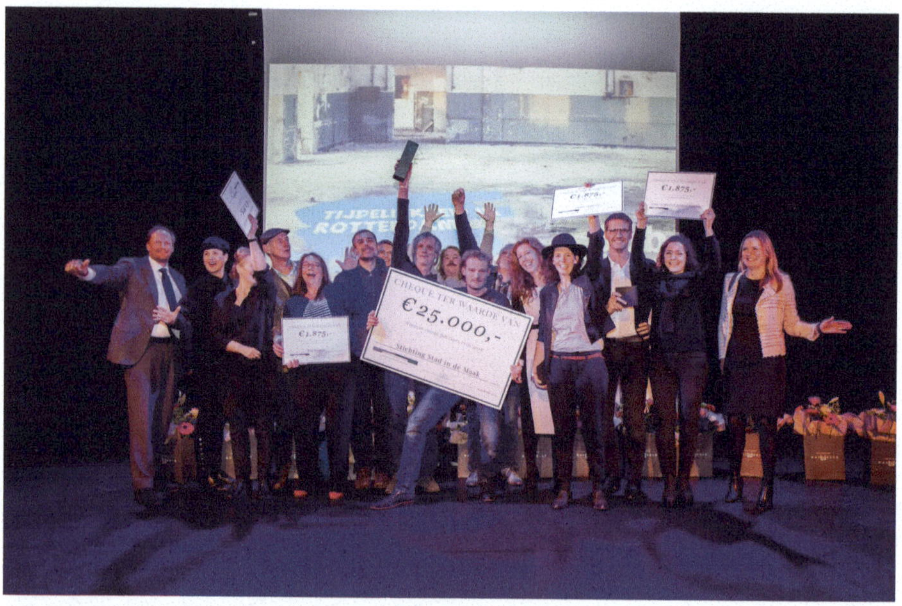

Fig. 4.4 City in the Making won the Job Dura Prize in 2016. (Available via Stad in de Maak: https://www.stadindemaak.nl/stad-in-de-maak-winnaaR-van-de-job-dura-prijs-2016/#more-762. Accessed 10 Aug 2020)

[3] Unpaid labor is the work that does not receive direct monetary remuneration. Here it refers to the work that is achieved by labor payoff in exchange for the right of residence in City in the Making properties, which is not within the traditional market labor scope.

2013, 81). In other words, scaling out and scaling up can reduce the understanding of governing urban commons to the rules of management, endanger the participatory nature of the commons and end up neglecting the "act" of commoning itself (Kousoulas and Sohn 2015, 4).

3.3 Scaling Deep

On this train of thought, scaling up and out are observable and even commensurable and both face certain limitations. Therefore, we return to the nature of urban commons as a social commons and attend to the intangible and incommensurable factors that might influence urban commons' sustainment and resilience. The intangible factor can be traced back to Ebenezer Howard's Garden City idea (Howard 1965) where he already recognized that the value of urban spaces is not from the intrinsic worth of the land and buildings but from the people and activities that take place there. In *Governing the Commons,* Ostrom mentions "norm" as an important part of the institutional change of the commons. Norms, unwritten social behavior rules, can evolve entirely internal to an individual. In the commons, norms can be developed in co-forming actions that are frequently not represented at all (Ostrom 2014, 11). Moore, Riddell and Vocisano (2015, 79) echo this observation and further develop the idea of scaling deep as shifts in "beliefs, ideas, and narratives of dominant social structures" occur. The earlier work by Van den Bosch and Rotmans (2008) also touched upon the requirements for durable changes, including the transformation of people's hearts and minds, values and cultural practices, as well as the quality of relationships.

The scaling deep of urban commons can thus be defined as a learning and regenerative process. Through the activities of urban commons, new norms, worldviews, beliefs and social relationships are generated. The stronger and more interactive social relations generated by urban commons are essential to the resilient governance of urban commons (Stavrides 2016, 39). It entails the mutual re-support and learning from peers and knowledge generation through collaborating on common resources. Meanwhile, scaling deep reinforces the notion of commoning in Peter Linebaugh's definition – the act of (re)producing in common (2009, 132). Additionally, scaling deep is ubiquitous but invisible. It is always ongoing, but organizations do not intentionally notice it, and thus fail to track, articulate, and actualize it. As a result, we rarely see an urban commons articulate its future vision with a goal to scale deep.

Let me take *R-Urban* in France as an example, the set of resident-run facilities that aims at becoming a model of socially and ecologically resilient forms of production and consumption (Bradley 2015, 100). The founders, Doina Petrescu and Constantin Petcou, have identified the necessity to find supportive tools to help existing projects to self-manage and for experiences and knowledge to be shared across locations. They believe that for R-Urban to become sustainable in the long term, generating new iterations through replication and multiplication (scaling out) and operating at levels beyond the local to influence policy and development strategies (scaling up) are essential (Baibarac and Petrescu 2017, 231). The creation of collective hubs and co-design networks in three European cities was the pioneer of the scaling-out process. The partnership with the city of Colombes, the art and architectural practice Public Works, and a local social bank foreran the scaling up process. Although R-Urban in Colombes was eventually evicted due to the conflict with the municipal authority after the local election in 2015 (Petrescu, Doina, Constantin Petcou and Corelia Baibarac 2016; Tribillon 2015), it is worth noticing that the scaling-deep process has already taken root and served as the foundation for the project to be reborn in Gennevilliers, France (Figs. 4.5 and 4.6). In a lecture by Petrescu and Petcou, they noted that participants who had been involved in urban gardening for years have become more politically motivated. In the practices of gardening and collective management, participants learnt from each other and further developed the awareness of social and political engagement (Baibarac and Petrescu 2017; Bradley 2015; Petrescu et al. 2020). In this sense, scaling deep through knowledge-generation and worldview-shifting synchronizes with scaling out and scaling up processes.

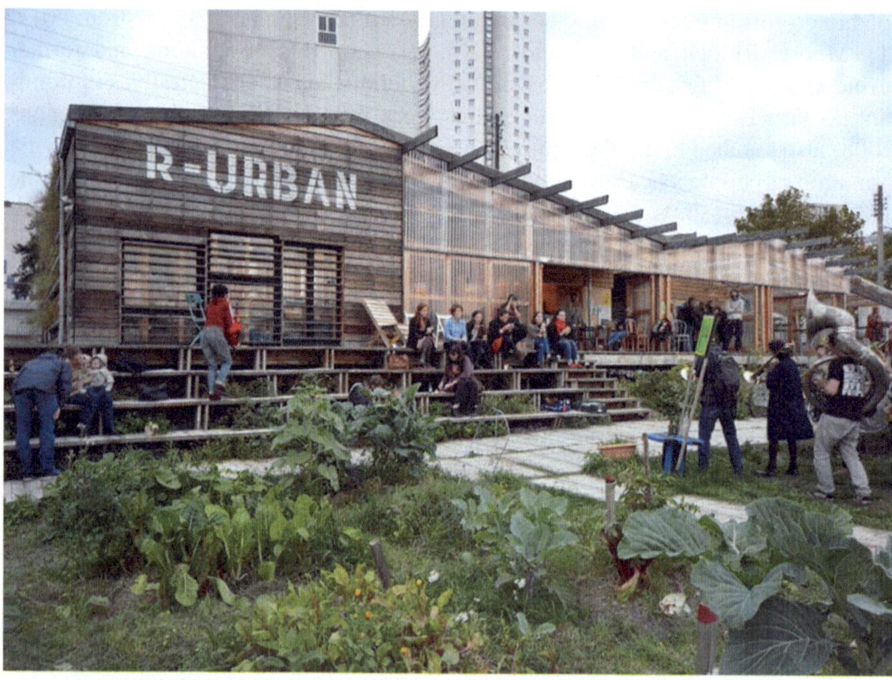

Fig. 4.5 R-Urban Agrocité in Colombes, France in 2013. (Available via Atelier d'Architecture Autogeree: http://www.urbantactics.org/projets/agrocite/. Accessed 10 Aug 2020)

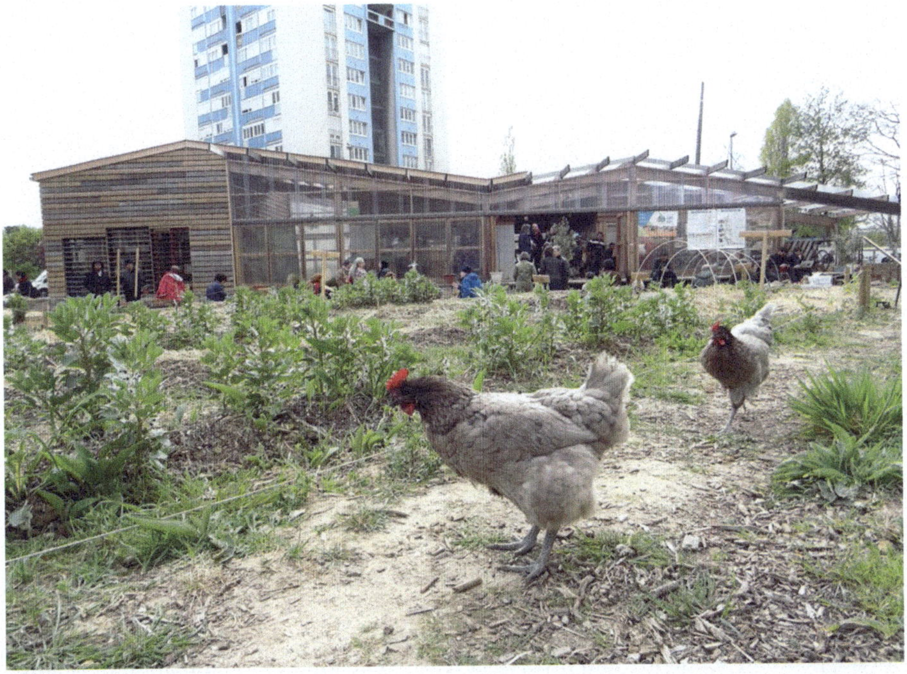

Fig. 4.6 R-Urban Agrocité in Gennevillier, France in 2019. (Available via Atelier d'Architecture Autogeree: http://www.urbantactics.org/projets/agrocite–gennevilliers/?fbclid=IwAR1HZ_BDwwfnAl–itIOQDqsMNlM2gfXmzGP_n4dErhXhrRHYkCUVlueV0. Accessed 10 Aug 2020)

4 Conclusion

I borrow the transition model by Geels (2005) to illustrate the three types of scaling in Fig. 4.7 and also summarize a synthesized explication of scaling out, up and deep for urban commons:

1. Scaling out: the geographical and horizontal expansion and duplication of urban commons from one site to a larger site or multiple satellite sites in different locations;
2. Scaling up: the vertical expansion through an accumulative process of power from a local small-scale urban commons to a higher regime or landscape level of engagement with a wider socio-political agenda.
3. Scaling deep: the internal development of urban commons to influence participants' worldviews and quality of relationships, and ultimately form a reciprocal self-learning and self-growing mechanism in the organization itself.

Moreover, the three examples discussed so far suggest that the three types of scaling do not exist separately. A linear method of either scaling out, up, or deep will not be able to sustain the momentum of the urban commons. The study by the J.W. McConnell Family Foundation in Canada also shows that scaling attempts do

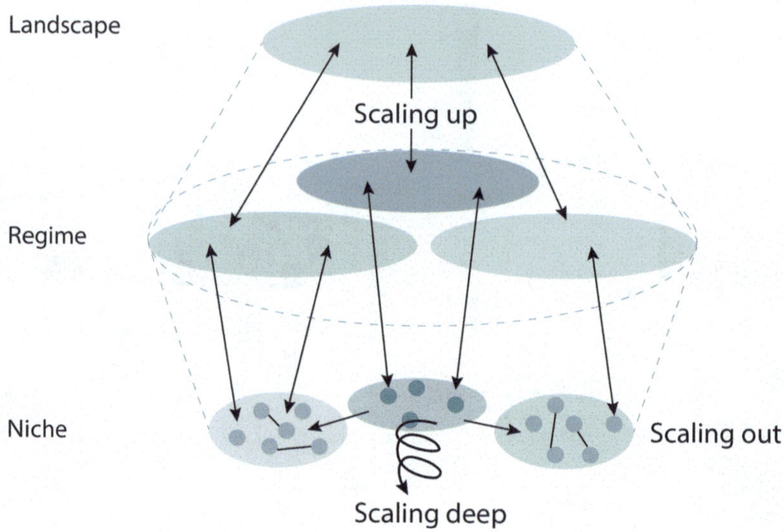

Fig. 4.7 Conceptual model of scaling out, scaling up and scaling deep (Based on Geels 2005; Van den Bosch and Rotmans 2008)

not happen independently and involve at least some kind of combination of scaling out, scaling up, and scaling deep (Moore et al. 2015). In this study, the observation of the co-existence of the scaling processes is as follows: "No single participant immediately jumped to scaling up or scaling deep, and thus, we believe that large systems change involves at least a combination of the three types of scaling." (p. 80) That is, an urban commons cannot simply expect to scale out and up – to sustain itself and remain resilient in the long run – without building up the capacity of value-shifting when scaling deep. The combination of the scaling strategies creates synergies which increases the opportunity of success of each scaling (Manzini 2015, 187). Connecting holistically to involve regional agendas promotes the vertical expansion of the commons and also provides more agency for participants to voice their needs (scaling deep). Although a formula does not exist for how exactly their combination performs the best, we can make the assumption that the three types of scaling co-exist, mutually advance and collectively contribute to the sustainment and resilience of urban commons.

If we stick to the definition of design by Herbert Simon (1988, 67), that is, "to design is to devise courses of action aimed at changing existing situations into preferred ones," then maintaining the resilience of urban commons is a design proposition in which designers' contribution is needed to steer the surfacing of scaling deep in combination with other strategies. In regard to the design of "things," Thorpe and Chapman also expanded the notion of "sustainable design" to a design for the psychological well-being and human experience through formed relationships with artefacts (Thorpe 2010; Chapman 2009; Vissonova 2018). Inferring from this logic, because scaling deep of urban commons happens implicitly and is impossible to intervene in directly, the design provides an entry point where tangible material and spatial forms and narratives can be manipulated to serve as the carriers for intangible, non-material purposes. When visions and ideas on sustaining urban commons are inadequate, designers need to make choices regarding which strategies to apply or not, and become bridges with other social actors to achieve collaborations that feed into the common goals (Manzini 2015, 179–195).

The purpose of this chapter is to examine the three scaling dimensions, what they mean, how they interact, and their implications for making urban commons sustained and resilient. My aim is to offer a set of clarifications that is consistent with the theoretical and empirical evidence of sustained and resilient urban commons, and which, I hope, will be useful to practitioners and researchers in urban commons and relevant fields, offering a more comprehensive angle of the governance of urban commons. The considerations of scaling in this chapter open up more theoretical and practical questions yet to be resolved. What is needed is a further theoretical development of a set of criteria of judgment and a system of weighing to indicate the scaling-deep processes without falling into the cliché of converting everything into monetary values. One could perhaps then imagine design interventions that facilitate the co-existence and co-evolution of scaling processes of urban commons.

References

Baibarac, Corelia, and Doina Petrescu. 2017. Open-source Resilience: A Connected Commons-based Proposition for Urban Transformation. *Procedia Engineering* 198: 227–239. https://doi.org/10.1016/j.proeng.2017.07.157.

Bauwens, Michel. 2017. The History and Evolution of the Commons. *Commons Transition*. https://commonstransition.org/history–evolution–commons/. Accessed 27 Aug 2020.

Benkler, Yochai. 2006. *The Wealth of Networks: How Social Production Transforms Markets and Freedom*. New Haven/London: Yale University Press.

Bollier, David. 2014. *Think Like a Commoner: A Short Introduction to the Life of the Commons*. Gabriola Island: New Society Publishers.

Bradley, Karin. 2015. Open-Source Urbanism: Creating, Multiplying and Managing Urban Commons. *Footprint* 16: 91–108.

Chapman, Jonathan. 2009. Design for (Emotional) Durability. *Design Issues* 25 (4): 29–35. https://doi.org/10.1162/desi.2009.25.4.29.

de Angelis, Massimo. 2017. *Omnia Sunt Communia: On the Commons and the Transformation to Postcapitalism*. London: Zed Books.

Džokić, Ana, and Marc Neelen. 2015. Instituting Commoning. *Footprint* 16: 21–34.

———. 2018. *Upscaling, Training, Commoning: Constructing a Future that is Yet to Be*. Berlin: Jovis Berlin.

Farjam, Mike, Tine De Moor, René van Weeren, M.A.E. Anders Forsman, Amineh Ghorbani Dehkordi, and Giangiacomo Bravo. 2020. Shared Patterns in Long-Term Dynamics of Commons as Institutions for Collective Action. *International Journal of the Commons* 14 (1): 78–90. https://doi.org/10.5334/ijc.959.

Folke, Carl, S.R. Carpenter, Brian Walker, Marten Scheffer, Terry Chapin, and Johan Rockström. 2010. Resilience Thinking: Integrating Resilience, Adaptability and Transformability. *Ecology and Society* 15 (4): 20. http://www.ecologyandsociety.org/vol15/iss4/art20/.

Geels, F.W. 2005. The Dynamics of Transitions in Socio-Technical Systems: A Multi-Level Analysis of the Transition Pathway from Horse-Drawn Carriages to Automobiles (1860–1930). *Technology Analysis and Strategic Management* 17 (4): 445–476.

Hardin, Garrett. 1968. The Tragedy of the Commons. *Science* 162 (3859): 1243–1248. https://doi.org/10.1126/science.162.3859.1243.

Harvey, David. 2013. *Rebel Cities: From the Right to the City to the Urban Revolution*. New York: Verso.

Howard, Ebenezer. 1965. *Garden Cities of To-morrow*. Cambridge, MA/London: MIT Press.

Kousoulas, Stavros, and Heidi Sohn. 2015. Commoning' Practices and the Production of Urban Space: Understanding Urban Asymmetries in the Athenian Case. Originally published in Spanish as 'Prácticas de conexión común y la producción del espacio urbano: comprendiendo asimetrías urbanas en Atenas. In *Repensar la Metrópoli II: Políticas e instrumentos para la gestión urbana*, ed. R. Eibenschutz and B.R. Ramírez Velázquez, vol. I. Mexico: UAM–X.

Linebaugh, Peter. 2009. *The Magna Carta Manifesto: Liberties and Commons for All*. Berkeley/Los Angeles/London: University of California Press.

Manzini, Ezio. 2015. *Design, When Everybody Designs: An Introduction to Design for Social Innovation*. Cambridge, MA/London: The MIT Press.

Moore, M.L., Darcy Riddell, and Dana Vocisano. 2015. Scaling Out, Scaling Up, Scaling Deep: Strategies of Non–Profits in Advancing Systemic Social Innovation. *Journal of Corporate Citizenship* 2015 (58): 67–84. https://doi.org/10.9774/GLEAF.4700.2015.ju.00009.

Nooteboom, Bart. 1999. Innovation, Learning and Industrial Organization. *Cambridge Journal of Economics* 23 (2): 127–150. https://doi.org/10.2307/23599580.

Ostrom, Elinor. 1990. *Governing the Commons: The Evolution of Institutions for Collective Action*. New York: Cambridge University Press.

———. 2014. Do Institutions for Collective Action Evolve? *Journal of Bioeconomics* 16 (1): 3–30. https://doi.org/10.1007/s10818-013-9154-8.

Petrescu, Doina, Constantin Petcou, Maliha Safri, and Katherine Gibson. 2020. Calculating the Value of the Commons: Generating Resilient Urban Futures. *Environmental Policy and Governance* 2020: 1–16. https://doi.org/10.1002/eet.1890.

Petrescu, Doina, Constantin Petcou and Corelia Baibarac. 2016. Co-producing commons-based resilience: lessons from R-Urban. *Building Research & Information*44(7): 717–736. https://doi.org/10.1080/09613218.2016.1214891. Accessed 26 Feb 2020.

Petrescu, Doina, and Kim Trogal, eds. 2017. *The Social (Re)Production of Architecture: Politics, Values and Actions in Contemporary Practice*. New York: Routledge.

Simon, H.A. 1988. The Science of Design: Creating the Artificial. *Design Issues* 4 (1/2): 67. https://doi.org/10.2307/1511391.

Stavrides, Stavros. 2016. Expanding Commoning: In Against and Beyond Capitalism. In *Common Space: The City As Commons*, 31–61. London: Zed Books.

Sutton, Philip. 2004. What is Sustainability? *Eingana* 27 (1): 4–9.

Thorpe, Ann. 2010. Design's Role in Sustainable Consumption. *Design Issues* 26 (2): 3–16. https://doi.org/10.1162/DESI_a_00001.

Tribillon, Justinien. 2015. Why is a Paris suburb scrapping an urban farm to build a car park? *The Guardian*. https://www.theguardian.com/cities/2015/sep/11/paris–un–climate–conference–colombes–R-urban–urban–farm–caR-park. Accessed 15 Jul 2020.

Van den Bosch, Suzanne and Jan Rotmans. 2008. Deepening, Broadening and Scaling Up: A Framework for Steering Transition Experiments. *Knowledge Centre for Sustainable System Innovations and Transitions*. Available via RePub: http://hdl.handle.net/1765/15812.

Vissonova, Karina. 2018. Effects of Design and Sustainable Design of Technical Artefacts. In *Advancements in the Philosophy of Design*, ed. Pieter E. Vermaas and Stéphane Vial, 433–451. https://doi.org/10.1007/978-3-319-73302-9_20.

Walker, B.H., J.M. Nick Abel, and Anderies and Paul Ryan. 2009. Resilience, Adaptability, and Transformability in the Goulburn-Broken Catchment, Australia. *Ecology and Society* 14 (1): 12. http://www.ecologyandsociety.org/vol14/iss1/art12/.

Wall, Derek. 2014. *The Commons in History: Culture, Conflict, and Ecology*. Cambridge, MA/London: MIT Press.
Westley, Frances, D.J. Nino Antadze, Kirsten Robinson Riddell, and Sean Geobey. 2014. Five Configurations for Scaling Up Social Innovation: Case Examples of Nonprofit Organizations from Canada. *The Journal of Applied Behavioral Science* 50 (3): 234–260. https://doi.org/10.1177/0021886314532945.
Yang, Jian. 2020. Urban Home Gardening Takes Root Amid Epidemic. *Shanghai Daily*. https://www.shine.cn/news/metro/2002262834/?fbclid=IwAR1cDp6qNN7-Sv5dPHdnl2P-Kpcb2jtlt3dhGlW8vQvs3386aWQIEgMBT24. Accessed 26 Feb 2020.

Chapter 5
Alignments of Architecture and Commoning in Tai O Village Architecture Critique and Fields of Adversity

Daniel Elkin, Chi-Yuen Leung, and Xiao Lu Wang

Abstract This chapter examines alignment between commoning and architecture's disciplinary limits. The first section discusses Tai O Village, a settlement in Hong Kong where changing development patterns require contested negotiation. In Tai O's context, the chapter asks how commoning relates to architecture's disciplinary foundations. What consequences come of alignment between their conceptual fields? To investigate this, the second section genealogically analyzes the commons' and commoning's expansion in scope from Elinor Ostrom to later literature, defining the terms' conceptual inclusions as prerequisite for alignment. The section thereafter reviews a prevalent structure in architecture critique: each text examined constructs politics and ethics for architecture, privileging certain agencies that direct resources via technology. We contend that this critical writing structure is foundational to architecture, through historical permutations vary. When authors of later texts include architectural processes as critique objects, previously partial alignments to commoning solidify. These distinctions in architecture's assessment priorities have, we argue, significant consequences for architecture's foundations. With this, critical architectural scholarship differentiates from contingent practice through a commoning framework. If we self-assess our architectural research in Tai O to make decisions, then the foundational structure above remains, including permissions for examining architectural processes as objects and privileging agency equity

D. Elkin (✉)
Environment and Interior Design, The Hong Kong Polytechnic University, Hung Hom, Hong Kong
e-mail: daniel.k.elkin@polyu.edu.hk

C.-Y. Leung
Department of Applied Social Sciences, The Hong Kong Polytechnic University, Hung Hom, Hong Kong

X. L. Wang
Centre for Social Innovation, University of Cambridge Judge Business School, Cambridge, UK

G. Bruyns, S. Kousoulas (eds.), *Design Commons*, Design Research Foundations, https://doi.org/10.1007/978-3-030-95057-6_5

as a disciplinary concern. The ethics and politics of our casework prioritize the agencies they include, and the access technology they afford. This changes how we appraise architectural products: their ability to structure engagement and generate knowledge make them relevant to commoning, while architecture's foundational critique structure acquires characterizing differences.

Keywords Commoning · Architecture · Design Foundations · Community Development · Architectural Critique

This chapter concerns the commons, commoning, and how these premises contribute to architectural design's foundations. It examines how ideas within commoning's conceptual field affect disciplinary determinations, rather than distinguishing contingent permutations in practice. Our casework in Tai O, a peri-urban village in Hong Kong, motivates this question. The first section presents Tai O's urban redevelopment situation, in which the complex development transition makes collective decision-making contested, prone to intractable problems within Hong Kong's administrative context. Commoning's conceptual focus on participatory agency models for resource management appears relevant to this transformation, as the contestation of Tai O's future is shaped by which agencies can participate in decision-making.

To ask whether commoning pushes architecture to new foundations, however, is to question the degree to which commoning's conceptual field inflects upon architectural critique as a decision-making process: how can commoning's relevance to work in Tai O, conducted through architecture and assessment of its objects, reflect upon alignment between the two conceptual fields? Is this merely a contingent alignment of differentiated practice, or do alignments between architecture and commoning foundationally affect architecture's performance? This chapter suggests that architecture self-determines ethics and politics through critical writing, privileging certain agencies in their control of technology and resources as critical objects. This, we submit, is a foundational structure. Later permutations, however, admit objects that allow architecture and commoning to align more closely. This requires different assessment permissions for architecture. Assessed objects' timescale must broaden to permit agency equity as an object. Before returning to Tai O to discuss how these changes inflect our self-evaluation, we review commoning and architecture literature to support this argument.

The second section briefly defines the commons and commoning from literature outside architecture. The words differ significantly in use between public administration literature and as motivations for transgressive spatial occupations and social movements. This section describes how the commons in Elinor Ostrom's (1990) *Governing the Commons* relates to commoning in *The Politics of the Commons: From Theory to Struggle* (Erkin et al. 2018a, b, c, d). Both concepts relate to contests over resource control. Begüm Özden Firat's (2018) and Bülent Duru's (2018) chapters summarize movements which increase the scope of the idea from Ostrom's book, commoning natural, cultural, social, and technological resources. This brief genealogy relates commoning to multiple resource categories, seeking participatory models for production independent state or private control. In this section we describe how commoning's conceptual field expanded from Ostrom, to clarify how alignment to architecture may perform.

Alignment between commoning and architecture as a foundational change rather than a historical permutation of practice is possible only if architecture has disciplinary foundations. We contend that critique structure is coherent as a foundation for architecture if we examine how critical architecture writing builds fields of adversity to determine disciplinary limits (Grove 2018). The third section's background suggests that, as a genus containing many species, architecture uses these assessments to construct ethics and politics for resource use, encapsulated through privileged agencies' use of technology. This mode of rhetoric emerges as a foundation for Western architectural texts, though expressions of this structure change. We review Vitruvius' *The Ten Books on Architecture* (1914), and *Towards a New Architecture* by Le Corbusier (1986) to argue that this structure is coherent as a foundation for architecture. Viewed as constructed fields of adversity, later alignment between critical interpretation of architecture and commoning's concepts is conceivable, though dependent on differences discussed in the next section.

If architecture uses critique to self-determine in this way, some texts approach commoning's conceptual territory as permutations of this structure. Robert Venturi's (1977) and Kenneth Frampton's (1987, 1983) critical texts frame fields of adversity that approach commoning, though only partially. More critically, tests such as Awan, Schneider, and Till's (2011) appraise objects over larger ranges of time, arguably to include agency equity as a disciplinary concern. With this change, alignment between commoning and architecture in texts like Sanchez's (2021) becomes coherent under the same foundational structure, but with distinctions in how critical scholarship appraises objects. This, arguably, is the way in which commoning alignment impacts architecture's foundations: contingent permutations of the long-present structure solidify distinctions in which objects critical evaluations will prioritize and how they appraise these objects.

With this stated, we conclude the chapter with self-assessment of our work in Tai O to elaborate how architectural products perform. The import of commoning alignment to architecture in our work is that it distinguishes the purpose architectural works serve. We present discipline-normative artifacts such as drawings and prototypes, to emphasize that we justify these objects according to relationship structuring and knowledge generating purposes outside their immediate purpose, whether to communicate a design possibility or prototype a technical solution. Commoning aligns architecture's critical scholarship to privilege the participatory measure of agencies included in its processes, and access afforded to precisely incremental uses of technology. As such, each critique object in the foreground implies emergent relational objects of greater significance to commoning's aims, as our conclusion elaborates. This change in how architecture views evaluated objects contains, we argue, the most significant implications for the discipline's evolving foundations.

1 Tai O Village: Situation and Commoning's Relevance

We have worked in Tai O Village, a former fishing settlement on Lantau Island in Hong Kong, since 2017. Our collaborative work between architectural design, social sciences, tourism management, and land surveying research centers took impetus from changes in the village's demographic and economic character. Through this transition, Tai O's unique development pattern suggests that action research aligned to commoning's conceptual field, particularly regarding participatory decision-making models, may be relevant. This section details Tai O's situation and commoning's relevance to it, before problematizing the editorial question of this volume regarding architecture's disciplinary foundations.

Tanka fishing people settled Tai O's estuary and riverbed, while Hakka and Cantonese people settled its shores and nearby Yi O Tsuen before the Qing Dynasty period. The village played regionally significant roles since this time, serving as a Qing salt monopoly production center, and a colonial anti-piracy outpost for what became the Great Bay Region. Villagers produced fish and salt commercially up until the 1970s. Tai O and Yi O were a significant agglomeration until British colonization transferred Hong Kong's political center to Hong Kong Island and Kowloon Peninsula (Wong 2000) (Fig. 5.1). Since a ban on commercial trawling in the 1970s, Tai O has transitioned to a tourism economy (Elkin et al. 2021). Tai O's *pang uk* (literally referring to "scaffolding or "wood," and "house") stilt houses, an architectural type that evolved from sampan boats to the houses present today, are a major tourist attraction (Yeung 2007) (Fig. 5.2). The transition to tourism as the village's major revenue source gained statutory emphasis throughout the 2010s, culminating in 2017's *Sustainable Lantau Blueprint* (SLB). This document, from the government's Civil Engineering and Development Department (CEDD) designates Tai O as a tourism center, provisioning investment for this new role (Sustainable Lantau Office 2017).

With this re-designation, Tai O approaches a redevelopment transition with multiple, often mutually exclusive prospective futures. The decision-making processes in determining Tai O's future are, as Kevin Grove described in *Resilience* (2018), prone to be complex and consequently for there to be limitations on knowledge and ethics. Wicked problems, which Grove defines as intractable rather than ambiguous, are palpable in making decisions for Tai O. First, flooding, fires, and statutory limitations threaten stilt houses and their residents. Under the Squatter Control Policy for Surveyed Squatter Structures (SCPS), the Lands Department (LD) *tolerates* stilt houses, conveying no land tenure. This means they are subject to clearance, and residents may not renovate, expand, or repurpose them without expressed LD permission (Squatter Control Office 2020). This has caused alienation and work stoppages, even for changes that mitigate the threat of fire and flooding (Elkin et al. 2021). Government statements at the regional level stop short of a conservation strategy, despite the stilt houses' cultural value (Wong 2019). Many also remain unsewered, despite service extensions. Though infrastructure investments continue, only one private bus company serves the village with inadequate service to transit

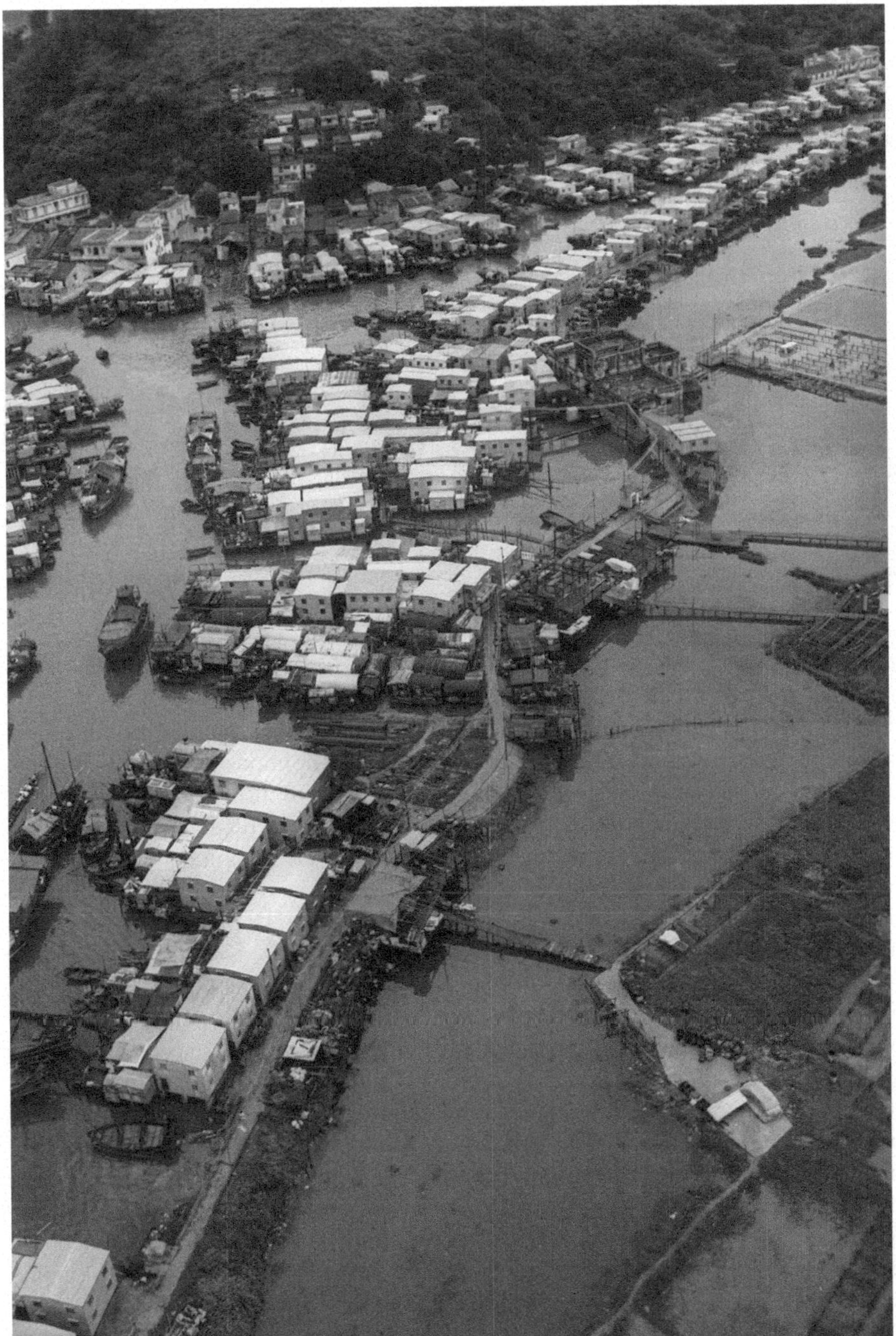

Fig. 5.1 Aerial image of Tai O Village taken in 1976, showing bigger population and greater numbers of Pang *Uk* houses. (Photo: *Census Image of Tai* O, Hong Kong Information Services Department Photo Library)

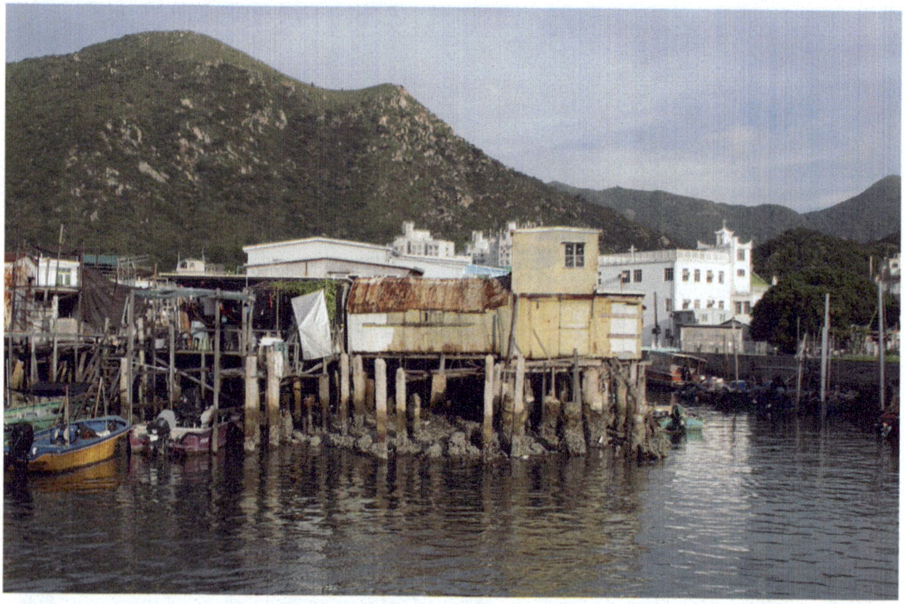

Fig. 5.2 Pang *Uk* houses in 2020. One of the oldest round-topped houses is visible in the foreground, and newer low-slope roof houses are visible in the background. (Photo: Daniel Keith Elkin)

centers forty-five minutes away. Speaking to the Tai O Village Rural Committee and stakeholders in spring of 2020, we heard ambivalent opinions on additional transport connections. Additional tourist traffic, they estimate, brings litter, congestion, and changes to their way of life as well as revenue. Tai O villagers we interviewed are acutely aware that government attention and research amount to complex mixes of opportunity and threats (Elkin et al. 2020). Covid-19 has made some threats more noticeable when visitors arriving in the village refuse to take precautions, exposing Tai O's elderly and underserved population.

Based on this situation, we argue that ideas in the conceptual field of commoning are relevant to work in Tai O. From Firat, we note that commoning movements respond to "intensifying commodification process[es] of urban space," (2018), a species of which is arguably taking place in Tai O. Hong Kong's Tourism Board and the CEDD foster Tai O's international commodity consumption through development strategy (Hong Kong Tourism Board 2020). More critically to commoning methods, Tai O's future transformation will likely depend on the "third way between the market and the state" that Aykut Çoban derives from Ostrom (Çoban 2018, 32). Rural Committee members note that residents hope to repurpose and expand stilt houses for commercial use rather than residence alone, even as many also emphasize their desire to maintain Tai O's rural character and quiet way of life (Elkin et al. 2020). In the presence of such wicked problems, it is possible that decisions made to respond to market imperatives for Tai O's space could destabilize the village's topological "form, function, and identity" (Grove 2018, 77). At the same time, the government explicitly delegates decision making on stilt houses to "non-governmental organisations for revitalisation and management" (Wong 2019). As such, commoning-oriented approaches for action research, seeking participatory, alternative models to make decisions about Tai O's future, may be relevant. These ideas influence collaborative governance and architectural work in Tai O, which we present at the end of this chapter.

Returning to the editorial question of this volume, however, means discussing architecture's disciplinary foundations, and our commoning-aligned work in Tai O as an object for assessment. We must ask how this work relates to architecture's limits as a discipline, and what relevance it has to architecture's conceptual self-determination. We can ask if commoning contributes ideas useful to architecture as evidenced in our work as practice, but interpretations of our work as such are historically contingent. To ask whether commoning contributes to architecture's *foundations* requires a different question: does alignment with commoning have consequences for how architecture self-determines, or is alignment to commoning only contingent on practice?

Before we return to our casework in Tai O to elaborate how foundational changes influence our work, we explore this question through literature review below. We argue that architecture self-determines through critical writing, in which texts construct fields of adversity containing critique objects. We contend that this structure, including certain categories, is *foundational*. Critical assessments construct ethics and politics for architecture that privilege certain agencies in their direction of resources using technology. More significantly, some critical writing aligns to commoning within this structure in a distinct way. They appraise objects over larger ranges of time, often to include agency conflicts as critique objects. This distinction in appraisal criteria, we argue, distinguishes architecture's foundational structure when aligned to commoning. At the end of this chapter, we elaborate the consequences of this distinction within our self-evaluation process for directing research.

2 The Commons and Commoning: Resources Concerned and Participation Models

This section defines the commons and commoning, describing their conceptual field regarding resource distribution and participatory administration models, such that alignments with architecture are perceptible. Elinor Ostrom's book *Governing the Commons* (2000) provides the concept's initial scope, while chapters from *The Politics of the Commons* (ed. Erdoğan, Nuran and Özdeş) progress from Ostrom to broader governance schema (Firat 2018; Duru 2018). As Aykut Çoban (2018, 29) states, "the term commons has recently been transformed into something similar to that of the parable of the blind man and the elephant… it is impossible to conduct a scientific discussion with a 'concept' to which everyone can attach a different meaning." Çoban, Duru and Firat discuss this definition problem adequately. We suggest commoning is an essentially contested concept, its precise definition less actionable than the genealogy of its use and themes (Grove 2018, 31). *The Politics of the Commons* and other texts transfer Ostrom's ideas to broader resource fields and models for control. This allows more challenges to the conceptual scope of commoning, while retaining Ostrom's focus on participatory, responsive resource administration.

Ostrom markedly confines the scope of her book. She argues that Garret Hardin's (1968) writing on the commons draws unsubstantiated conclusions about common-pool resource scenarios. Hardin claims that shared natural resource system failure is inevitable without privatization or government coercion. Ostrom argues that Hardin's models for behavior are not predictive and should not direct policy, and that components of Hardin's dilemma model lack verisimilitude. Ostrom describes cases where rational actors self-regulate shared resources without privatization or government intervention, further sketching characteristics of long-enduring management organizations. Of interest for this chapter is Ostrom's focus on resource administration and participatory legislative, monitoring and executive models. Contrary to Hardin, Ostrom notes that appropriator-governed, iterative resource management models create durable solutions outside the privatization or coercion dichotomy.

From Ostrom, discussion of the commons grew to larger scopes in commoning, leading to Çoban's parable and volumes of scholarship (Çoban 2018, 29; Bauwens 2019; Sanchez 2021). As commons, Bülent Duru enumerates, "firstly, *natural resources* such as air, water, soil, forests, and seeds; secondly *urban areas* such as roads, streets, parks, squares, and coasts; and lastly, *social and cultural values* such as science, internet, arts, languages, and traditions," (2018, 11, emphasis original), in recounting the concept's development. From Ostrom to Duru and others, the commons conceptual field evidently expands. Ostrom and Hardin deal with shared resource scenarios exclusively focused on shared *natural* resources. Contests over urban territory and other resources emerge later.

The Arab Spring contextualizes this expansion in *The Politics of the Commons*, including discussion of commoning in the revolutionary moment and its failure to "create sustainable organizational models suitable for the object of the movements and the object of their demands" (Dardot and Laval 2018, quoted in Firat 2018). However, commoning emerges in governance and technological literature outside this context. We suggest that failures within these revolutionary moments are contingent on expressions rather than endemic to the idea, given that scholars apply the concept in fields like technology and architecture, where conflicts do not include the revolutionary context (Sanchez 2021; Ramos 2016). Across these applications, we summarize commoning as: concepts and decision-making biases concerned with resource governance, whether natural, territorial, socio-cultural, and/or technological, motivated toward participatory agency in governance, especially exterior to privatization, commercial, and state-controlled models. As shown, commoning's conceptual field extends beyond Ostrom and social movements. This established, the following sections discuss how architecture asserts its conceptual territory, where and how alignments to commoning occur, and how they affect the discipline's foundations.

3 Architectural Critique as Foundational Structure

This section reviews two Western architectural texts that construct fields of adversity around similar categories of objects to determine the discipline's limits. We contend that this critique structure is logically foundational to architecture, though any single expression of it is not. Critical writing on architecture stages fields of adversity to self-determine disciplinary ethics and politics, in which privileged agencies direct resources using technology. In some alignments, objects in architecture's critical field of adversity consequentially relate to commoning's conceptual field. Despite changing appraisals of critique objects, we argue the critical structure is a foundational technique to discern the discipline's limits. We borrow this premise from Kevin Grove's discussion of Stephen Collier and Michel Foucault. This separates objects of critique from fields of adversity which

> [gesture] to a wider field of *strategic engagement* with the object of critique: what vision of proper forms of ordering and organizing social and environmental relations do critics promote, and *what stands in the way of realizing this vision?* … a field of adversity is that element that allows critique to identify limits in established ways of doing and thinking that inhibit other ways of doing things and show how these limits can be transgressed. (Grove 2018, 90–91, original emphasis)

The texts we examine construct fields of adversity around evaluated objects. This process suggests alternative visions for architecture, and the society surrounding it. As they contain strategic fields of adversity, critical architectural writing is not hermetic. Each depends upon actor and behavior categories *beyond* architectural products. Interactions between objects within these categories define the species of each text, within a genus of assessment structure. Each text shares the task to define ethics and politics for architecture, including aesthetic and moral rightness. Each entangles resources, technology, and agencies differently depending upon this assemblage. For this chapter, we analyze Vitruvius's *Ten Books on Architecture* (Pollio 1914) and Le Corbusier's *Towards a New Architecture* (1986), texts chosen for their age and prevalence in Western critique. We discuss each text as a permutation to this foundational structure, with alignments to commoning emerging with specific, consequential distinctions.

We examine Dr. Morris Hicky Morgan's 1914 edition of *The Ten Books on Architecture*, in which Vitruvius constructs his field of adversity for architecture under Caesar Augustus's patronage. He articulates signals for rightness and propriety in architecture using direct, detailed terms:

> Symmetry is a proper agreement between the members of the work itself, and relation between the different parts and the whole general scheme, in accordance with a certain part selected as standard. Thus, in the human body there is a kind of symmetrical harmony

> between forearm, foot, palm, finger, and other small parts; and so it is with perfect buildings. (Pollio 1914, 28)

Though anatomical politics are immediately present, Vitruvius positions other objects more neutrally. He most visibly defines architecture through ethics and politics when invoking authority, explicitly and in process of empire. The otherwise technical book on building materials makes a characteristic turn to defend the use of brick for architecture: Vitruvius describes the history of King Mausolus and Queen Artemesia. The Rhodians attack after the King's death, objecting to the Queen's rule. Artemesia subdues the Rhodians, whereupon Vitruvius uses this victory to defend the use of brick, because Mausolus built the Mausoleum at Halicarnassus of brick (Pollio 1914, 63–67).

Vitruvius endorses brick technology for architecture through authority, rather than consequential verification of its aesthetic or constructed performance. His parable justifies brick to architecture because of its validity to a victorious monarchy. Similarly, statements on Rome's continued population growth and the conquest-enabled discovery of larch timber intertwine technological norms for Vitruvian architecture with politics and ethics (Pollio 1914, 67, 76). Vitruvius defends technology and resource use depending on the agencies that direct them and their privilege. Architecture, like other uses of technology, is valid for Vitruvius when directed against Rome's adversaries. Anachronistic biases in this permutation of architecture become conspicuous over time, but from the contemporary viewpoint this text coherently displays the structure we present. Vitruvius, along with later authors, defines a field of adversity requiring a broad range of actors and value constructions to define architecture's ethics and politics.

Frederick Etchells's translation of *Towards a New Architecture* also displays this structure. Ethics and politics, signaling moral and interpersonal rightness for the society around architecture, as well as visual and experiential rightness for architecture itself, appear throughout the book. Le Corbusier does evaluate objects in his field of adversity according to compositional and formal criteria alone on occasion, as in, 'Primary forms are beautiful forms because they can be clearly appreciated,' (Le Corbusier 1986, 2). However, as in the passage below, ethical domains for appraising objects emerge promptly:

> Our eyes are made to see forms in light; light and shade reveal these forms; cubes, cones, spheres, cylinders or pyramids are the great primary forms which light reveals to advantage; the image of these is distinct and tangible within us and without ambiguity. It is for that reason that these are beautiful forms, the most beautiful forms. Everybody is agreed as to that, the child, the savage [sic] and the metaphysician. (Le Corbusier 1986, 29).

Statements like these entangle normative appraisal for compositional rightness with other assemblages. Along with long, accusatory passages in the second person, explicit references to morality within his evaluations suggest that, for Le Corbusier, the discipline should privilege certain agencies' uses of technology and resources, as well as compositional and experiential signals (Le Corbusier 1986, 114–16, 6, 13, 180). Later scholars note that his writing privileges resource direction according to the ethics of industrial capitalism (Banham 2002) (Padovan 2002; Frampton

2008; Cattaneo et al. 2016). Even without the text's aureate tone, he entangles architecture with workers' behavior as an evaluated object, creating a morality for labor in his field of adversity,

> The spirit of the worker's booth no longer exists, but certainly there does exist a more collective spirit. If the workman is intelligent, he will understand the final end of his labour, and this will fill him with a legitimate pride. When the Auto announces that such and such a car has reached 180 miles an hour, the workmen will gather together and tell one another: 'Our car did that!' There we have a moral factor which is of importance. (Le Corbusier 1986, 275)

Le Corbusier's writing, similar to Vitruvius's, discernibly interrelates categories of critique objects as a historically contingent field of adversity to articulate architecture's limits. Contemporary interpretation of this expression is less the issue, though the exclusively male architect and remaining colonial presences in his writing make it problematic (Le Corbusier 1986, 279–88). This structure, entangling politics and ethics with architecture depending on object categories, continues into later expressions of the discipline, such that we argue that this structure's continuing presence makes it a disciplinary foundation. In the following section we discuss later architectural texts in the same way, including distinctions in critical texts' appraisal criteria that we contend allow alignment with commoning.

4 Expanding Rhetorical Bounds, Partial Alignments

If we view the structure of criticism as foundational to architecture, questions follow whether permutations of critique approach commoning, and whether alignment is contingent on practice or reflects significant consequences for this foundation. This section suggests that *Complexity and Contradiction in Architecture* by Robert Venturi (1977) shares this structure but privileges a broader range of agencies as a permutation within postmodernism. In similar literature, we find that Kenneth Frampton's (1983, 2002) Critical Regionalist argument contains ideas partially aligned to commoning. Frampton constructs a field of adversity sympathetic to commoning's conceptual field within a critical architectural text. Importantly, however, Frampton appraises objects within this field differently than other critics, particularly those who align their work directly to commoning. Texts like *Spatial Agency* (Awan et al. 2011) and *Architecture for the Commons* (Sanchez 2021) appraise processes surrounding completed buildings as critique objects, using this mode to problematize agency equity within the discipline. This, we argue, is a significant difference in assessment structure as architecture approaches commoning, as it distinguishes how architecture appraises objects within its fields of adversity. This section discusses manifestations of this difference within this evaluation structure, with subtle distinctions emerging in how it performs.

Along with tonal differences to the texts above, Venturi states his ambition: to "talk about architecture rather than around it" (Venturi 1977, 14), noting the failure of Sir John Summerson's criticism of architecture to construct a hermetic assessment of architecture. He self-consciously restricts the scope of his commentary, working to validate critique objects without constructing ethics and politics beyond the discipline. Still, he follows a similar structure, constructing complexity and contradiction as epistemological, as well as architectural, concerns via August Heckscher (Venturi 1977, 16). However, Venturi privileges broad and pragmatic agencies as compared to the texts above. Along with rhetorically emphasizing his appraisal's subjectivity by using the first person, Venturi allows wider spectra of agencies to validly make decisions within architecture. He includes the Philadelphia Savings Fund Society building within architecture *according to* the urbanistic and programmatic complexity it faces, rather than in spite of it (Venturi 1977, 30, 33). He also admits strategically banal exigencies from real estate development:

> Flat plate construction consists of concrete slabs of constant depth and varied reinforcement, with irregularly placed columns without beams or caps... This permits, in apartment houses especially, a constant ceiling profile for the spaces below in order to accommodate partitions. Flat plates are structurally impure: their section is not minimum ... Form follows function here in a contradictory way; substance follows structural function; profile follows spatial function. (Venturi 1977, 36)

Within compositional limits including "incoherence or arbitrariness," (Venturi 1977, 16), landlords, tenants, contractors, and their values contribute validly within Venturi's field of adversity for architecture. However, as elaborated in similarities to Frampton's arguments below, a remaining characteristic of these agencies' performance as critique objects is that Venturi appraises their contributions to completed buildings with the importance of their participatory agency in creating that architecture not considered. The agent party in Venturi's commentary is the architect. She or he may accommodate other parties' concerns to manifest more complex practice, but Venturi does not inherently privilege other agencies' contributions.

The distinction in structure held by critical writing fully aligned to commoning is more apparent compared against a partial alignment. We argue that Kenneth Frampton's (1983) commentary on critical regionalism is a case where architecture aligns yet closer to commoning's conceptual field, but without a distinction prevalent in full alignments. As before, Frampton structures his argument similarly, building a field of adversity to place critique objects within architecture. First, he quotes Paul Ricoeur to stage his appraisal ethically and politically within a conflict between "world civilization" and regionally distinct culture (Foster and Frampton 1987, 16). Afterward, Frampton privileges objects which synthetically mediate between universality and regional distinctiveness. Here he appraises Jørn Utzon's Bagsvaerd Church, speaking of

> a revealed conjunction between, on the one hand, the rationality of normative technique and, on the other, the arationality of idiosyncratic form. Inasmuch as this building is organized around a regular grid and is comprised of repetitive, in-fill modules... we may justly regard it as the outcome of universal civilization... However, the universality of this productive method... is abruptly mediated when one passes from the optimal modular skin of the exterior to the far less optimal reinforced concrete shell vault spanning the nave. This last is obviously a relatively uneconomic mode of construction. (Foster and Frampton 1987, 22).

Frampton stages this object in a field of adversity that contains contests increasingly aligned to the commoning concerns discussed above. Along with claiming similar literature, Frampton criticizes positivistic urban planning and the rapacity that characterizes capitalist development (Foster and Frampton 1987, 17). However, he makes no causal connection between universal civilization, expressions of which architecture must mediate, and globalization. This is in spite of privatization, wage growth repression, and internationalization of capital that took place between the 1970s and 90s, within Frampton's scope of experience (Akçay 2018, 54–56). Frampton's assessment stands between a hermetic disciplinary definition arguably addressing architecture consequent from these economic and political contests, but without specifically including those events as objects.

Rather than to criticize Frampton, we suggest this shows alignment between architecture and commoning is historically tangible but requires specific, intentional distinctions in how critical architectural scholarship appraises its objects, a foundational difference. Notably Frampton constructs resistance through aesthetic signals in interpreted architectural works, rather than interpreted practice. Frampton's currency for resistance is experiential, as in the Bagsvaerd Church, or phenomenological, as in his assessment of Aalto's Säynatsalo Town Hall:

> The main route leading to the second-floor council chamber is ultimately orchestrated in terms which are as much tactile as they are visual. Not only is the principal access stair lined in raked brickwork, but the treads and risers are also finished in brick. The kinetic impetus of the body in climbing the stair is thus checked by the friction of the steps, which are 'read' soon after, in contrast to the timber floor of the council chamber itself. (Foster and Frampton 1987, 28)

He endorses this experience as resistance via Heidegger's writing on phenomenology, and noticeably the *user experience* of the object is what connects architecture to resistance. His argument does not explain the consequences of resistance through practice's qualities, examining processes that created the work and their placement within the conflicts sketched at the start of this chapter. If we argue that Frampton's critique partially aligns to commoning, this characteristic of its structure foundationally differs from texts that align more closely.

In texts like *Spatial Agency* (Awan et al. 2018) and *Architecture for the Commons* (Sanchez 2021), further alignment between architecture and commoning conceivably depends on validating architectural products throughout their genesis, including *processes* of this genesis as objects in a field of adversity. As Sanchez's commentary explicitly aligns itself with commoning, we first suggest that Awan, Schneider and Till's text is similarly aligned. As evidence, we refer to their justification for titling the book spatial, rather than architectural, agency:

> The third limit of the word 'architectural' is that it suggests that only architects are involved in the creative production of the built environment… New ways of working and behaving are demanded if we are to avoid being impotent passengers on the rollercoaster of boom and bust cycles. Clues as to these other ways are given by the examples in the book, most of which prioritise values outside the normal terms of reference of the economic market, namely those of social, environmental and ethical justice. (Awan et al. 2011, 28)

Their introduction also mentions several conflicts relevant to the commoning movements described above. More critically, *Spatial Agency* relies on the difference in structure we have suggested, in the objects they choose to critique and in explicit claims. The book refers to "the dynamic context of social space, rather than within the static context of architecture as building," (Awan et al. 2011, 28) suggesting that, to address commoning-related conflicts, processes and relationships *around* architectural works will more often enter this field of adversity. This is borne out in the rest of the book, which reviews network visualization practices, institutions like Community Design Centers, and open-source technical publications to construct the response of agency-oriented practices to commoning-related conflicts (Awan et al. 2011, 76, 126–27, 216). Their critical writing scarcely justifies such objects based on signals in users' experience, reflecting *Spatial Agency*'s distinct timescale for assessment. Consistently, the ethics and politics for architecture in this criticism validate resource and technology use *during* production, as distinct from the texts above.

As such, it is not surprising that *Architecture for the Commons* includes processes and tools as critique objects more than completed works of architecture. Along with explicitly claiming alignment to commoning in the title, Jose Sanchez discusses the ethics and politics of "the post-2008 perspective," (Sanchez 2021, 7), which situates his commentary toward conflicts related in *The Politics of the Commons*. To orient architecture towards these conflicts, he proves objects in his field of adversity almost exclusively according to their methods and practice as they come into being, even when they contribute to completed works. For example, he analyzes Patrik Schumacher's work according to practical economic inequities in the parametric practice model, rather than in completed buildings (Sanchez 2021, 3). His writing is coherent in this aspect. Sanchez rarely connects exegesis of completed architecture to his interpretation, instead describing how open-topology building kits, parametric design software, and spatial scenario planning games allow broad and agent participation in architecture, framed as a response to conflicts also included in commoning literature (Sanchez 2021, 63–93, 99–107). Instead, the emergent roles architectural objects play in relationships between potential agent parties defends them to commoning to the extent that they foster participation in architecture. When critical assessment performs this way, architecture appraises physical objects, especially discipline-normative products like drawings, models and mock-ups, differently. This, we argue, is a significant difference in interpretive structure as a foundation for architecture, and a significant consequence for how the discipline aligns to commoning. Given this argument, we return to our work in Tai O Village in the last section. Reviewing our work, we discuss the consequences of this change in self-appraisal as a process for research decision-making, and its impacts on the interplay between architectural and other research.

5 Architecture in Tai O Village: Critical Appraisal of Architectural Products

We conceptualize our work in Tai O as fostering a collaborative governance regime, following Emerson, Nabotchi, and Balogh's article integrating the concept. The researcher's role in this framework is to build capacity and relationships for stakeholders and administrators to solve problems (Emerson et al. 2011; Elkin et al. 2020). As discussed, Tai O's unique situation suggests this commoning-related framework, between state and private enterprise, as a response. However, questions remain of what architecture or design bring to this approach, and its reciprocal consequences for architecture. We suggest that our architectural research progresses through self-evaluation, appraising products according to methods distinguished in commoning-aligned writing above. As we decide to accept costs in directing resources or technology, we appraise our architectural products according to how they interface between agencies in Tai O. An architectural product serves to structure engagement, create knowledge, and articulate politics, even while it serves other technical, spatial, or compositional objectives as a discrete object. With these dual purposes, our architectural products are often fugitive, contributing to consensus and dissensus interactions along spatial activist schema (Fuad-Luke 2017). To elaborate this assessment, we present three architectural products: a footing mold produced in collaboration with contractors in Tai O, a pedestrian traffic-monitoring initiative, and a predesign package for strategic development initiatives. We review these to elaborate the distinction commoning alignment makes to how we evaluate our architecture and discuss implications this makes for the discipline's foundations.

We used digital and computational design techniques to prototype a template which Mr. Chou, a fiberglass boat repair contractor in Tai O, used to build a two-sided mold for stilt house footings (Fig. 5.3). At a technical level, this design marginally improves upon existing stilt house building technology: contractors cast *pang uk* house footings in fifty-gallon drums, or in two-sided cylindrical molds. We designed the mold to distribute load over a wider area and to release more easily from cured footings. While this upstreaming generated knowledge within an architectural research subset, we additionally appraise this product's ability to engage us with Tai O's construction technology network, to use John F.C. Turner and Robert Fichter's (1972) term, or communities of practice, after Lave and Wenger (2018). This prototype prompted intimate knowledge exchange between fabricators and researchers. We know more about Chou's norms of practice, mold-making skills, and priorities. This includes design projects he led himself, including a portable fiberglass toilet design. Along with our unrealized truss prototype (Elkin et al. 2021), these products gather information and build relationships, with orientation toward technical improvement partly an initiating conceit. While we could conceivably address issues with *pang uk* construction independently, we appraise these products according to how they demonstrate, emphasize, and connect situated agencies in the research scenario, similar to Awan, Schneider and Till (2011).

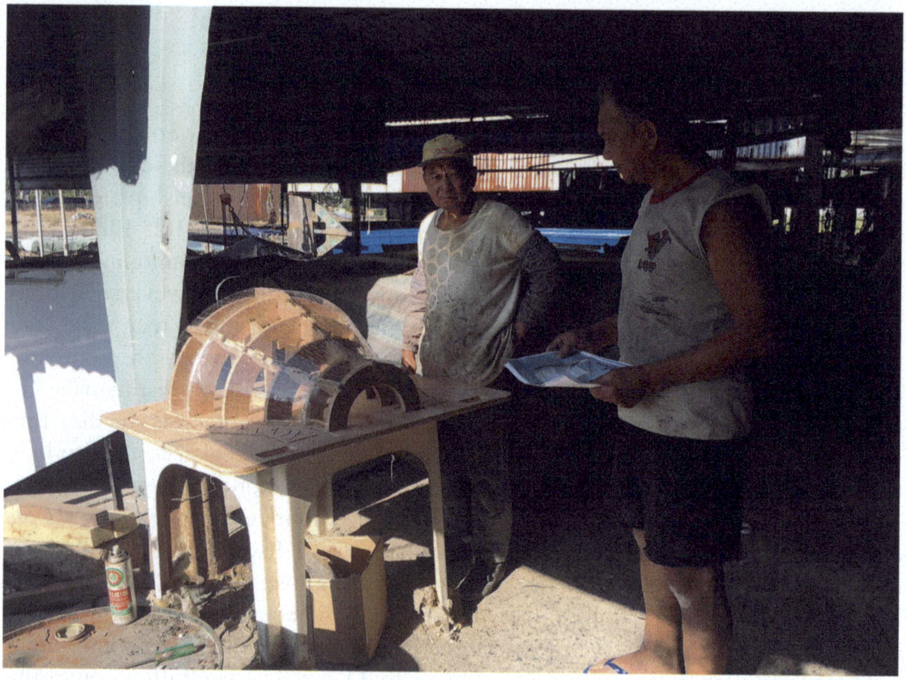

Fig. 5.3 The digitally fabricated footing mold prototype in Mr. Chou's repair shop in Tai O. (Photo: Daniel Keith Elkin)

We will soon start work oriented toward creating knowledge and structuring relationships. Between October and December 2020, we will install six infrared monitors throughout Tai O (Fig. 5.4). We will enclose each USD 80 device in a 3D-printed enclosure for easier mounting, tamper-proofing, and to carry an information card. These cards explain that the monitors neither connect to networks nor collect identity information and display current traffic information as part of our engagement strategy. The monitoring locations create a perimeter around Tai O at locations where visitors enter or turn toward destinations. Over a year of monitoring, we anticipate pedestrian count data will make flows in and out of Tai O more visible. While the architectural product in this effort is limited to drawings and small technical objects, we expect it, along with other engagement work, to orient this project toward a politics, rather than a policing, of Tai O's space. Up-to-date, public and situated presentation of this data aims to alleviate residents' concerns, but also to structure greater awareness of change as an alternative politics in the village.

Along with commoning literature, Alistair Fuad-Luke's (2017) writing on design activism contributes to our self-appraisal. Of import is his elaboration of the alternatives premise, also found in commoning's conceptual field, regarding consensus and dissensus. Fuad-Luke distinguishes between social design and design activism in the latter's "history applying practices designed to provoke (antagonise) and to create dialogue and positions of contestation (agonism)" (Fuad-Luke 2017). We

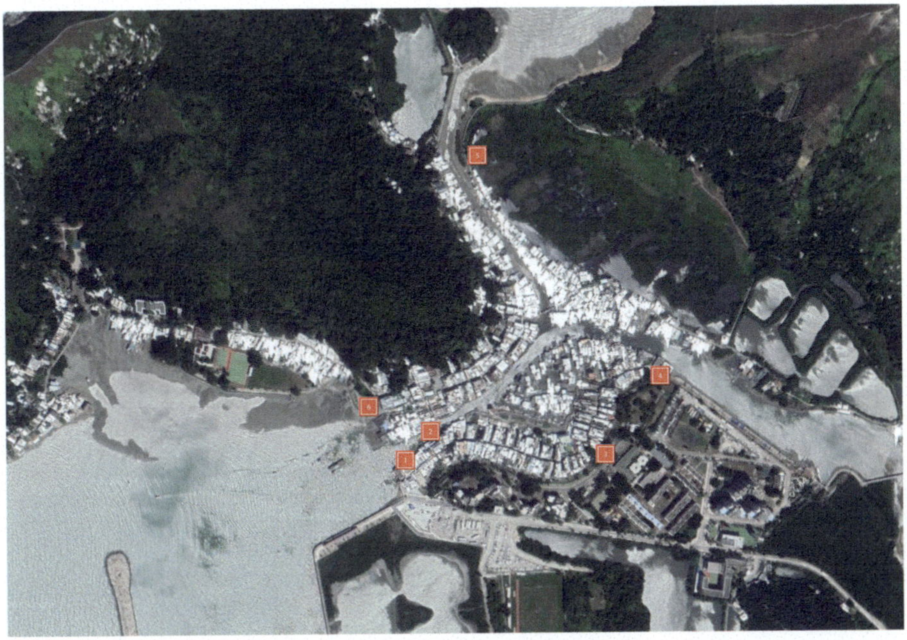

Fig. 5.4 Aerial photograph of the six pedestrian monitoring locations in Tai O. (Source: Diagram by Daniel Keith Elkin and Gideon Yeung Chi Hang. Photograph by Google Maps)

validate our pre-design project according to how it creates or reveals contestation in ways suited to architectural products, even if closer to the "weak agonism"-end of Fuad-Luke's spectrum (Fuad-Luke 2017). The predesign project includes renovation schemes for the Tai O Rural Committee Historic and Cultural Showroom, and for a community hall on Tai O's western edge (Figs. 5.5 and 5.6). The drawings shown reprogram the showroom as a crowd-control and visitor information gathering facility to contribute to knowledge production. While our relationship with the Rural Committee makes consensus on this project conceivable, the community hall project is, somewhat deliberately, more fraught.

The CEDD owns the parcels shown, which are designated government, institutional, or community zoning, G/IC. This designation reserves land for meeting halls, cultural institutions, public clinics and infrastructure. This specific investment via the Sustainable Lantau Blueprint includes an outdoor Cantonese opera theater and event space. We suggest this contributes to commodification of Tai O's space within regional strategy which, while not malign, is insensitive to village residents' concerns. The new facility nearly doubles capacity for private cars and tourist coaches in Tai O by removing an underutilized park. Even without the specific dangers of Covid-19 to Tai O residents, the frustration reported by village stakeholders when the regional government "just does things" may stem from projects like this one, as was the case during the expansion of the Tai O bus terminus. With the caveat that these products have yet to enter our structured engagement process, we suggest

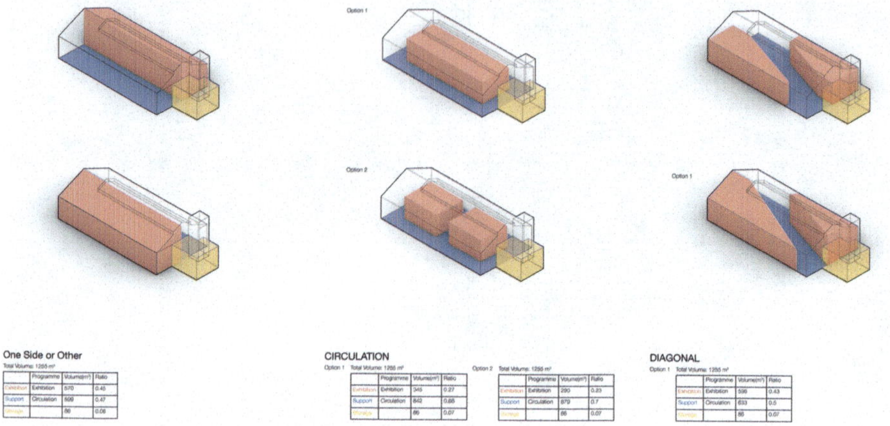

Fig. 5.5 Predesign schemes for a renovated Tai O Rural Committee Historical and Cultural Showroom addressing crowd control, wayfinding, and tourist visitor data collection objectives. (Design team: Daniel Elkin, Lee Chung Pan, Yeung Chi Hang)

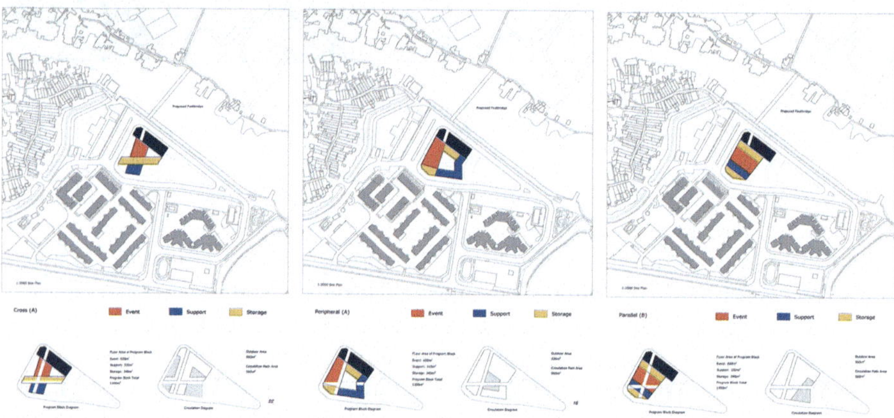

Fig. 5.6 Predesign schemes for a Tai O community hall facility, including storm shelter and coach visitor check point. These expand upon the anticipated program for a new facility by the CEDD, given feedback from villagers on flooding, storm, and Covid-19 threats. (Design team: Daniel Elkin, Lee Chung Pan, Yeung Chi Hang)

that commoning-aligned biases in our self-assessment inflect how we appraise and shape our predesign work.

The stakeholders we spoke to, contributing agencies in Tai O, recommended crowd control, storm and flood shelter, and visitor education facilities as exigent needs. Each predesign scheme presented for the community hall provides spatial and budgetary arrangements for these programs on the site, elaborating critical program at a schematic level. For example, we provide published architectural precedents for bathroom and utility planning that accommodates family groups staying overnight or washing clothing during storm events. We recommend ways for the new rural committee meeting hall to accommodate tour group gatherings during off-hours. The site planning of each scheme aims to re-integrate the expansive site with its surroundings and provide wayfinding. We suggest that, after Fuad-Luke, these architectural products clarify contestation in a case where *specifying difference* between possible futures makes their consequences more discernible. We suggest that this potential is one thing architecture contributes to commoning initiatives. Returning to distinctions in architecture critique caused by alignment with commoning, we contend that our work's movement from predesign to design, deliberately delayed, reflects our distinct self-appraisal criteria. We evaluate our work on its ability to specify to the amount of design resolution needed to clarify contestation, but not to suggest its resolution without consequential engagement. As we do so, we reflect on a stakeholder's feedback during earlier research. An activist participating in a feedback workshop on community projects commented that he supported cultural exhibition facilities in Tai O but would not record his support for fear that others would use it to endorse projects not suitable to the Village. Especially when the biases of commoning drive our work, appreciation of such epistemological complexity distinguishes how architecture critical assessment performs: critique must justify its objects for expanded ranges of time, coordinating methods and processes to produce architecture in accordance with the ethics and politics it claims, such that alignment to commoning can be coherent. We suggest that this change is the most foundational consequence between architecture and commoning's conceptual field.

Acknowledgements Research contained in this chapter received funding support from The Hong Kong Polytechnic University School of Design and The British Council in Hong Kong's Social Innovation and Entrepreneurship Fund.

Research Assistants contributing to this work include Yu Chui Sang, April Chan Ying Shan, Joan Yan Ho Siu, Ben Lee Chung Pan and Gideon Yeung Chi Hang.

We extend special thanks to Mr. Leslie Ho of the Tai O Community Work Office for his support.

References

Awan, Nishat, Tatjana Schneider, and Jeremy Till. 2011. *Spatial Agency: Other Ways of Doing Architecture*. London: Routledge.

Awan, Nishat, Tatjana Schneider, and Jeremy Till. 2018. *Spatial Agency: Other Ways of Doing Architecture*. Routledge.

Banham, Reyner. 2002. *Theory and Design in the First Machine Age*. Oxford: Architectural Press.

Bauwens, Michel, Vasilis Kostakis, and Alex Pazaitis. 2019. *Peer to Peer: the Commons Manifesto*. London: University of Westminster Press.

Cattaneo, Daniela Alejandra, and Jimena Paula Cutruneo. 2016. "The Outside Is Always an Inside": The Idea Of Space and Its Theoretical Heritage in "Toward an Architecture". *Journal of Architecture and Urbanism* 40: 250–258. https://doi.org/10.3846/20297955.2016.1210049.

Elkin, Daniel, Norah Xiaolu Wang, and Chi-Yuen Leung. 2020. International Social Innovation Research Conference 2020. Regional and Place-Based Approaches to Resilience in Hong Kong: Nested Administrative Visioning and Social Innovation Initiatives as Mediator in Tai

O Village. In *Proceedings of the International Social Innovation Research Conference 2020*. Sheffield: University of Sheffield.

Elkin, Daniel, Norah Xiaolu Wang, Chi-Yuen Leung, and Wantanee Suntikul. 2021. International Congress of Architects 2020 Conference. Inequality in Development Futures: Tourism Economies and Construction Technology in Tai O, a Village near Hong Kong. In *Proceedings of the International Congress of Architects 2020 Conference*. Rio de Janeiro: International Congress of Architects.

Emerson, Kirk, Tina Nabotchi, and Stephen Balogh. 2011. An Integrative Framework for Collaborative Governance. *Journal of Public Administration Research and Theory* 22: 1–29. https://doi.org/10.1093/jpart/mur011.

Erkin, Erdoğan, Yüce Nuran, Özbay Özdeş, Carol Williams, and Ümit Akçay. 2018a. The Crisis of Capitalism and the Commons. In *The Politics of the Commons: From Theory to Struggle*, ed. Erkin Erdoğan, Yüce Nuran, and Özbay Özdeş, 53–65. İstanbul: Sivil ve Ekolojik Haklar Derneği.

Erkin, Erdoğan, Yüce Nuran, Özbay Özdeş, Carol Williams, and Aykut Çoban. 2018b. Ecological Commons and Enclosure Policies in Turkey. In *The Politics of the Commons: From Theory to Struggle*, ed. Erkin Erdoğan, Yüce Nuran, and Özbay Özdeş, 29–52. İstanbul: Sivil ve Ekolojik Haklar Derneği.

Erkin, Erdoğan, Yüce Nuran, Özbay Özdeş, Carol Williams, and Bülent Duru. 2018c. What are the Commons? On Natural, Urban, Social Commons and Their Effects on Urban Social Movements. In *The Politics of the Commons: From Theory to Struggle*, ed. Erkin Erdoğan, Yüce Nuran, and Özbay Özdeş, 11–28. İstanbul: Sivil ve Ekolojik Haklar Derneği.

Erkin, Erdoğan, Yüce Nuran, Özbay Özdeş, Carol Williams, and Begüm Özden Firat. 2018d. Global Movement Cycles and Commoning. In *The Politics of the Commons: From Theory to Struggle*, ed. Erkin Erdoğan, Yüce Nuran, and Özbay Özdeş, 66–79. İstanbul: Sivil ve Ekolojik Haklar Derneği.

Foster Hal, Frampton Kenneth (1987) Towards a Critical Regionalism: Six Points for an Architecture of Resistance. In: *The Anti-Aesthetic: Essays on Postmodern Culture*, 16–30, NY: New Press.

Frampton, Kenneth. 1983. Prospects for a Critical Regionalism. *Perspecta* 20: 147–162. https://doi.org/10.2307/1567071.

Fuad-Luke, Alistair. 2017. Design Activism's Teleological Freedoms as a Means to Transform our Habitus. *Agents of Alternatives* (blog). 4 January 2017. http://agentsofalternatives.com/?p=2539

Grove, Kevin. 2018. *Resilience*. London: Routledge.

Hardin, James. 1968. The Tragedy of the Commons. *Science* 16: 1243–1248.

Lave, Jean, and Etienne Wenger. 2018. *Situated Learning: Legitimate Peripheral Participation*. Cambridge: University Press.

Le Corbusier. 1986. *Towards a New Architecture*. Translated by Frederick Etchells. Mineola: Dover Publications.

Ostrom E. 1990. *Governing the Commons: The evolution of institutions for collective Action*. Oxford: Cambridge University Press.

Padovan, Richard. 2002. *Towards Universality: Le Corbusier, Mies and De Stijl*. London: Routledge.

Pollio, Marcus Vitruvius, Morris Hickey Morgan, and Herbert Langford Warren. 1914. *The Ten Books on Architecture*. Cambridge, MA: Harvard University Press.

Ramos, José Maria. 2016. *The City as Commons: A Policy Reader*. Melbourne: The Commons Transition Coalition.

Sanchez, Jose. 2021. *Architecture for the Commons: Participatory Systems in the Age of Platforms*. Abingdon, Oxon: Routledge.

Squatter Control Office. 2020. *Squatter Control Policy on Surveyed Squatter Structures*. Hong Kong: The Government of the Hong Kong Special Administrative Region.

Sustainable Lantau Office. 2017. *Sustainable Lantau Blueprint*. Hong Kong: The Government of the Hong Kong Special Administrative Region.

Tai O: Hong Kong Tourism Board. 2020. *Discover Hong Kong*. Hong Kong Tourism Board. https://www.discoverhongkong.com/eng/interactive-map/tai-o.html. Accessed Oct 7 .

Turner, John F.C., and Robert Fichter. 1972. *Freedom to Build: Dweller Control of the Housing Process*. New York: Macmillan.

Venturi, Robert. 1977. *Complexity and Contradiction in Architecture*. New York: Museum of Modern Art.

Wong, Michael. 2019. LCQ1: Conservation of the stilt houses at Tai O. Press release posted online, 22 May 2019. https://www.info.gov.hk/gia/general/201905/22/P2019052200546.htm

Wong, Wai King. 2000. *Tai O: Love Stories of the Fishing Village*. Hong Kong: Concern Group About Tai O's Culture and Antiquities.

Yeung, Gary. 2007. Practicing the Built Tradition in Tai O Hong Kong. *HKIA Journal*: 2–7.

Part II
Design, the Commons and Culture

Chapter 6
Persistent Modeling of the Built
A Collective Experiment Merging Structural Preservation and Digital Design Between Academia and Industry

Frank Bauer and Lasse Sehested Skafte

Abstract With an interest in the dynamics of structure and decay, this experiment challenges prevalent logics of building preservation by developing a hybrid strategy between digital geometrical data and physical manipulation. It informs an intervention to an exemplary construction, with the intention of adding structural redundancy and thereby converging model and building. In this regard, the chapter discusses the results of several on-site workshops with partners from business and industry, conducted at and on a derelict brick barn in Brandenburg, Germany. It merges 3D scanning data with industry practices for foundation injections and an external tensile structure, exemplifying an extension of prevalent strategies on building and the built, of what is finished and what may be persistent. Overlaying the obtained point cloud with an ideal model construed upon evidence from observed detailing and constructional history, it approaches the multiplicity of forces that over time have acted upon the brickwork. A translation of this deviation, traditionally conceived of as material and structural failure on a path towards collapse, functions as a qualitative representation of global deflection and allows for quantitative assessments of prospective strategies for intervention.

Keywords Persistent Modeling · Structural Preservation · Common Practices · Digital-Physical · Systemic Brick Bond Behavior

F. Bauer (✉)
Department of Digital and Experimental Design, Berlin University of the Arts, Berlin, Germany and Cluster of Excellence Matters of Activity. Image Space Material, Humboldt Universität zu Berlin, Berlin, Germany
e-mail: frank.bauer@hu-berlin.de

L. S. Skafte
Department of Digital and Experimental Design, Berlin University of the Arts, Berlin, Germany

G. Bruyns, S. Kousoulas (eds.), *Design Commons*, Design Research Foundations, https://doi.org/10.1007/978-3-030-95057-6_6

It was just when theoretician Manuel DeLanda openly embraced emerging technologies in their potential "to think about the origin of form and structure not as something imposed from the outside on an inert matter" (2004, 21) that his colleague Antoine Picon (2004, 19) would caution against "the loss of all political and social bearings" in their affirmative, self-contained innovation towards efficiency and optimization. Even years after, controversies around the fetishization of digital design and making still prevail (Sheil 2014; Hughes 2014). As such, it is slowly beginning to show how the epistemic threshold posed by the 'digital turn' forces designers to fully reformulate the shapes and realities of their practice at large.

Through the merging of conventionally disparate disciplinary cultures, this chapter probes novel understandings of agency emerging along and with this process. It proposes to conceive of designing, modeling and building not as isolated, sequential steps, but as intertwined conditions of a holistic approach. Based upon an iterative experiment in the context of building heritage preservation, the described measures, workshops and interventions seek to challenge the prevalent logic of architectural practices at the convergence of digital representation and built reality. It furthermore substantiates how digital designers have to re-figure their stake between individual and collective work, and along this, their stance towards 'commons'. By doing so, it explores how designers can extend both the scope and interest of such translations in order to overcome dominant obsessions with quantitative narratives of precision and optimization, instead pursuing the unexpected, qualitative merits – and how such more diverse, concurring approaches may serve ecology, economy at once while extending our understanding of design.

Conceived as an experimental crossover of fields of knowledge with relation to more 'common' definitions of the practice, this first survey of the project asks for conceptual prospects of 3D modeling while pushing current understandings and conceptions of structural preservation simultaneously. Choosing a derelict brick barn in the Prignitz region of Western Brandenburg in Germany as subject, the inquiry begins by gathering and processing digitally recorded data on building deformations. While such deformations are usually conceived as mere material and structural failure on a path towards collapse, here they are used as a catalyst to innovate an existing technology of foundation injection – a widespread strategy of soil stabilization through ground injections of an expanding resin foam. Establishing moments of reciprocal feedback between model and building along, the chapter then exemplifies an extension of prevalent strategies on modeling – one which negotiates the threshold of the result and its process, of the realized and the represented, of what is terminated and what may be persistent. Very openly, this inquiry reconsiders the systemic behavior of brick bonds between equilibrium and eventual collapse in new ways. It consciously conceives of building as an always dynamic and flexible process, where previous actions and effects as well as prospective strategies find manifestation through recursive, iterative interventions.

1 Context in "Commons"

In relation to the described inquiry, this approach probes the threshold of design and commons in several ways. In relation to the broader argument of the present publication, three concurring perspectives and backgrounds may be outlined.

1.1 Digital Agencies

Along the digital revolution of building industries, ongoing transformations and stratifications continue to challenge conventions and procedures across trades and disciplines. At this threshold, new actors and agencies (material, operative and numerical among others) are addressing, diversifying and reallocating questions of agency, activity and authorship (Carpo 2009). Such bias towards more collective processes expands known roles and responsibilities usually deemed to be static, just as much as it scrutinizes prevalent understandings from an individual authorial act of design.

In the field of Science and Technology Studies (STS), various approaches from actor-network theory (ANT) after anthropologist Bruno Latour have illuminated these circumstances; and even though it appears neither viable nor necessary to cover their range here at large, they hint us towards the roles of manifold, human and non-human design agents such as materials, processes or technologies (Latour 2011) – how they manifest themselves in increasingly distributed authorial networks between collaborative processes, theoretical scholarship and experimental work (Rheinberger et al. 2010, 146). While one may assume *that* such innovative processes do relocate agency, it appears crucial to get a more thorough understanding of *how*, by *whom* and *where to* such relocations take place. To fully understand digital design from its actors and networks will then inevitably incorporate an acknowledgement of how all this "embraces a complex conglomerate of many surprising agencies" (Latour and Yaneva 2008, 86).

1.2 Novel Design Perspectives

With regard to more collective and shared forms of design, new concepts of authorship and 'commons' are opening up just as new realities for the practice at large. As a result, designers may embrace the epistemic turnover induced by the 'digital turn' as an exceptional opportunity to reshape and reconfigure design both in research and practice. While they are disclosing future issues, forms and applications of their work, this may offer exceptional potentials to forge new and bespoke processes within collective setups involving specialists from equally diverse fields of expertise.

Such crossovers stand related to similar trends in the realms of digital design and fabrication, which are just as much off the beaten path: manifesting themselves in trends towards "hacking" (Witt 2011) or "unmaking" (Ibach et al. 2019) one the one hand, or in tendencies towards more practice-based forms of architectural research like 'research through design', 'design through making' or 'design-based research' (Mareis 2016) on the other hand. Both developments offer new perspectives to rearticulate given disciplinary realities. For the involved researchers and practitioners, this then allows them to hypothesize on, to qualify and to experiment with cooperative planning processes to new extents. It is this merging of so-far unlinked areas of computational technologies and trade-specific knowledge, which raises yet unknown potentials to pose and answer emerging questions in unexpected ways.

1.3 Preservation and Commons

It may be too simple to claim that building heritage, in its very essence, can claim a 'common' position for the simple reason of its bare existence, potency and efficacy in modern societies. Such economical reductions of heritage to "cultural commons" have been contested for good reasons, namely their neglect of symbolic, material and abstract values. (Barrère 2019).

Due to their naturally long, complex and often conflicting histories, heritage buildings are, however, exceptionally apt examples to investigate broader, truly 'common' standpoints surrounding the multiplicities of agency and collectivity mentioned above. Imagine, for instance, the manifold of human (owners, craftsmen, architects, contractors) and non-human (materials, tools, building codes, site conditions, machineries) agents affecting the generations of their existence. But how to relate to the complexity formed by such multiple agencies, how to treat related conflicts – such as precedence versus originality or historicity versus innovation – remains anything but resolved. In order to problematize such conflicting interests, Alois Riegl's systematization of *Denkmalwerte* as an "eternal display of cycles of growth and decay" (Riegler 1995 [1903], 72, this and all following translations by the authors unless otherwise noted) has been re-read with growing interest in recent years (Dolff-Bonekämper 2010). This inquiry takes on this discussion of the political and social relevance of Riegl for the evolving forms of more 'common' practices between activists, researchers, designers and preservationists. Doing so, it follows preservationists Niklaus Kohler and Uta Hassler in understanding this largely also as a methodological issue:

> The need to understand the value, the composition and the long-term dynamic of the building stock is vital for the definition of future practice for architects and engineers. The required knowledge can only be produced by multi-disciplinary scientific research and must be related back into the professional practice. Many of the methods of this research do not yet exist and will have to be designed by architects and engineers, who are familiar not only with the traditional skill and art of building, but also with the various scientific methods required (2002, 234–35).

Beginning from this assumption, the present chapter is a tentative approach towards such methods. It probes an approach to merge heterogeneous knowledge and technologies to preserve and innovate heritage building between partners from academia and industry. As an inherently collective endeavor, it requires to not treat the 'digital chain' as an isolated entity, but as an arena open for investigations through external discourses and references.

2 Method

Just as much as the starting point of the project is informed from various disciplinary backgrounds, so is its methodological approach. Its conception stems from the conviction that the instruments, media and processes of design mutually influence each other in an inherently intertwined production of knowledge – and thus clearly stands in the tradition of a cultural technique perspective on design:

> Separating the act of design from its media misses to recognize the actual productivity of cultural techniques in designing, misses also their adequate discussion in the age of digital technologies. As opposed to that, the agency of cultural techniques within design becomes clear. In the architectural realm, cultural technologies and designing constitute and inform themselves mutually and hence define each other always anew. (Gethmann and Hauser 2009, 9)

In a more recent contribution, cultural historian Susanne Hauser extends these thoughts further, as she refers explicitly to the generative potential of disciplinary crossings in this context. Borrowing from a wider set of concepts and instruments, the present inquiry is consciously performing design as such a "transcending practice" (Hauser 2013, 376), a cultural technique dissociating implicit from explicit knowledge to develop unexpected, interdisciplinary responses to epistemic voids – such as the 'digital turn'.

Operationalizing this general stance, the given approach follows a working concept by design educators Hannah Groninger and Thomas Schmitz: the distinction of tools or devices for making ("Werkzeug") from devices for thinking ("Denkzeug"). More precisely, it is based upon the hypothesis that it is especially within instances of creative applications, eventually abandoning given programming towards unexpected purposes, when the former are rendered innovational in their transformation to the latter: "A device for making [Werkzeug], which is used as opposed to its conception, for other purposes and thus gets deconstructed, may at any time become a device for thinking [Denkzeug]." (2012, 20). While specific methods and tool sets are explained below upon their application, their reprogramming or reinvention through disdiplinary crossings – an interdisciplinary *re*-tooling towards unexpected merits, if you will – sums up the general methodological approach of this inquiry.

2.1 Persistent Modeling

At the intersection of simulation and building, this project puts one fundamental concept of architectural practice, the *model*, at its core. The working process may be read as an open attempt to exemplify an augmented understanding of this concept – the model is not only to be understood as *medium* to verify and represent, but as a processional and speculative *instrument* for generating knowledge through its capacities for abstraction. Notably, this conception builds upon the concept of the "persistent model" as proposed by Phil Ayres:

> And the persistent model is starting to look at the opportunities that digital representation admits relative to the physical world, or physical constructs.. So essentially what the persistent model is, is a form of nested feedback system, which couples the environment of representation to the artifact;.. but then that information is being fed back, to actually begin to temper the ideal, and so what we start to have, is actually the actual re-informing the ideal. And so it's a cybernetic loop actually, where what we are talking about is a form of control that is being passed between the represented and the actual, and continually oscillating between the two, informing them, and governing each other. (2010, min. 9:47 and 13:32)

The project takes further practical inspiration from existing strategies surrounding the development of Ayres's concept: notably the research group CITA (Center for Information Technology and Architecture) at The Royal Danish Academy promotes an understanding of practice-based architectural research openly "Merging Digital and Physical Enquiries" (Ramsgard Thomsen and Tamke 2015). Across various instances of the project, this installs what Ayres and his colleagues at CITA have framed and experimented with as a *persistent* workflow intertwining digital simulation with a physical intervention; and in doing so, actively mediates knowledge from domains such as construction, traditional craft and computational modeling into the digital realm.

In doing so, the present approach experiments with prevalent dependencies of modeling and building, meaning that models (the virtual or physical simulations or representations) conventionally only appear as preliminary drafts, subordinated to their later realization on site (that which is built). By upsetting this workflow, it can approach a more symbiotic convergence of structure and model as a hybrid system.

Intersecting two usually very disparate fields of knowledge (foundation injections and 3D scanning) by concomitant means of construction analysis, intervention strategies and modeling tools, the working method explicitly does not attempt to revolutionize the actual preservation of building stock. Rather, it is to be conceived as an experimental realization of what Ayres would refer to as a nested feedback system between the systemic behavior of the brick building and its persistent visualization (2012); a practice-based attempt to converge structure and model in an aesthetic and functional hybrid existing somewhere between analogue and digital, between ideal and real, between model and building.

2.2 Foundation Injection

Soil stabilization techniques based on injections are by no means a novel invention. After its first introduction in the 1980s, a patented soil lifting process from 1998 still forms the basis of most competing strategies on the market today, including the deep injection method by Uretek, a major stakeholder in this area, who was willing to cooperate within this experimental setup as project partner. This method, "which from a geotechnical perspective is at the border between underpinning and grouting" (Buzzi et al. 2008, 1013), employs an intervention under and around building foundations to address settlements and stabilize surrounding soil layers. It is meant to work against differential soil settlements, which are caused by factors such as vegetative drainage, softer water-responsive clay soils, subsurface erosion, and soil consolidations due to decay, among others.

Unlike conventional strategies – involving excavation and fortifying the foundation – the deep injection method injects an expanding two-component polyurethane resin through 16 mm drill holes under the building foundation. Individual injection depths are determined through preliminary drill sample analysis of the surrounding soil. Along with the chemical activation of its components, the injected resin is distributed and expands in its transition to a rigid high-density foam, gradually compacting neighboring soil layers (vertical fracking). As soon as adjacent soils turn dense enough that the lateral pressure exceeds static ground and building loads, the resin expands and densifies upwards (horizontal fracking). This leads to a gradual lifting of the upper soil levels, eventually the sub-foundation soil and finally consolidates the building's foundation.

This method is used in various scenarios, when signs of soil settlement are evident through cracks in floors, walls or other parts of construction, or are to be prevented (such as in the case of floor additions). While injection points under load bearing elements are determined by experience, reference point levels are continuously monitored with tape markers and laser sensors, where injection conventionally stops as soon as the foundation is supported anew. Usually, the intervention is conceived to be the consolidation of the current state, as restorations and veritable liftings involve unpredictable dynamic behavior, as Uretek engineers repeatedly remarked throughout initial discussions. However, it became exactly this unexpected dynamic behavior which we then operationalized within the current inquiry, to innovate the process by utilizing information from a 3D deviation analysis, providing information of the effect on the entire structure, rather than just singular reference points. This augmented understanding could allow us to more precisely guide the injections and render them more feasible, efficient and predictable, but first and foremost – by reducing the required material consumption – more economically and ecologically sustainable. Also, it extends the scope of the technology beyond the consolidation of the already present, towards an active dynamic manipulation of the structure, and ensuing structural sustainability.

3 Study

In a first workshop on site in August 2017, a thorough building survey acquired necessary information for an initial assessment of the structure. Applying a traditional architectural approach of analogue measurement and observations of structural detailing, that is, brick formats and bonds, as well as knowledge of construction history, an idealized 3D model of the building was constructed. Building upon this diverse evidence, the aim was not to draw a precise representation of the structure in its current state, as this would be digitally recorded in a following step. Rather, it was the abstraction to its ideal geometric model, that is, the digital representation of some sort of proto-structure. This model would then be used as reference in order to align and quantify assessed deformations from the following 3D scans, as well as serve as the base layer for the progression towards the persistent model.

3.1 Initial Survey and Ideal Model

Throughout a second workshop on site, 3D surveys of the structure and its surroundings were recorded with the collaborating partner Plan3D Berlin using a LIDAR (Light Detection and Ranging) scanning device by manufacturer FARO (Fig. 6.1).

This widespread method, with applications in various industries, illuminates the target object with laser, and senses the distance of returning laser reflections through

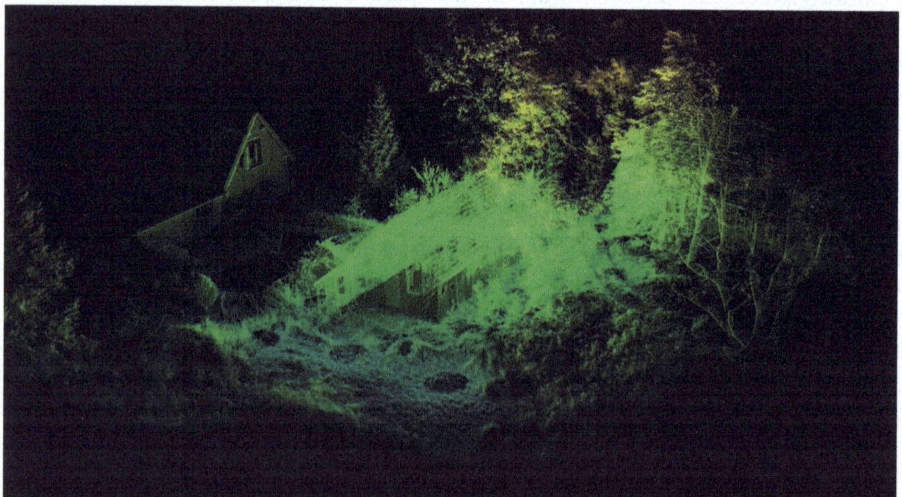

Fig. 6.1 Point cloud of pre-intervention record of the structure in its environment. (Source: Frank Bauer and Lasse Skafte)

their individual speed and wavelength. The resulting cloud of registered points can then be aligned through reference points and further manipulated towards a 3D model of the scanned object. In this case, the scans covered both the interior and exterior of the structure and were merged, using reference spheres for alignment. For reasons of clarity and reference, the scans were further augmented with a photographic layer, assigning color information to each point of the cloud. After culling and compiling the recorded point clouds with the Faro Scene software and trimming with the open source Cloud Compare to isolate only the structure in focus, they were translated into Rhinoceros 3D for further processing.

3.2 Deviation and Translation Analysis

In the following step, the two resulting layers of the recorded point cloud and the ideal model were mapped in one 3D representation. The data was aligned through reference points on rather undeflected facade sections, which had been identified throughout the recordings. Utilizing a point deviation analysis routine with Cloud Compare and the CITA Grasshopper plugin Volvox to calculate the deviation between measured objects and target geometry, it was then possible to visualize the global geometrical deflection of the structure; that is, to get a qualitative understanding of the extent to which the respective parts of the building in the recorded state differed from what had been reconstructed as the ideal model of the structure. The resulting analysis graphically illustrates the sum of the multiplicity of dynamic forces that have acted upon the brickwork, the resulting geometry representing the altered state and conditions of the structural system. This step also generated the first feedback of the persistent model, feeding substantial assessed geometry into the digital model, thereby establishing a connection between the actual building and the digital model – which then no longer serves only the purpose of representation, but becomes an active generator of knowledge on the current state of the structure.

On this basis, two parts of the southeastern and northwestern facades, notably two opposed building corners, were identified as potential exemplary intervention sites. In order to quantify the assessed information from the analysis on global deflections, the translation of randomized brick sets was traced and calculated in order to determine aggregate deformation vectors, and therefore boundary values to each injection spot. As an excerpt from the overall analysis, it may suffice here to discuss this along individual exemplary bricks at the two intervention locations – measuring the three-dimensional translation of determined locator points (upper edge facing the building corner) with a custom Grasshopper routine, from their idealized position (grey) to their deflected positions (yellow) (Fig. 6.2). As for our exemplary bricks, breaking down the calculated translation vector to its components resulted in deformation boundary values of -85 mm in x-axis, +26 mm in y-axis and +73 mm in z-axis at intervention location I and +43 mm in x-axis, -22 mm in y-axis and +41 mm in z-axis at intervention location III.

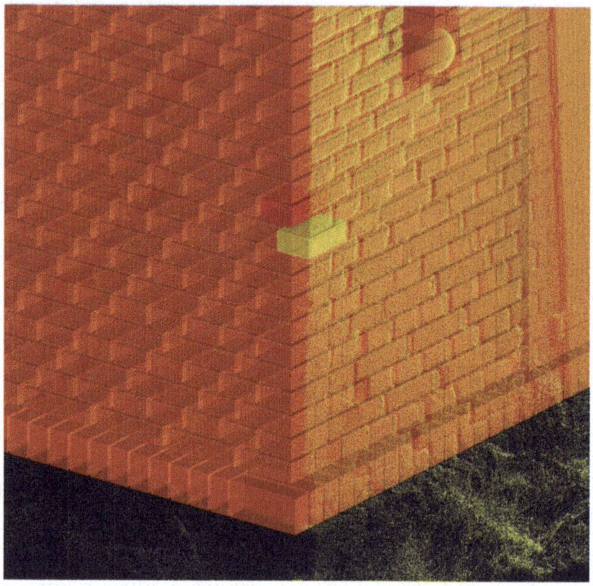

Fig. 6.2 Quantitative deviation analysis of sample brick at injection spot I.1, overlaying of pre-intervention point cloud (yellow) and ideal model (light red), estimate of individual brick translation from construction time (red) to 2017 (yellow). (Source: Frank Bauer and Lasse Skafte)

3.3 Intervention

Within the wider context laid out above, a specific aim of the intervention and motivator for the industry partner was to act as a prototype of a more qualified application of the given foundation injection method in cases of minimally invasive restorations of heritage buildings. Juxtaposing knowledge and experience of the involved staff technicians by means of contemporary scanning and modeling technology should render this usually rather practice-based process of compacting the ground by literally inflating it, more feasible, predictable and sustainable.

Regarding the specific conditions on site, our interest was in approaching the deteriorating brick structure as a dynamic constructive system, where individual bricks are held together in bonds, thus reacting dynamically to all forces influencing or applied to the structure. To reinforce the partially failing bonds and to reinstate a layer of structural redundancy, a minimal external supportive structure was designed and installed around the barn. In applying nothing but external, both compressive and tensile forces, the actual intervention refrains from enforcing a definite state of the building's appearance, but sets boundary conditions for the system, restricting the outward movement of the load-bearing walls to avoid collapse.

To do so, the assessed information from the digital model on existing building deformations was operationalized in order to inform each individual injection set. This formed the next step in the evolution of the persistent model, feeding back information from the digital model, including the nested data on the current state of the deformation, to the physical structure onsite. The quantified boundary values and additional structural information were prepared for prior evaluation and planning with project engineers from Uretek. Injection spots I.1 to I.7 and III.1 to III.4 (Fig. 6.3) were then distributed along the facades under load bearing wall elements (obviously sparing lintels and openings). Following preliminary ground analysis,

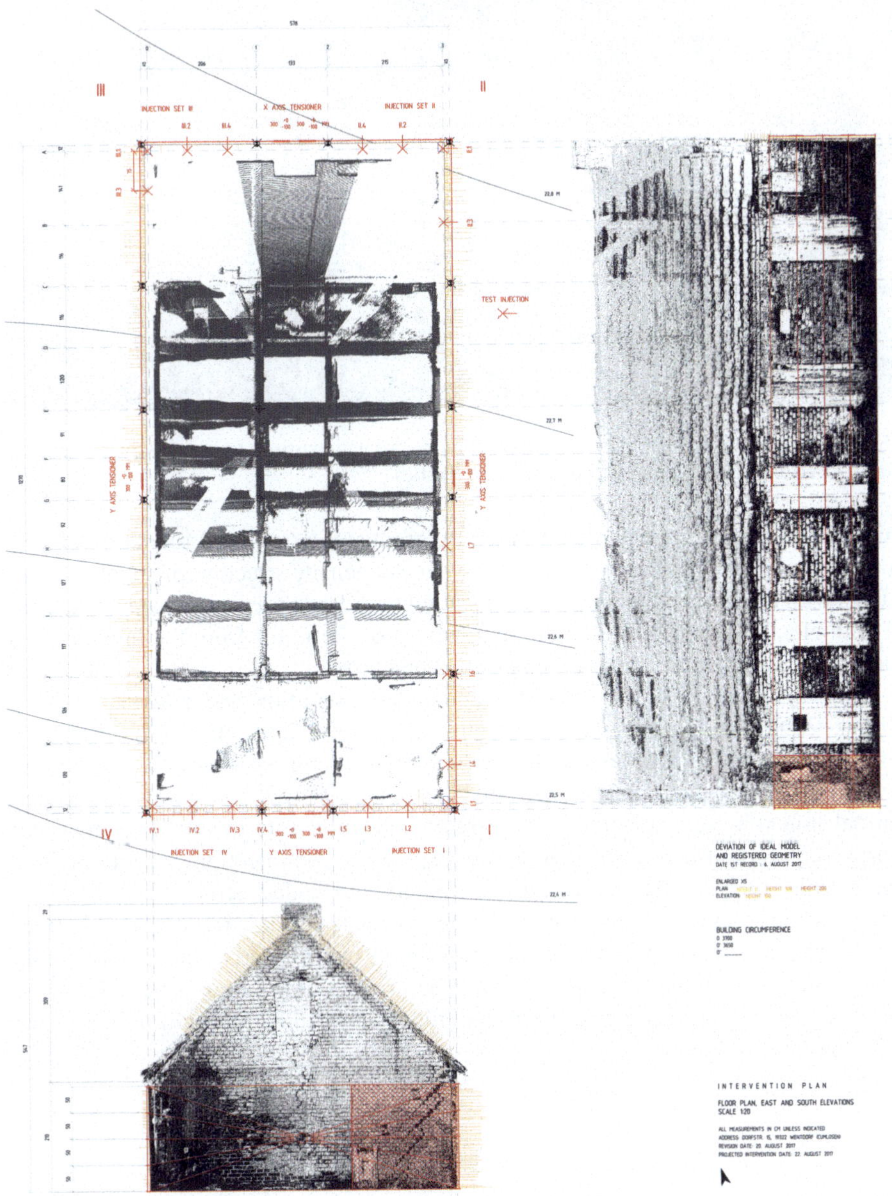

Fig. 6.3 Intervention plan, floor plan and elevations with graphic representation of deformations (x5) and individual injection sets. (Source: Frank Bauer and Lasse Skafte)

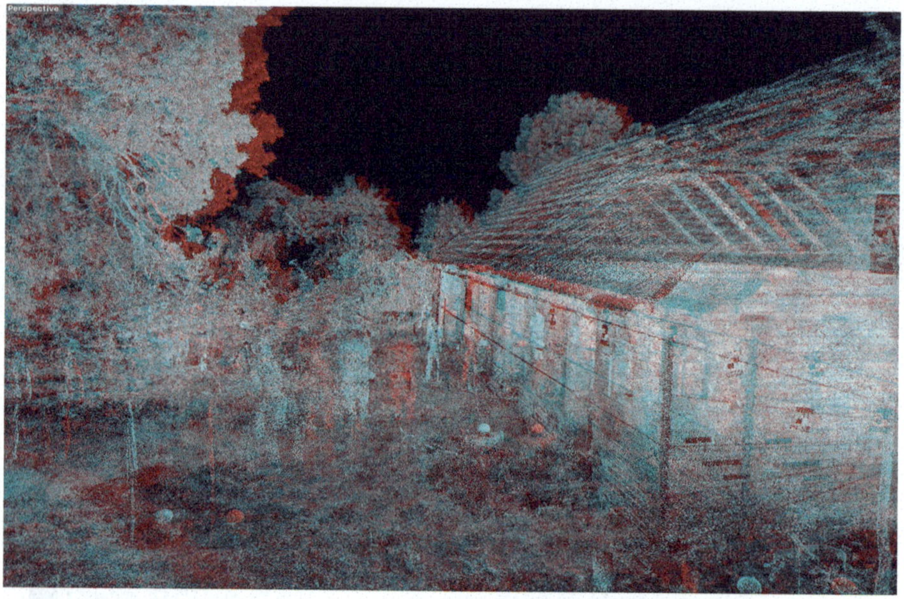

Fig. 6.4 3D visualization of injection intervention record with partners from industry, Uretek and Plan3D Berlin. (Source: Frank Bauer and Lasse Skafte)

injections were set to begin at 3,5 m depth, gradually rising until foundation level. Above, encompassing the brick walls, the devised tensile structure of stainless steel cables with adjustable tensioners was installed in order to prevent dynamic shearing of wall elements, and to utilize the momentum of the uplift from the injections to further consolidate the structure along its x- and y-axis.

The injections progressed fluently and under execution and guidance of two Uretek technicians in attendance, a technical engineer and other staff (Fig. 6.4). Ongoing issues with dispersion due to the poor soil composition of a rather large area around the foundation (through contact cracks, soil zone transitions, burrowing animal holes) were countered with iterative injection sets to minimize material waste. Regarding the brick structure, the response to the injections was immediate and visible. Only rarely did various sediments and added cement mortar layers (from various lackluster repair attempts, for instance) of existing cracks and fractured bed joints, restrict or hinder the movement of wall parts, thus complicating their behavior as dynamic system. While such impediments may have to be taken into account or eliminated prior to intervention, they did not seem to influence the actual process at large.

3.4 Post-Intervention Recording and Analysis

In a second scanning workshop on site in September 2017, a point cloud of the post-intervention building geometry was recorded with recording staff from Plan3D Berlin. Using markers that had been preserved at several parts of the manipulated building, as well as reference markers in its immediate, unaffected surroundings, alignment of the new point cloud with the previous layers of the model was possible. On this basis, a further qualitative deviation analysis was conducted, between ideal model, pre-intervention and post-intervention state, suggesting at least noticeable success in reverting the previous deformations of the structure 6.5. Forming the last iteration of the persistent model to be discussed here, feeding data on the post-intervention geometry back into the digital model space, it further augmented the hybrid recursive connection of digital simulation and physical artifact.

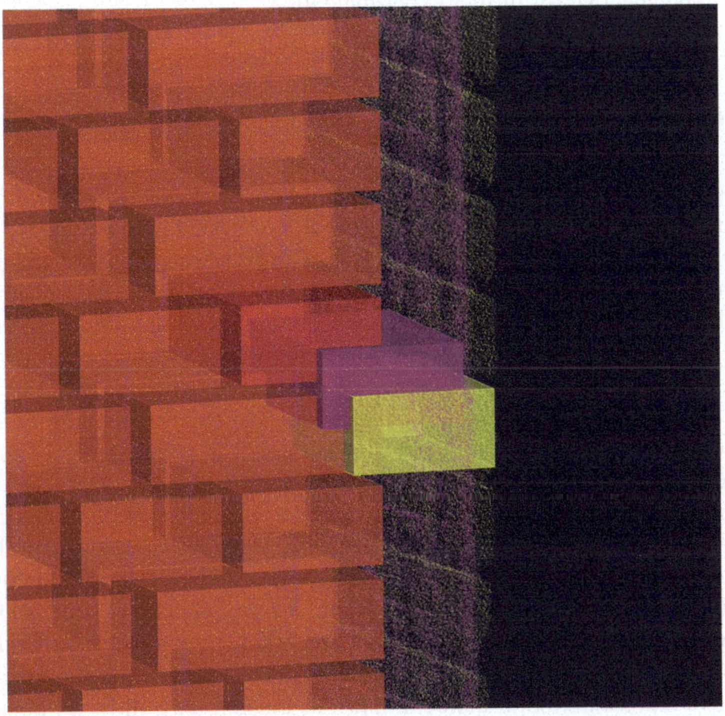

Fig. 6.5 Quantitative deviation analysis of sample bricks at injection spot I.1, overlaying pre- (yellow) and post-intervention (magenta) point clouds and ideal model (light red) as visualization of sample brick translation from construction time (red), until 2017 (yellow) and through the 2017 intervention (magenta). (Source: Frank Bauer and Lasse Skafte)

A subsequent second quantitative analysis showed major translations, at times even approaching the set boundary values. Looking at our sample bricks again, translations amount to (boundary values in brackets) −27 (−85)mm in x-axis, +13 (+26)mm in y-axis and +43 (+73)mm in z-axis at injection spot I.1 and +18 (+43) mm in x-axis, -19 (-22)mm in y-axis and +31 (+4142)mm in z-axis at injection spot III.1 (Fig. 6.5). To all involved partners, the transformations in z-direction were especially significant, as they showed how the brick structure could be raised profoundly towards its original level during the intervention. Seeing that the industry partner would usually only work towards closing cracks of several millimeters, the range of these translations appear exceptional.

The concomitant translations along the x- and y-axis have to be attributed to manipulations through fastening of the installed tensile structure, which could make use of the dynamics of the vertical momentum, literally drawing back sheared-off wall elements. While this measure was primarily installed in order to provide boundary conditions for the structure and add structural redundancy to avoid its collapse, the unexpected success in contracting the brick bonds may form another interesting perspective for further experimentation.

4 Conclusion

If one agrees upon the "digital moment in architectural practice" (Zardini 2017, 11–12) being an epistemic threshold to the realm of design, that is, one which fundamentally changes the conditions and realities of whole segments of society, then the given project may be understood as an attempt at reaction. Speaking with art and cultural historian Susanne Hauser, it is precisely in these situations that one has to cultivate novel understandings of design generating unexpected, interdisciplinary responses (2013, 376).

With an interest in the dynamics of structure and decay, this inquiry seeks to challenge prevalent logics of architectural design at the convergence of modeling and building, which is not conceived as something to be completed, but continuously conducted. Inspired by existing strategies experimenting with more persistent forms of the model (Ayres 2012), the established working process exemplifies an augmented understanding of modeling, fully embracing this cultural technique in its speculative and productive potentials by taking it from *medium* to *instrument.*

The investigation of such surveyed and simulated deviations, traditionally conceived of as material and structural failure on a path towards collapse, functions as a qualitative representation of deflection, just as much it may allow for quantitative judgments towards prospective strategies. The ambition of this first insight into the intervention was to highlight conceptual ramifications of a symbiotic convergence of structure and model in a more persistent relation. The intervention reinstates and reinforces the former while informing the latter, thus pursuing an aesthetically and functionally hybrid character. In applying nothing but external forces, it does not enforce a definite state of the building's appearance. Rather it produces new structural redundancy and subsequent degrees of freedom by adding constructional layers to an already complex system of manifold agents affecting the structure. Authoring an altered graph between equilibrium and collapse, further relations to existing research on dynamic masonry behavior (Van Mele et al. 2012) appear an interesting perspective. As a preliminary result, the intervention establishes a persistent feedback system between the actual building and its digital representation, which may be subject to further analysis as well as inspire more process- and time-based simulations in preservation contexts, notably.

In a broader sense, the given approach confronts linearities in an industrial division of labor towards more 'common' perspectives, as it attempts to re-feed diverse modes of experience and learning, both by man and machine into the digital chain. If one follows Norbert Palz in assuming that it is "rather through a synthesis of heterogeneous and even competing findings from which innovative, surprising and conceptually complex design solutions can emerge" (2016, 16), then our project may be read as one possible proof-of-concept. It invites us to reconsider how only cross-fertilizing knowledge streams and concepts on materials, processes and technology in unexpected ways will allow for such solutions. It substantiates that one may take a simple brick as a starting point to discover new perspectives on larger contexts, that one may allow the unexpected field of preservation to teach digital designers to extend both the scope and interest of their instruments.

While Mario Carpo recently argued that "digital technologies seem to be nurturing a new pervasive style of digital making – but not the one architects had in mind, and not one all architects may like" (2011, 41), he was suspecting innovation largely among more open formats and collective logics of the discipline. If this holds true, the innovation to computational tool sets may also mean to overcome their obsessions with quantitative narratives of precision and optimization towards their unexpected qualitative merits, thus pursuing the crossover from devices for making [*Werkzeug*] to devices for thinking [*Denkzeug*]. So perhaps now more than ever, the time has come to learn about the digital tool sets of the future from our treatment of relics from the past.

Acknowledgements The authors would like to acknowledge the support of the Department of Digital and Experimental Design, the Department of Design and Building Construction and the Department of History and Theory of Architecture at Berlin University of the Arts. Also all efforts of our collaborating partners from business and industry, namely the firms and all involved staff from Uretek, Plan 3D Berlin, NowLab at BigRep and Carl Stahl for their material and personnel support to the project are greatly appreciated. Working towards publication was in part made possible through the support of the Cluster of Excellence »Matters of Activity. Image Space Material« funded by the Deutsche Forschungsgemeinschaft (DFG, German Research Foundation) under Germany's Excellence Strategy – EXC 2025 – 390648296. Special gratitude goes to the facilitating owner of the historical brick barn, without whose support the project could not have been realized.

References

Ayres, Phil. 2010. Persistent Modelling. Lecture at the Digital Material Seminar, Oslo. https://vimeo.com/11615361. Accessed 14 September 2020.

———. 2012. Introduction. Persistent modelling – reconsidering relations. In *Persistent Modelling. Extending the Role of Architectural Representation*, ed. Phil Ayres, 1–10. London: Routledge.

Barrère, Christian. 2019. Cultural Heritage: Capital, Commons, and Heritages. In *The Oxford Handbook of Public Heritage Theory and Practice*, ed. Angela M. Labrador and Neil Asher Silberman, 139–154. Oxford: University Press.

Buzzi, Olivier, Stephen Fityus, Yasumasa Sasaki, and Scott Sloan. 2008. Structure and properties of expanding polyurethane foam in the contextof foundation remediation in expansive soil. *Mechanics of Materials* 40 (12): 1012–1021.

Carpo, Mario. 2009. Revolutions: Some New Technologies in Search of an Author. *The Log* 15: 49–54.

———. 2011. Digital Style. *The Log* 23: 41–52.

DeLanda, Manuel. 2004. Material Complexity. In *Digital Tectonics*, ed. Neil Leach, David Turnbull, and Chris Williams, 14–21. Chichester: Wiley.

Dolff-Bonekämper, Gabi. 2010. Gegenwartswerte: Für eine Erneuerung von Alois Riegls Denkmalwerttheorie. In *DENKmalWERTE: Beiträge zur Theorie und Aktualität der Denkmalpflege*, ed. Hans-Rudolf Meier and Ingrid Scheurmann, 27–40. Berlin: Deutscher Kunstverlag.

Gethmann, Daniel, and Susanne Hauser. 2009. Einleitung. In *Kulturtechnik Entwerfen: Praktiken, Konzepte und Medien in Architektur und Design Science*, ed. Daniel Gethmann and Susanne Hauser, 9–18. Bielefeld: transcript.

Hauser, Susanne. 2013. Verfahren des Überschreitens. In *Wissenschaft Entwerfen*, ed. Sabine Ammon and Eva-Maria Froschauer, 363–384. Munich: Fink.

Hughes, Francesca. 2014. *The Architecture of Error: Matter, Measure, and the Misadventures of Precision*. Cambridge, MA: MIT.

Ibach, Merle, Michaela Büsse, Felix Gerloff, Viktor Bedö,Shintaro Miyazaki and Jamie Allen. 2019. Unmaking: Against General Applicability. In *The Critical Makers Reader: (Un)learning Technology*, ed. Loes Bogers and Letizia Chiappini, 47–60. Amsterdam: Institute of Network Cultures.

Kohler, Niklaus, and Uta Hassler. 2002. The Building Stock as a Research Object. *Building Research & Information* 30 (4): 226–236.

Latour, Bruno, and Albena Yaneva. 2008. 'Give Me a Gun and I Will Make All Buildings Move'. An ANT's View of Architecture. In *Explorations in Architecture: Teaching, Design, Research*, ed. Reto Geiser, 80–89. Basel: Birkhäuser.

Latour, Bruno. 2011. Why Do Architects Read Latour? An Interview with Bruno Latour. *Perspecta* 44: 64–49.

Mareis, Claudia. 2016. Doing Research: Design Research in the Context of the "Practice Turn". In *Design as Research: Positions, Arguments, Perspectives*, ed. Gesche Joost et al., 35–41. Basel: Birkhäuser.

Palz, Norbert. 2016. "Personal Media Practice: Collapsing the Manifold." Unpublished manuscript, UdK Berlin.

Picon, Antoine. 2004. The Ghost of Architecture. *Perspecta* 35: 8–19.

Thomsen, Mette Ramsgaard, and Martin Tamke. 2015. Prototyping Practice: Merging Digital and Physical Enquiries. In *Rethink! Prototyping: Transdisciplinary Concepts of Prototyping*, ed. Christoph Gengnagel, Emilia Nagy, and Rainer Stark, 49–62. Berlin: Springer.

Rheinberger, Hans-Jörg, Karin Krauthausen, and Omar Nasim. 2010. Papierpraktiken im Labor: Interview mit Hans-Jörg Rheinberger. In *Notieren, Skizzieren: Schreiben und Zeichnen als Verfahren des Entwurfs*, ed. Karin Krauthausen and Omar Nasim, 139–158. Zurich, Berlin: Diaphanes.

Riegl, Alois. 1995 [1903] Entwurf einer gesetzlichen Organisation der Denkmalpflege in Österreich: Wesen und Entstehung des modernen Denkmalkultus. *Kunstwerk oder Denkmal? Alois Riegls Schriften zur Denkmalpflege*, ed. Ernst Bacher, 49–120. Vienna: Böhlau.
Sheil, Bob. 2014. High Definition: Negotiating Zero Tolerance. *Architectural Design* 84 (1): 10–13.
Witt, Andrew. 2011. Design Hacking: The Machinery of Visual Combinatorics. *The Log* 23: 17–25.
Zardini, Mirko. 2017. Eight Million Stories. In *When Is the Digital in Architecture?* ed. Andrew Goodhouse, 9–22. Berlin: Sternberg.
Van Mele, Tom, James McInerney, Matthew DeJong, and Philippe Block. 2012. Physical and Computational Discrete Modeling of Masonry Vault Collapse. In *Proceedings of the 8th International Conference on Structural Analysis of Historical Constructions*, ed. Jerzy Jasieńko, 2552–2560. Wroclaw: DWE.

Chapter 7
The Commons in African Spatial Production: A Critical Review of Geographies of Power

Gert van der Merwe

Abstract This chapter situates the South African policy discourse within global theoretical concerns regarding *Surveillance Capitalism* (Zuboff. The Age of Surveillance Capitalism: The Fight for a Human Future at the New Frontier of Power. New York: PublicAffairs, Perseus Book, a subsidiary of Hachette Book Group, 2019) and contemporary manifestations of neo-liberalism. It posits that the commons be understood as part of indigenous systems of spatial production, viewed as an ongoing and relational process in a geography of external power dynamics, and that power/capital will seek divisive tactics that obscure its operations. Analyzing South African precedents it recommends working from a place of knowledge, inclusionary practices and the relationship between bottom-up organization, while critically testing these operations against the Weberian state. It calls for deep solidarity and practices that go beyond charity or a welfare-based approach in a "politics of compassion" which evoke the literal meaning of the word to suffer with. It casts suspicion on practices of representation and the consolidation of heterogeneous spatial rights and practices, and advocates for openness.

Keywords Commons · Politics of compassion & solidarity · Surveillance capitalism · Spatial production · Weberian state

The commons corresponds with the South African indigenous value system of *ubuntu*. Described as a humanist philosophy embedded in African thought, it reveals a fundamental aspect of African society (Da Costa and Van Rensburg 2008). It is communal in nature, and in contrast with Western traditions, has strong non-individualistic characteristics of inclusiveness and participation (Da Costa and Van Rensburg 2008). However, although *ubuntu* is a pre-existing framework from which understandings flow, this is not descriptive of all particularities of cultural understandings and spatial conditions. It serves only as a generalized characteristic of

G. van der Merwe (✉)
Independent Scholar, Windhoek, Namibia
e-mail: gert.merwede@gmail.com

G. Bruyns, S. Kousoulas (eds.), *Design Commons*, Design Research Foundations, https://doi.org/10.1007/978-3-030-95057-6_7

African spatial production from which reciprocal and relational spatial understandings may be unpacked. It is therefore important to acknowledge that deeper cultural and spiritual dimensions dependent on cultural specificities should not be disregarded.

The danger in falling back on *ubuntu* without being rooted within the culture lies in the abstractions that have existed ever since Europeans first set foot in the Cape, and this chapter retains the perspective of an outsider to the tradition. It is not an attempt to provide ethnographic explanations yet acknowledges that *ubuntu* has a profound impact on how space is understood and negotiated in the South African context.

This contribution takes a cynical position on the commons, not to discount it, but rather, to acknowledge the darker dimensions of power and how it is exerted, in order to incorporate it into the discourse. Furthermore, this perspective represents a particular experience and is not a report or truthful map of consensus from the global South, but a conjecture against such homogenized polarization of theory (Watson 2014). It suggests complexities that unsettle taken-for-granted assumptions and focuses attention on new circumstances and challenges (Watson 2014). This is important when considering the Weberian state (here understood as the bureaucracy through which political theory is generally framed), but which remains at odds with practices in the global South (Watson 2014).

The emerging modernity, described by Shoshana Zuboff (2019) as "surveillance capital", colonizes the interior world of an individual's private life, and therefore, "southern theory" provides some guidance in unpicking these operations and providing mechanisms for resistance. This chapter shows that space is created through social reciprocities, deeply rooted in social relations, and thus, African spatial production has been undermined by abstractions which negate these relations in favor of quantifiable metrics rooted in reductionism. I shall refer to it as a materialist worldview. To clarify what is meant by materialism I employ a comparison between Marx, who measures social relations through the production of material goods, and Lefebvre, who implicitly recognizes social relations that do not necessarily manifest in material relations, but nonetheless describe power dynamics.

1 Tracing the Origins of Materialism and Instrumentalization

Materialism harks back to Mesopotamian cuneiform (3500 BCE) which first recorded human transactions, effectively creating a rendering of society through debt exchanges (Graeber 2011, 38), and reemerged as a dominant mode of thinking in the science of the Enlightenment.

With it came a darker side. It is easier to skew numbers than to rob a woman in person, and thus the inevitable suspicion of "money men" was naturally ever present. In the Southern context, Beinart et al. (2018, 101–2) report on a "master narrative for loss and restoration" suggesting that Africans lived in coherent egalitarian communities in which everyone, since time immemorial, held land in commons, until European colonization wreaked havoc and stole it by sword and pen. This is an oversimplified view, and in reality, disposition took place over long periods (Beinart et al. 2018, 101–2), calling for an approach that takes account of the many particularities. It is indicative of contradicting world views fundamental to this discussion, in which the materialism undermines indigenous socio-spatial relations that aim to affirm all humans within a matrix of socialization, and instead reduces the individual from a social being to being simply an economic actor, free from social relations.

1.1 Clashing Systems

Discussing co-production, Watson (2014) shows that the prevailing goal of administrators remains cost-effective/efficient state service delivery instead of upliftment, inclusion or empowerment as a goal. It suggests that the focus remains fixed on the management of material goods as the measure of fairness and remains at odds with African spatial conceptions which affirm social equity, aimed not at material equity, but each individual's place within social relations.

The pragmatic approach (bureaucracy) is the core legacy of modernism and colonial subjugation, in which representation is instrumentalized. Watson (2014) indicates that co-production is a highly political means to redistribute power among stakeholders, blurring the boundaries between voluntary, public and private sectors, diluting public accountability and violating the principles of the Weberian state (Bovaird 2007, 856).

1.2 The West Awakens

The growing interest in the commons over the past two decades seems to coincide with a recognition that the neo-liberal project failed to deliver the promises made by pundits. Less than a decade after Fukuyama (1993) declared "the end of History," the 9/11 attacks shattered his dream, giving rise to a slew of books on the failure of capitalism to deliver its emancipatory promises. Coinciding with the War on Terror, titles such as *The Shock Doctrine: The Rise of Disaster Capitalism* by Naomi Klein (2007) and Richard Sennett's *The Culture of the New Capitalism* (2006) expanded on how (following the financialization and globalization of the preceding three decades) the rise of a new form of capitalism distinct from industrial capitalism and geared towards new forms of subjugation was emerging as a global force, penetrating political power. It seemed that instability, since Marx's day, was capitalism's only constant, and Sennett (2006, 16) argues that black-and-white contrasts should be viewed with suspicion, particularly when suggesting progress as a measure of success. It indicates a recognition that capitalism has become divorced from the social functions, and highlights that the focus on growth was instrumentalized at the expense of social upliftment and inclusion.

After the 2008 credit crisis this became more pronounced as too-big-to-fail corporations were bailed out at the expense of the most vulnerable sectors of society (Graeber 2011, 16–17, 381). In the Global North this gave rise to the general discontent with a failing system. It culminated in the 2011 Occupy Wall Street protests, squarely focused on wealth disparity, and the London protest, which highlighted dimensions of exclusion such as race and a sense of injustice as central political themes (Zuboff 2019, 36–37). If 9/11 signaled the end of a unified globalized political hegemony led by the West, the credit crisis of 2008 signaled a loss of legitimacy of the prevailing capitalist system and the social contract which held it in place.

1.3 Finding Alternatives – Comparing Contexts

The purpose of discussing this global context is twofold. Firstly, the discontent with failing democratic systems marked by anti-government, pro-democracy movements signals the loss of faith in hegemonic power structures and resulted in a number of bottom-up social movements which sought to rectify the system. This explains not only why Ostrom was awarded the Nobel prize in Economics in 2011, but also the explosion of literature on the commons. The assumption is that while the left has traditionally resisted capital and the right resisted the "coercive state," the galvanization of the commons as a way out seems to undermine this divide (as all forms of power and the status-quo become scrutinized). The commons provides a new sphere of solidarity politics which embraces diversity. The second reason is to situate the South African narrative within a global context so we may read the specificity of experiences from this context in relation to these global shifts in discourse.

At the time in question, the South African government had finally acknowledged the HIV/AIDS pandemic and the economy performed comparatively well (when measured against the rest of the world) since the 2010 FIFA Soccer World Cup ensured an organic form of Keynesian stimulus, and the effects of "load shedding" (political spin for electricity blackouts due to an aging electricity grid) had not yet set in. With its inclusion in the BRICS block, as the "gateway to Africa" and "industrial powerhouse" of the continent, South Africa seemed to provide a glimmer of hope, and was expected to achieve reasonable economic growth, on condition that strategic assets such as water and electricity supply be maintained, and that the state sustained programs of economic transformation and social upliftment.

This was not to be realized, as what is now called "the lost decade" under the reign of President Jacob Zuma saw the systematic hollowing out of key state functions (to capture state resources) in a systematic process of looting and patronage which enriched the political elite. Deep systems of corruption were however a pre-existing threat to democracy rooted in the concentration of power within the African National Congress (ANC) – they held a two-third majority. This resulted in a situation where loyalty to the party and political allies within the party superseded the oath of office and thus fundamentally undermined the Weberian state.

To explain the clash of the Weberian system, rooted in materialism, with indigenous practices, it is helpful to consider what preceded the nation-state as formalized in the Union of South Africa in 1910. Beinart et al. (2018, 133–34) describe spatial practices in which different cultural-linguistic groups sharing geographical areas lived side by side. These territorial conceptions were not rooted in measuring physical space or managing material resources, but rather one's ties to particular social groupings based on shared ancestry and highly localized practices which took account of specificities. The primary nexus of these relations lies in the family, nested in further social structures.

Colonial administration, on the other hand, was highly centralized and aimed at managing material resources with very little concern for local customary practices, but what is interesting to note in this context is that this came after 250 years of systematic colonization spearheaded by capital in the form of the VOC (Dutch East India Company), only later to be displaced by British imperial rule in the early nineteenth century.

In David Graeber's account of colonialism, he argues that it did not introduce a concept of the soul through Christianity (Graeber 2011, 244). Rather, it introduced the idea that the body was separate from the soul, as a mere material collection of tissues and nerves, and the prison of the soul (Graeber 2011, 244). This situates the body in the material world, in which, following Foucault, power is exerted through processes and technologies – surveys and maps – which Watson (2014) suggests should be appropriated by those affected by it to restore equilibrium and dignity. By stripping the individual of their relational value, based on the intrinsic humanity, the body becomes an object subjected to external power (similar to the Panopticon Prison Plan described in *Discipline and Punish – Foucault, 1975*), rather than a person deriving value from social relations irrespective of the body's relation to Cartesian space.

1.4 Theoretical Landscape

Academic discourse in South Africa is largely "pro-poor," up to date with recommendations of international bodies. "Pro-poor" policy is taken as the foremost agenda to achieve transformation, but the concern remains framed in a materialist model (rather different from how transformation is promoted in politics where it is not distilled to a single understanding). Here equitable distribution is measured through Rawlsian concepts of justice, often called spatial justice.

When developing theory "from the South", we need to be wary of presenting "southern theory" as homogeneous (Watson 2014). The theoretical distinction between the global South and global North remains superficial (Watson 2014). The dehumanization experienced through materialist insistence on negating human socio-spatial dimensions as experienced in the global South parallels the colonization of the interior world of an individual's private life experienced in the global North as described by Zuboff (2019).

Watson (2014) underscores the importance of place in shaping thought and action. This conception of place is more than cartesian geography and is rooted in social relations and conditional negotiation. It incorporates an implicit understanding of the commons, providing opportunities for developing theories for resistance. It insists on the recognition of human values that precede material welfare and regards the common good as the fundamental metric.

1.5 Situating the Argument

The expropriation of the inner private realm which poses an existential threat to Western democracy (Zuboff 2019) is therefore not dissimilar to the clash between a system which insists on measuring space from a materialist perspective and renders social relations that govern them illegitimate. The sum of social interactions cannot be reduced to material relations which do not account for social meaning and are unable to affirm human value in itself. Wealth, and material goods, only serve as an approximation. This implies that there is a need to theorize social relations that operate outside these dimensions, while still recognizing their existence and the force they exert on social structures.

What is enthralling about the emergence of money is that it holds more value than the material itself; suggesting that it is derived from trust, rooted in social reciprocity (Graeber 2011, 245–46). What was previously rooted in social relations, which were then recorded as a catalogue of debts, gradually shifted, finally transferring this trust to a centralized state (Graeber 2011, 245–46). Any materialist philosophy, therefore, negotiates a fundamental contradiction between substance and shape, form and content (Graeber 2011, 245–46). Similarly, centralized administrative systems, aimed at managing resources, fail to consider the social dimensions. Theorizing the commons allows this gap to be closed.

A framework that recognizes heterogeneous and overlapping perspectives and honors the "origin narrative" (Watson 2014) is important to situate the particularity of these observations. The hypothesis provides an account which suggests that the Weberian bureaucratic view is inherently at odds with indigenous spatial systems. Tracing the origins of liberal democracy, we need to interrogate the origins of liberalism and capitalism, which lent legitimacy first to the colonial rule, followed by the apartheid state and subsequently a "democratic" South Africa (all of which are concerned with the bureaucratic management and not real agency and inclusion).

Western tradition separates the person from the system and liberates them from culturally defined relations, organizing the division of labor into a rationalized order. Western individualism is born from this separation, and it is dangerous to suggest that it should be applicable in a context where meaning is socially generated because it disenfranchises people from the production of socially constructed meaning and the affirmation of their humanity within a reciprocal social structure.

To highlight the contradiction between bureaucracy and indigenous systems it is important to recognize the resemblances between the hierarchies of state, research institutions, and corporations, concerned with efficiency, at the expense of a human-centered approach (Watson 2014). The bureaucratic mode is pervasive across the political spectrum, with the right favoring individual agency through choice, regulated by the invisible hand (delivered by a free market, but hierarchically organized corporations); and the left, where the state utilizes its scale to gain efficiency in order to provide opportunities broadly.

Both center-right and center-left, despite political differences, maintain the market faith as primary instruments for achieving the public good and defining merit, and should be examined when concerned with the common good (Sandel 2020). The conceptions remain focused on providing material goods or services and Western individualism. This fails to recognize far more fundamental needs for dignity and social affirmation as understood from the perspective of social/relational conceptions of space. The attempt here is to relate South African experiences to theory from the global North while acknowledging that this mode of knowledge production remains part of the Western/bureaucratic traditions and that the limitations should therefore be recognized (Watson 2014).

When Zuboff (2019) reports on interviews with 270 participants in the London riots of 2011, there are remarkable resemblances with more recent South African discourse running concurrently with protests such as the #BlackLivesMatter movement in the global North. She explains that for some it was broadly social, not just material, but hinged on how respondents felt they were treated compared with others (Zuboff 2019, 36). The "sense of being invisible" and "the denial of dignity" (Zuboff 2019, 36) go to the heart of legitimacy and are the conceptual boundaries between tyranny and democracy.

Similarly, Michael Sandel, reflecting on the subject of race in an interview on his latest book *Tyranny of the Masses* (2020) suggests that we separate racial injustice and class too severely, undermining both (Sandel 2020). He argues that if all we did to address racial injustice was to remove barriers to meritocratic competition so that everyone, regardless of racial background, can have the same chance to rise up the economic ladder, we miss the social fact that inequality is increasing, meaning that even if there is upward mobility in a material sense, it does not mean people are gaining more agency (Sandel 2020).

As Zuboff (2019, 36) observed, for some, this is economic in nature (the lack of a job, money, or opportunity), but for many there is a deeper social dimension that is equally important. Sandel (2020) points to Martin Luther King who stated that the person who picks up your garbage is as important as the physician. He describes it as a radical theme for reconfiguring the economy to acknowledge that "all labor has dignity," when seen as a contribution to the common good (Sandel 2020). This indicates that value is socially ascribed, rather than being a utilitarian supply-demand model, and it provides an alternative point of departure for theorizing inclusivity.

Sennett (2006, 110) warns against meritocracy where a bureaucracy measures something deep inside the individual, punishing her for a lack of "usefulness" and only strengthening those who succeed. This results in a person's value becoming an impersonal procedure, stripping it of its human dimensions (Sennett 2006, 110), rather than affirming social relations. Zuboff (2019, 29) goes one step further, describing our relationship with knowledge, authority, and power as having no end, and that each generation must assert its will and imagination to resist subjugation. She points to Thomas Piketty to argue that values of social justice, and therefore democratic societies, are threatened; as a neo-feudalism, in which power and wealth are consolidated, and inherited wealth rather than merit determine achievement (Zuboff 2019, 29). She continues that the system seems to have reverted to preindustrial feudal patterns, but people know themselves to be worthy of dignity and the opportunity to live an effective life – and once they have realized this it is like an "existential toothpaste" which, once liberated, cannot be squeezed back into the tube (Zuboff 2019, 29) resulting in an irreversible revolutionary fervor.

In Ostrom's account, community members and households gain equal access to services, while exclusionary factors such as income, gender and ethnicity play an insignificant role (Watson 2014). This world is held in place by a fair, uncorrupted, non-politicized, consensual state (Watson 2014). It neglects to sufficiently theorize power dynamics, which when designing the system to integrate with the Weberian state may redistribute power among stakeholders in asymmetrical ways (Watson 2014), often along the lines of political patronage.

The challenge here is to recognize that when considering the commons, the emphasis should not be on the management of material goods, but rather should place human dignity and inclusiveness at the center of discourse. The bureaucratic systems of knowledge production remain theoretical and fundamentally contradict highly localized spatial conceptions rooted in negotiated practice and may be instrumentalized to legitimize further abstraction. Just as the pyramid shape of Weberian bureaucracy enables a corporation to add ever more people at the lower ranks, and capitalism seeks to create markets, the state apparatus may become obese for the sake of social inclusion in order to galvanize power (Sennett 2006, 30). In the South African system where patronage systems are endemic, there is often a conflation of the Weberian separation between the political post and the person or party. Sennett (2006, 30) warns that this leads to an ever-increasing bureaucracy, fattening up institutions as a form of pacification by "giving everyone a place." Indeed, Watson (2014) points to Benit-Gbaffou and Oldfield (2011, 446) to suggest that the Weberian state should be understood as an ideal, that in some conditions never emerged in the first place.

2 Case Studies

Studying South African case studies, this chapter shows that similar dynamics exist in the commons and that while it serves as a model for bottom-up governance, it needs to be theorized as a dynamic process of resistance against hegemonic power rather than a universal panacea. The case studies illustrate that the theorization of the commons is taken to be a theoretical starting point for policy makers in the

South African context, but that these do not necessarily address the inherent nature of power or capital to obfuscate its operations by manipulating the surrounding geography of power (exerting influence that is difficult to sanction since it is not bound by the internal rules that governs the commons, but rather exerts influence from the outside). The chosen case studies are presented as a reflection of the status-quo, while also providing some conceptual springboards for extending theoretical approaches, especially with respect to researchers and policy makers operating from within the Weberian State (potentially part of a process of 'decolonizing'). This contrasts with the sections on land that follow where I shall demonstrate how the concept of the commons is manipulated and the instrumentalization of the 'community' are an instance of patronage not dissimilar to the bloating of the state Sennett (2006, 30) warns against.

Underscoring an emergent, dynamic process of sustained mobilization rather than organizational design, Watson (2014) prescribes a departure from the Weberian State. What is to replace it should therefore be rooted in an understanding of social production. Following Da Costa and Van Rensburg (2008) this chapter is theoretically grounded in a Lefebvrian concept of the "production of space," and by extension the "right to the city". Here the production of space is conceived as an expression of a society's collective mind and expression of culture (Da Costa and Van Rensburg 2008).

To affirm and support an indigenous spatial ability, conceived as the ability to organize spatial information and interpret knowledge, where interrelations constituted through the social interactions define spatial production (diverging from more static Eurocentric models of spatial definition), space may be understood as a realm of possibility (Da Costa and Van Rensburg 2008), remaining unbounded by physical space. This spatial concept is expansive and generated through social signification, not external abstractions in pursuit of measurement and quantification.

In contrast, the corporation, once an innovative means of organizing collaborative investment, has enjoyed many of the rights of personhood but is now free of democratic obligations, moral calculations, and social considerations (Zuboff 2019, 206). Similarly, in relation to the Weberian state, the commons may be instrumentalized by powerful individuals acting in their own interest or on behalf of capital to evade legal constraints, if left unchecked.

To explore the South African condition, I shall discuss two interdisciplinary research projects published in 2015 (halfway through the "lost decade"), one focused on homelessness in Tshwane (capital of South Africa), and the other focused on street trading in Johannesburg. Although both of these examples are the product of the Weberian bureaucratic system (academics and other partners organized hierarchically), they both make recommendations for bottom-up systems which conform to Ostrom's principles for the commons: "Those who are affected by rules need to have agency to change them; in a system of nested tiers" (Ostrom 1990, 101–2).

These case studies both implicitly acknowledge the failure of the state to deliver a socially just system and seek to find alternatives for integration, thus supporting the core argument put forward that the commons is part of an indigenous mode of spatial production, and is implicitly recognized in policy, but is not guarantee to be immune from external coercion. From this, the discussion will turn to land, tracing how the commons, the common good and community have been instrumentalized for personal acquisition of wealth and political power at the expense of those it was meant to compensate for historical dispossession, before considering possible tactics of resistance.

2.1 *Johannesburg*

In the aftermath of Operation Clean Sweep (November 2013); the AFTRAX project (Alternative Formalities, Transnationalism and Xenophobia in Johannesburg Inner City), a research commission aimed at assessing existing knowledge of the informal economy, reflected on failures to take into account existing practices. Based on these investigations and discussions, Claire Bénit-Gbaffou, a researcher at the Center for Urbanism and the Built Environment Studies (CUBES) at the University of the Witwatersrand, compiled a report in 2015.

Core to the diagnosis of the issues put forward by AFTRAX is a restrictive approach where policies and practices are based in fear and the assumption that "the number of trading sites is limited" and perceptions that inclusive approaches would lead to the "invasion" of the streets (Benit-Gbaffou 2015, 11–13). It suggests that the city remains concerned with control and management of material resources rather than upliftment and affirmation of the citizenry that constitute the life of the city and is characteristic of the Weberian materialist mode of thinking about space. Here we observe how the Weberian state comes into conflict with bottom-up organization in its inherent need to control and dominate space, superseding human-centered negotiation, and management of space. Bénit-Gbaffou asserts a positive and affirmative approach as an activation of the street and heterogeneous land uses which facilitates buy-in from traders (Benit-Gbaffou 2015, 11–13). Because informality is endemic (inherently functioning outside the boundaries of regulation), the impossibility to enforce a restrictive approach is reflected onto the Weberian state itself as having failed to meet the mandate of government (when conceived as a territory to be controlled rather than taking a catalytic approach to space making), while oppressing the very traders it is supposed to uplift through illegalization, as well as creating insecurity of tenure and of status (Benit-Gbaffou 2015, 11). While the language of service delivery is used, the notion of service is wholly missing in the current approach since the city is viewed not as generative, but as a finite resource for which people need to compete. This breeds cultures of secrecy and abuse of power, displacing responsibility and resulting in deception to avoid accountability, corrupting and delegitimizing the system, while an inclusive approach could benefit the city in various ways to build solidarity and trust (Benit-Gbaffou 2015, 13).

Affirmative practices would end the mass of illegal traders, making it easier to manage and allowing fees from thousands of traders (valuable resources for urban management) to flow into the municipal accounts, while limiting opportunities for corruption (Benit-Gbaffou 2015, 13). It would be a pragmatic acceptance of reality and social needs, and addresses the management issues with the necessary power, legitimacy and resources (Benit-Gbaffou 2015, 13). These are negotiated through a multi-scalar approach, designed to be context-specific and include stakeholders (Benit-Gbaffou 2015, 11). Where security of tenure for street traders is recognized, control and management works (Benit-Gbaffou 2015, 13).

Where the private sector is involved, innovations from retail improvement districts (RID) suggest that coordination with cleaners and security could solve issues between the informal and formal sector quickly (problems typically arise from streets being perceived by the formal sector as dirty or cluttered due to the presence of the informal traders – it is fundamentally a classist view) (Benit-Gbaffou 2015, 13). Similarly, lending legitimacy to the informal sector and utilizing the cleaners and security as intermediaries allows city managers (whether private or municipal) to find flexible solutions from problem-solving, rather than revenue-generating, approaches (Benit-Gbaffou 2015, 13). From a municipal perspective the affirmative approach put forward by the researchers suggests enhanced participation of traders and consolidation of institutions, to build ongoing partnerships and to frame agreements, while the report indicates that the provision of dedicated funding and administrative capacity for the inner city should be recommended (Benit-Gbaffou 2015, 13). The literature critically views municipal repression of street trade as a legacy of colonial practices reproduced or reinvented in the post-colonial era, while also warning against laissez-faire approaches where tolerance deteriorates into apathy, and political competition between tiers of local government (who collect fees) devolves into collusion and a different kind of power struggle (Benit-Gbaffou 2015, 15).

In keeping with Watson (2014), who emphasizes location-specific approaches and starting from a place of indigenous knowledge, the researchers first conducted a survey of all existing traders and included them in formulating the survey (Benit-Gbaffou 2015, 14). In addition, the researchers call for a suspension of policing of "illegal traders" in order to establish an accurate picture of the situation as experienced on the ground (Benit-Gbaffou 2015, 14). This affirms the map as a mechanism of the Weberian state but puts the power of decision-making in the hands of those affected by it (but risks it ultimately being turned against them should commitments to affirmative and inclusive policy be reversed). Based on the survey, they suggest that a multi-stakeholder committee should decide on policy, by-laws and subsequent implementation (Benit-Gbaffou 2015, 14). Simultaneously, a street trader organizations forum, funded by the city and facilitated by an independent party, could empower street traders to effectively participate in the multi-stakeholder committee and other processes (Benit-Gbaffou 2015, 13–14). These are to be rooted in area-based management that use consensual and incremental definitions of the "street carrying capacity" not technical definitions or absolute terms, but rather a negotiated common understanding (Benit-Gbaffou 2015, 10).

2.2 Tshwane

In Tshwane, the *Pathways Out of Homelessness Research Report* (2015) goes one step further in questioning the relationship between researchers and what they call a "new kind of politics" expressed in compassion, generosity and justice. Explicitly rejecting neutrality, it views prevailing socio-spatial arrangements as remnants of the apartheid city and a severely unequal society. De Beer and Valley (2015), 1–2) argue that this new kind of politics departs from "mutual respect" towards "working collaboratively towards the common good, and replacing a culture of violent charity, handouts and crumbs, with a culture of generous investment and reciprocity."

They explain that a "politics of compassion" must go much deeper than charity or a welfare-based approach, to evoke the literal meaning of the word "to suffer with," and that researchers, officials, and practitioners must operate from a position of deep solidarity, without patronizing the socially vulnerable or treating poverty as a pathology (De Beer and Valley 2015, 1–2).

The purpose of this collaboration, despite the various independent institutions (with diverging goals) involved, was to work towards a social contract where various role players collectively accept responsibility for implementing a strategy which addresses the structural causes of homelessness and supports the essential social service infrastructure (Mashau 2015, 60–61). In collaboration with other parties, researchers from the University of South Africa (UNISA) are developing a theoretical framework for the concept of a social contract or social covenant which goes beyond the socio-spatial dimensions to interrogate the history, political and legal frameworks of the concept and incorporates spiritual/religious/theological dimensions and a health perspective (Mashau 2015, 60–61).

From a spatial perspective, the researchers note a recent development. In the past, street homelessness had been concentrated in the city center and while still more visible in these regions, current research indicates that a shift has occurred into higher-income suburban areas (Pathways out of Homelessness Research Report 2015, 63). The same applies to parks, areas next to water bodies, bushes and vacant land, which also play an increasingly important role (Pathways out of Homelessness Research Report 2015, 63). This indicates that inclusive practices around the commons become ever more important and that natural commons must be included in discussions on the urban commons.

As in the Johannesburg case study, consensus points to contextually defined collectives and collaborative action, according to the unique characteristics of street homelessness in every region (Pathways out of Homelessness Research Report 2015, 67). De Klerk (2015) posits that design needs to accommodate for transience to help people survive on the street. She provides examples such as amenities and basic services such as ablutions, electricity and cooking facilities in commercial buildings such as malls, or in community facilities such as churches, parks and community centers (Pathways out of Homelessness Research Report 2015, 17). Her contribution to the report further suggests that there is a need for mobile survival kits in the form of backpacks (Pathways out of Homelessness Research Report

2015, 17). This might be of particular importance to informal waste recyclers who provide a vital service where the state failed.

Spatial analysis suggests that the homeless typically congregate and sleep in deteriorated open spaces and areas where built fabric is incoherent or vacant, while well-maintained open spaces are places of refuge and provide access to amenities (Pathways out of Homelessness Research Report 2015, 20). This suggests that invisibility is often a means of survival. Vacant and therefore uncontested space is used to congregate and sleep away from the harassment of police or security services, while the ability to disappear in highly controlled spaces can provide safety and resources. Having agency over when one wants to be visible or invisible is therefore an indicator for self-determination and dignity. This echoes issues related to surveillance capitalism, where the agency to choose which aspects of a person's private life may be seen, and under which conditions, must remain in the control of the individual and may not be infringed upon by an external party. It is a fundamental right to ensure self-determination and sanction against undue influence from external parties who can otherwise turn this against the unknowing individual through subversive manipulation. The underlying questions here are what information is known and how it may be used, suggesting that any information about a person belongs to the individual person and ensures their agency.

3 Observations

The IsiXosa proverb *Umunthu ngumunthu ngaBanthu,* meaning "I am a person by reason of other people" (Lloyd 2003, 113; Da Costa and Van Rensburg 2008) underpins the theoretical matrix of this chapter which suggests African spatial production is ongoing, reciprocal and relational. Against this background of generative spatial exploration that acknowledges multiplicity and heterogeneity, or "differential space" in Lefebvre's language (1991, 52), socio-spatial differences are valued and emphasized in order to restore the precedence of the human body and the social condition over "abstract space" (Da Costa and Van Rensburg 2008). This divergence from physical conceptions of spatial formulation in favor of the non-physical undermines the Weberian state by denying the abstracted map as absolute or representative. It acknowledges the map or survey only as a tool for building consensus and claims the power of ownership over how information about the person may be utilized in order to resist the imposition of power and control through surveillance as first conceptualized in Foucault's panopticon (Foucault 1995, 195). Paradoxically, as is often the case in the global South, this does not deny the usefulness of such tools, and formalized ownership remains a critical goal for upwards mobility.

The reciprocal and relational concept, along with the right to decide how information may be utilized in the context of *ubuntu* and the African modes of spatial production shows that "empowered participatory governance" and the specific design of institutions to deliver transformative democratic strategies should be taken seriously to ensure affirmative practices (Watson 2014). Beyond producing data, it encourages a common understanding and dialogue between community members, from which a political basis is formed to support ongoing and repeated processes of engagement with government (Watson 2014). This should, as the research suggests, go beyond socio-spatial dimensions to affirm common humanity (Mashau 2015, 60–61), a politics of compassion and of deep solidarity (De Beer and Valley 2015, 1–2). This is operative in a landscape where *ubuntu* and the understanding of the *munthu,* the unique life force vested in each individual, is situated within a collectivized *bunthu*, and challenges individualism and the bureaucratic hierarchies that support them.

3.1 Land and Divisive Representation

For the final analysis of the South African landscape, we turn to land restitution processes to uncover power dynamics at play when representation and other abstractions interfere with historical spatial production and management patterns. In the introduction to this chapter we have already considered the misconception which holds that dispossession took place as a single event or in a linear process (and can therefore be easily reversed), rather than acknowledging that these processes diverge widely, playing out over long periods where African families remained on white-owned farmland (surviving sometimes as tenants and shopkeepers), well after conquest up until the beginning of the apartheid era (Beinart et al. 2018, 101–2). Different spatial understandings and different rights coexisted therefore until 1948, when one world view and spatial understanding was effectively criminalized, and bodies in space became regulated by race.

The 1913 cutoff date for land claims was chosen, since this coincides with the Natives Land Act of 1913, which is the first nationwide, state-run, dispossession and denial of rights systematically based on race. However, it was also introduced partly because the African National Congress (ruling party since the end of apartheid in 1994) under President Nelson Mandela recognized that earlier claims were likely to be tribal in nature, and hence avoided recreating the old chieftaincies of pre-colonial South Africa, since it could be the basis of ethnically based and divisive politics (Beinart et al. 2018, 99). Gradually, however, the government came to accept the role of traditional authorities in rural administration, formalizing their position in the House of Traditional Leaders through the Traditional Leadership and Governance Framework Act of 2003 (Beinart et al. 2018, 99).

Historical research found that the tribal and community claims are often grossly inflated or, in some cases, lack any support whatsoever, since it is clear that chiefs did not have as much authority over the area as is often claimed (Beinart et al. 2018, 106–7). This challenges the concept that all Africans lived in tribes ruled by chiefs that "owned" the land on behalf of communities and posits that the notion is rooted in segregationist and apartheid ideology (Beinart et al. 2018, 106–7). Instead, the researchers paint a picture where even in territories where particular chiefs held the nexus of power through a dense cluster of followers, there were other people, with diverging or no political allegiances (Beinart et al. 2018, 106–7). It is a geography characterized by high levels of mobility between diverse, interpenetrating groups with multiple forms of political allegiance, living in the region (Beinart et al. 2018, 106–7). It illustrates that the idea of exclusive tribal communities with effective control, especially in relation to land allocation, is not supported by the evidence, and this was further disrupted by labor tenancy agreements with white farmers (Beinart et al. 2018, 106–7).

Interestingly, most individuals did not receive land directly from chiefs, but by allocation from groups formed around a core kin and through inheritance, with custom granting "a woman exclusive right to cultivate any area which she had once turned over, no matter how long it was kept lying fallow" (in specific cases) and she kept the right to pass it on to further generations (Beinart et al. 2018, 98). Communal pastureland, which formed the majority of most chiefdoms, was open to all who owned land parcels, and was subject to local grazing practices (Beinart et al. 2018, 98). This places labor at the center of the socially produced space and suggests that a Lefebvrian perspective might be appropriate, while it could also form the basis on which women may seek greater recognition. The authors point to A.J. Kerr (author of the only extended analysis of customary laws of property) who considers the word "ownership" appropriate to describe these rights since they are "good against the world," thus serving as an appropriate baseline from which to understand customary rights (Beinart et al. 2018, 98).

To describe how the commons is situated within a larger geography of power and may therefore be instrumentalized, I draw on the well-publicized South African example when President Jacob Zuma encouraged the House of Traditional Leaders to seek legal advice in lodging large claims "on behalf of the people" (Beinart et al. 2018, 131–32). He saw the chiefs as valuable political allies, and when the Restitution of Land Rights Amendment Act was passed in 2014, he explicitly encouraged chiefs to lead restitution claims (Beinart et al. 2018, 99), while a parliamentarian scolded an audience of disgruntled public participation members on the matter for asking for title deeds, because they were not part of "our culture … we want our land, but our chiefs there must be our eyes, in our land" (unfortunately the researchers did not provide further particularities as to who this individual is or which community participation process they are referring to but it serves to illustrate how political agendas often use rhetoric to supersede legitimate stakeholder claims) (Beinart et al. 2018, 131–32). In one case, members of the public objected: "chiefs should make their own claims like other citizens and not launch their claims on behalf of their communities," while others lamented that "people were urged to respect their chiefs, as they are the "custodians of culture" (Beinart et al. 2018, 131–32).

Without any supporting evidence, the Minister of Rural Development and Land Reform argued (at a parliamentary hearing on the amendment bill) in February 2014, stating that "land had been removed from traditional leaders as custodians of the land, so the position could not be discounted" and that concerns about land being returned to traditional leaders "reflected 'colonial' viewpoints" (Beinart et al. 2018, 132). This divisive appropriation of the narrative was also recognized at other hearings where speakers suggested that the government was exacerbating racial divisions between whites and blacks, when in fact government officials were capturing the benefits of land restitution at the expense of claimants (Beinart et al. 2018, 142). The encouragement of tribal claims were in direct opposition to the intentions of the first Restitution Act and threatened an ethnicization of local landholding and politics (Beinart et al. 2018, 132) along tribal lines. By choosing a singular representative, a dominant group may claim a particular area, thus justifying the exclusion of those "undeserving" because they do not belong to those who "own" the land, while opening the system up for corruption since it is easier to bribe or exert pressure on an individual than it is to subjugate a heterogeneous population with highly localized spatial negotiation and autonomy.

Some social clusters were critical of the role that chiefs had played in the homelands during apartheid (since it was then, as now, a tool of division and subjugation through mock representation) and wanted to establish alternative, local democratic structures, pursuing opportunities under the government's pilot land redistribution program (Beinart et al. 2018, 111–12). These claimant communities were keen to keep traditional authorities at the margins, but this became increasingly more difficult over the years after apartheid (Beinart et al. 2018, 111–12). Ironically, this happened concurrently with the systematic dismantling of the checks and balances that the state apparatus had in place (essentially attacking the Weberian state) yet also resorting to apartheid racial tactics to disenfranchise minorities and deny indigenous spatial practices that allow heterogeneous populations to administer localized solutions.

Land claims over large areas, led by chiefs, suggests a supposition that, based on their origin, some communities have a deeper connection to particular areas of land than others, and that they are the "rightful owners" of the land (Beinart et al. 2018, 133–34). The authors cite an example in Mpumalanga province, where Tsonga people are considered by some to be "immigrants" from Mozambique, despite having immigrated into the area before the colonial boundaries were entrenched (Beinart et al. 2018, 133–34). In a further exceptionally divisive example, they illustrate how the same Tsonga speakers were described as "'imported on a large scale' to serve whites," and other wording which systematically dehumanizes these minorities (Beinart et al. 2018, 133–34). Most chillingly, Beinart et al. (2018, 133–34) report the comparison with insects, an echo of xenophobia elsewhere, and even the Rwandan genocide.

The Land Claims Commission was legitimizing the idea that land restitution is about returning land back to the "original" settlers, "indigenous" people, rather than respecting the Restitution Act, which explicitly states the claimants should qualify on the basis of disposition through racial discriminatory practices after 1913. The logical extension of neglecting such clear provisions become even more alarming in the Western Cape where Khoi peoples claim that the Western Cape, or even the entire country, "rightfully" belongs to them as the first people, with deep implications of a shift towards what Beinart et al. (2018, 133–34) call "indigeneity." It might even be considered counter-constitutional because the South African Constitution (1996) makes the declaration that "South Africa belongs to all who live in it, united in our diversity, whilst recognizing the injustices of our past" (Constitution of South Africa 1996). This is not to deny that we also have the deep responsibility to cherish and protect these cultures, threatened by cultural-linguistic extinction, and should make special use rights operable to protect the cultural heritage. Here theory developed out of the commons could prove very useful.

Less belligerent, yet equally divisive, is the danger that a pervasive rhetoric of "community" can shroud the concept in a dense fog which obscures cleavages, conflicts and identities within local society in order to obfuscate networks of patronage and political power that monopolize resources, whatever their intended destination (Beinart et al. 2018, 100–101). The reality is that family claims and community claims are often merged into large overarching claims by the Land Claims Commission (sometimes by claimants themselves – sometimes by coercion), resulting in traditional leaders being privileged, at the expense of individual or family claims (Beinart et al. 2018, 117–18). In some cases, based on the merger, entirely new fictional "communities" (this is a rhetorical device used to reallocate resources along patronage lines) are created, even if they have no rules of land allocation, shared histories, or experience of disposition (Beinart et al. 2018, 117–18). The justification provided is that the commission is spared the task of adjudicating between conflicting claims and only has to deal with one claimant group rather than several (Beinart et al. 2018, 117–18). This is a good example where bureaucratic efficiency trumped justice and illustrates the key mechanism on which this critique is founded.

The instrumentalization of representation on behalf of a particular "community" (used as a rhetorical smokescreen), supposedly homogeneous, serves the agendas of politicians and chiefs and the complicit Land Claims Commission, and is therefore in direct contravention of the Restitution Acts intentions. This happened despite a Constitutional Court judgement handed down in 2007 which stated that the task is made easier by the key term "community" being defined in the land restitution legislation which limits it to a community dispossessed of land after 1913, and which, unless the context otherwise indicates, is a group of persons whose rights in land are derived from *shared rules* determining access to land held in common by such a group (Beinart et al. 2018, 119). The judgement reiterates that shared rules, and not the presence or absence of a traditional authority, was the crux of the matter, and that none of these attributes are requirements in themselves or collectively, yet the commission ignored this and maintained this divisive mode of operation (Beinart et al. 2018, 119).

An earlier ruling on the issue affirmed that dispossession and some element of commonality between the claiming community constitutes "a sufficiently cohesive group" (Beinart et al. 2018, 120). It ruled that labor tenants evicted by white farmers in the context of generalized racial discrimination did not need to have been removed by direct state action or resulting from specific legislation to have suffered dispossession (Beinart et al. 2018, 120). This offers a vast number of individual labor tenant rights, while rejecting most amalgamated community claims based on tribal groupings, yet the commission continued to favor the latter interpretation for processing redistribution (Beinart et al. 2018, 120), and suggests that the definition of "community" itself is a territory that may be colonized by external political powers.

4 Conclusion

This chapter posits that interactions between state apparatus and the commons needs to be critically evaluated as a geography of power. It suggests that existential threats to freedom in the global North is not dissimilar to the subjugation experienced through an abstraction of physical space which negates social signification in the South African context.

Policy in South Africa recognizes that dignity and inclusiveness need to be primary measures, but that this is not sufficient because it does not necessarily provide agency and does not yet close the gap between hierarchical bureaucracy and bottom-up organization. The chapter highlights that the whole is bigger than the sum of the parts, since social production incorporates spatial and social dimensions that are highly localized and cannot be accounted for in totalizing abstractions. It illustrated that territory was never divided into neat geographical administrative regions, and therefore, the idea of having a centralized representative political system is inherently foreign, undermining indigenous practices. The mechanisms utilized by a powerful elite to "represent" a "community" are often framed as a means of restoring a pre-colonial order focused on tribal claims for the "rightful owners," but in fact obfuscates the reality that this is a continuation of apartheid rhetoric and fundamentally undermines heterogeneous spatial practices and diverse, coexisting peoples.

With planning and academia being rooted in the Weberian hierarchical order it is often difficult to escape these pitfalls of "representation" or expert bias. Alternatives may be found in affirmative policy and a practice of deep solidarity or "politics of compassion" (De Beer and Valley 2015, 1–2) which primarily affirms humanity and takes account of local specificities, to avoid an oversimplified focus on resource allocation. What is required is a theoretical framework which recognizes the commons as a pre-existing framework, and agents operating from within the hierarchical structures of the modern nation state should not seek to formalize it from the outside, but rather should pay attention to how it may be undermined by patronage and corruption. It requires practices that work from a place of knowledge and consider existing structures, which in dialogue with the affected community seek ways to integrate with the formal state system. In such scenarios, researchers and practitioners must avoid being experts, but rather carry only a specific set of knowledge that should be viewed on equal footing with that which is indigenous and highly localized. This also relates to other forms of representation, with the case study questioning the legitimacy of oversimplified use of phrases such as "community" as homogenizing groupings and suggests that it should not be based on any one identity, but rather on shared rules, negotiated to suit a particular geography and circumstances.

To develop tactics of resistance, three potential avenues may be investigated. The first points to linguistics and textual analysis which takes rhetorical devices seriously. This suggests that key indicators may signal that terms such as "community" may be totalizing, obfuscate true intentions, and organize access along patronage lines that seek to agglomerate power rather than being truly inclusive. The second focuses on structural analysis, in particular indicators of openness and the right to information. It is evident that the map or survey are instruments of power in their own right, and how we organize knowledge systems is a territory for contestation. Who decides how this knowledge is produced, and to what end, is a central theme in both the global North and the global South with the possibility for theoretical cross-pollination.

Lastly, it may be conceptually fruitful to distinguish between "natural commons" and "constructed commons," where the former is defined by what pre-exists in the natural world and is the purview of all organic life, thus rendering it a pre-legal concept to which socially constructed organizations such as corporations and the state hold no claim, protecting natural resources from exploitation (particularly from powerful individuals who may utilize these vehicles for their own gain), while the latter pertains to all forms of socially constructed domains. This socially constructed commons presupposes that corporations and powerful individuals benefit from a socially constructed inheritance and suggests a need for theorizing systems of reciprocity.

References

Beinart, William, Peter Delius, and Michelle Hay. 2018. *Rights to Land: A Guide to Tenure, Upgrading and Restitution in South Africa*. Johannesburg: Jacana Media.

Benit-Gbaffou, Claire. 2015. *In Quest of Sustainable Models of Street Trading Management: Lessons for Johannesburg after Operation Clean Sweep Claire*. Johannesburg: CUBES, University of the Witwatersrand.

Benit-Gbaffou, Claire, and Sophie Oldfield. 2011. Accessing the State: Everyday Practices and Politics in Cities of the South. *Journal of Asian and African Studies* 46: 445–452.

Bovaird, Tony. 2007. Beyond Engagement and Participation: User and Community Co Production of Public Services. *Public Administration Review* 67: 846–860.

Constitution of the Republic of South Africa, Act 108 of 1996. *Archive*. https://www.justice.gov.za/legislation/constitution/SAConstitution-web-eng.pdf. Accessed 26 Oct 2020.

Da Costa, Mary-Anne, and Rudolf van Rensburg. 2008. Space as Ritual: Contesting the Fixed Interpretation of Space in the African City. *South African Journal of Art History* 23 (3): 30–42.

De Beer, Stephan, and Rehana Valley. 2015. Finding Pathways Out of Homelessness: An Engaged, Trans-disciplinary Collaborative in the City of Tshwane. In *Pathways Out of Homelessness Research Report 2015*. Jointly published by Tshwane Homelessness Forum, the City of Tshwane, University of South Africa and University of Pretoria.

De Klerk, Marianne. 2015. Policy Recommendations for Pathways Out of Homelessness: A Spatial Perspective. In *Pathways Out of Homelessness Research Report 2015*. Jointly published by Tshwane Homelessness Forum, the City of Tshwane, University of South Africa and University of Pretoria.

Fukuyama, Francis. 1993. *The End of History and the Last Man*. New York: Perennial, Harper Collins.
Foucault, Michel. 1995. *Discipline and Punish: The Birth of the Prison*. New York: Vintage Books, Random House.
Graeber, David. 2011. *Debt: The First 5,000 Years*. New York: Melville House Printing.
Klein, Naomi. 2007. *The Shock Doctrine: The Rise of Disaster Capitalism*. New York: Metropolitan Book – Henry Holt and Company.
Lefebvre, Henri. 1991. *The Production of Space*. Trans. D. Nicholson-Smith. Oxford: Blackwell.
Lloyd, Rod. 2003. Defining Spatial Concepts. *Urban Design International* 8 (3): 105–117.
Mashau, Thinandavha Derrick. 2015. Thinking Globally, Acting Locally: Pathways Out of Homelessness in the City of Tshwane. In *Pathways Out of Homelessness Research Report 2015*. Jointly published by Tshwane Homelessness Forum, the City of Tshwane, University of South Africa and University of Pretoria.
Ostrom, Elinor. 1990. *Governing the Commons: The Evolution of Institutions for Collective Action*. Cambridge University Press.
Sandel, Michael. 2020. *Event: Darren Walker and Michael Sandel Discuss "The tyranny of merit" (video)*. YouTube: Ford Foundation. Archive. https://www.youtube.com/watch?v=ldLYP7unW4A&t=3527s. Accessed 20 Sept 2020.
Sennett, Richard. 2006. *The Culture of the New Capitalism*. New Haven and London: Yale University Press.
Watson, Vanessa. 2014. Co-production and Collaboration in Planning: The Difference. *Planning Theory and Practice* 15 (1): 62–76.
Zuboff, Shoshana. 2019. *The Age of Surveillance Capitalism: The Fight for a Human Future at the New Frontier of Power*. New York: PublicAffairs, Perseus Book, a subsidiary of Hachette Book Group.

Chapter 8
Expressing Urban Commons: Architectural Ambiguity in the Construction of an Improvisational Future

Nicholas Frayne

Abstract Using ambiguity as both a conceptual framework for understanding architecture and as a social practice for living together in an increasingly urban world, I argue that by utilizing uncertainty to guide the stories we create from our lived environments, ambiguous architecture can turn our cities into transformative spaces. In light of growing inequality, continued exploitation, mass urbanization, and climate change, architecture can no longer remain a practice of ordering and disentangling urban life. Instead, architecture needs to support the fertility that exists within uncertainty and entanglement, enabling people to form new relationships, new practices, and new tools for addressing our collective challenges. This chapter engages in three investigative plateaus: (1) living with ambiguity, (2) composing ambiguity, and (3) encountering ambiguity; with the focus narrowing from the urban scale down to the architectural. At each plateau, ambiguity is used in different ways, describing modes of urban sociality, expressive structures, and communication through bodily experience, to argue that architecture can shape not only our actions, but the foundational attitudes that guide social dynamics and underlie urban commons. Throughout this exploration, I draw on emerging scholarship on decolonialization, urbanism, and design to propose an architectural philosophy that could lead us towards an enabling, improvisational future. Ultimately, this chapter offers ambiguity as an explorative tool for imagining the ethical responsibilities and roles of designers in our urban futures.

Keywords Architecture · Identity · Violence · Commons · Urbanism · Decolonization · Ambiguity

---

N. Frayne (✉)
School of Architecture, University of Waterloo, Waterloo, ON, Canada
e-mail: nffrayne@uwaterloo.ca

G. Bruyns, S. Kousoulas (eds.), *Design Commons*, Design Research Foundations, https://doi.org/10.1007/978-3-030-95057-6_8

On a hill overlooking the Alabama River you can encounter a transformative architecture. Hundreds of rusted metal figures hang above and around you, vivified as you walk through the space. This space tells the stories of thousands of named and unnamed victims of terror lynching, siting violence as part of our familiar, everyday lives. This architecture forces us to rethink the familiar, destabilizing the foundations of existing social dynamics. In doing so, it opens up space for new dynamics, new ways of living with each other, and new ways of thinking.

While the National Memorial for Peace and Justice in Alabama is focused specifically on the challenges of memorializing mass violence, the way the architecture deals with its subject matter can help us confront the unprecedented issues that we face today: namely, the global challenges of city-building in a future defined by mass urbanization, increasing inequalities, and a rapidly changing climate. Aligned with emerging critiques of neoliberal individualism, spaces like the National Memorial for Peace and Justice work towards social and political practices that resist violence and the limiting definitions of identity that it relies on (Butler 2020, 24–25; Sen 2006; Waller 2007). These practices are enabling, focused on expanding our capabilities to realize our own futures by destabilizing the terms of recognition by which we define who is part of "our group" (Butler 2020; Sen 1999). Such unprescribed ways of living with each other open up paths for new forms of exploring, new connective rhythms, new fertile ensembles that can help us confront and alter the trajectory of exploitation, enmity, and oppression that we are on today (Mbembe 2019; Smith 2016). An architecture that supports these destabilizing practices is one that engages the potential of the built environment to actively engender a future that is more open, more sustainable, and ultimately more humane. It is on such an improvisational future that ambiguous architecture finds its foundation.

Yet urban development today can largely be described as a practice of ordering urban space, creating what Richard Sennett has called the overdetermined "brittle city" (Sendra and Sennett 2020, 24). This enclosed ontological conception of urban life is enacted and supported through contemporary practices of building that destroy rather than adapt, that discourage experimentation, restrict extra-normative behavior, and prescribe use (Sendra and Sennett 2020, 25–30, 75; Simone and Pieterse 2017, 9, 14–16, 22–25). In turn, these environments perpetuate existing identities and entrench normative ways of living that have led us to the harsh realities of extreme poverty, oppression, and exploitation amidst what Doug Saunders has called the "great, and final, shift of human populations" to the city (2010, 1). Since our conceptions of how we relate to each other are formed socially, through relational, lived experience, it follows that spaces which inhibit our ability to speak, act, and create together scaffold the narrow worldviews that support violence and limit the ethical assessment of justice (Arendt 2018; Butler 2020, 16, 147; Sen 2009; Sen 2006). The common spaces where social relations are held, captured, or enabled become instruments for building either a limiting or an open future (Amin and Thrift 2016, 17; Simone 2019). To borrow from Ash Amin's definitions of the "good city," we can see that "brittle" urban practices work against the "habit of solidarity" on which our future cities need to be built (2006, 1011). This foundational habit is a practice of equality, working towards "an increasingly *avowed* interdependency" between human and non-human life (Butler 2020, 45). Our urban spaces can begin supporting this practice, operating to expand our capabilities for connection.

Structured around ambiguity as both a conceptual framework for understanding architecture and as a social practice for living together in an increasingly urban world, I argue that architectural ambiguity can help build such a connective future. It is not a checklist, but a way of thinking of architecture in tandem with the challenges of our urban future. This chapter engages in three investigative plateaus: (1) living with ambiguity, (2) composing ambiguity, and (3) encountering ambiguity. Together, these suggest that a philosophy of architectural ambiguity is a way to conceptualize and make cities that enable vital processes of connection and alliance on which our futures can be built. Within each plateau, ambiguity is explored at different scales and from different angles. *Living with ambiguity* describes improvisational modes of urban sociality and development at the district scale of the city; *composing ambiguity* looks at the challenges of making and experiencing architecture in relation to our lived histories; and *encountering ambiguity* offers figural modes of architectural expression as a way to transform our senses of normalcy. Positioning architecture as a framework for understanding who we are in relation to other people, I argue that architecture influences identity primarily through the affective manipulation of bodily movement and memory. This philosophy sees architecture as a potent force in building a more sustainable, adaptive, and enabling future. Architectural ambiguity is an urban practice that can expand our capabilities to openly engage with each other in and through our urban commons.

1 Living with Ambiguity: Cities Today, Cities Tomorrow

At the 2020 World Urban Forum in Abu Dhabi, over 168 countries agreed that cities are undoubtedly the sites where today's futures will take shape, and that cultural heritage and identity in cities are essential to peoples' empowerment and to the development of a sustainable social, political, economic, and environmental future (Amin and Thrift 2016, 11–15; UN-Habitat 2020). This declaration highlights the entanglement of "urbanization, culture, and innovation" through which the complexities of our global challenges can be assessed and addressed (UN-Habitat 2020).

However, not all urban culture works toward creating a more sustainable, innovative or just future. There is a violent dimension to the preservation and enunciation of culture and origins that we see today in rising ethno-nationalist, neofascist, and other xenophobic movements that mobilize a warped communal ideology as a moral justification for violence (Arendt 1976; Butler 2020, 53). In these exclusive senses of identity, victims of violence can be falsely separated from what we consider the self, eliminated from perception and perceptibility, meaning that they have "already died a 'social death' in the eyes of the perpetrators" and so any violence against them does not register as such (Arendt 2006; Butler 2009; Butler 2004; Mbembe 2019, 76–78; Sen 2006; Waller 2007, 197). The construction and systematization of such violent ideologies happens through both formal laws, policies, and regulations, and the informal social practices that underly them (Arendt 1976; Lemkin 1944). While the policies and codes that govern architecture and urban development are undeniably integral to practice, I focus here on the latter – the attitudes that guide those governing regulations and the role of the built environment in shaping them. Given that the lives lived within cities are inextricable from the city itself, we need to understand what urban practices and attitudes could work against these violent cultures.

Often rooted in on-the-ground investigations into informal urbanism, recent literature characterizes cities as places of inconclusive vitality and improvisation that form the "social, physical, and political terrain or our collective lives" (Amin 2004; Amin and Thrift 2016; Cuff et al. 2020, 1; Simone 2019; Simone and Pieterse 2017). As such, they are relational networks of transient, overlapping flows that extend far beyond the local geographies and histories of a place and constantly reform the collectives that collide and connect within them (Cuff et al. 2020, 44–52; Amin 2004; Amin and Thrift 2016; Mabogunje 1990).

Yet in practice urban development is largely concerned with fixing urban space, opposing the fluid nature of urban life (Sendra and Sennett 2020, 21; Simone and Pieterse 2017, 31–59). Formal cities are dominated by plans, visions, and regulations that value order, control, and the individual over unpredictability, contestation, and interdependency. The global project of urban development continues to extend a philosophy of overdetermination and "speculative destruction," turning cities into sites of control in the service of a status quo rooted in exploitative social dynamics (Lowe 2015; Sendra and Sennett 2020, 23; Simone 2019, 24; Toušek and Strnadová 2016). While the need for safer spaces, less vulnerable structures, and less precarity often underlies these efforts to order and organize urban life, this philosophy disregards the vitality that exists within uncertainty and incompleteness (Sendra and Sennett 2020, 32; Simone 2019, 98; Simone and Pieterse 2017). This uncertainty is, I argue, a key characteristic of urban life that development should aim to support. Aligned with major shifts beyond the field of urbanism, ambiguous urbanism is not focused on making a questionable perfect city, but rather on expanding our capabilities to realize the things we have reason to value (Sen 2009, 231; Sen 1999).

While on the surface disentanglement (ordered, controlled urban form and function) may seem beneficial, the removal of "blind spots" also removes spaces for undercover operations, where one can work outside of the formal structures of power that are, despite some good intentions, hardly benign. Moving people into more sturdy housing does not necessarily mean more opportunities or choices; such relocation or displacement can actually break the makeshift relations on which urban life relies (Simone and Pieterse 2017). The lack of clarity that entanglement provides allows for contestations, for making claims, and for residents of urban districts to remain "in play" in their own lives (Simone and Pieterse 2017, 13). By unravelling and simultaneously controlling urban space, building practices can redefine so-called democratic life in limiting and oppressive terms, muting our voices and halting our actions. In so doing, such controlling practices can constrain our ability to conceive of the possibility that we are connected with even the strangest among us – a possibility required if we are to confront and resist injustices (Mbembe 2019, 9; Sen 2009, 124–51; Sendra and Sennett 2020, 39; Simone 2019, 24).

Yet these destructive practices of regulation, control and order are not the only forces at play. Operating in the shadows cast by expanding bubbles of detached enclosure, we find ways of living that engage strangers in new and fundamentally open ways (Simone 2019, 127). The exploration of such vital practices is a frontier for urban research, forming what Vanessa Watson at the University of Cape Town has called a "southern theorizing project" that looks to "outside" spaces for ways of operating with complexity (Watson 2016, 36–40). Featuring prominently in this project is the work of AbdouMaliq Simone (2019), who examines "uninhabitable" districts that operate in the urban peripheries produced in the wake of systematized enclosing – the process of making the urban "habitable." Instead of problematizing these districts as something to be fixed (although they have innumerable problems that desperately need fixing), Simone argues that they offer a way of *living-with* the urban "as something strange, seemingly impermeable to calculation or figuring" (Simone 2019, 93). These urban districts are characterized by unfixed rhythmic improvisation, a politics of opportunistic "strange alliances" in an atmosphere unfettered by a fear of failure (Simone 2019, 136).

Places like Seelampur in Delhi or Kalibata City in Jakarta suggest the catalytic potential of interacting with each other in fundamentally open ways: an urban culture that, despite its oppressive hardships, strives to creatively overcome new challenges through the uncertain formation of unexpected alliances. Spaces like the informal markets that persist across much of Asia and Africa support these "unsettling" interactions, laying a foundation for residents to remain active in realizing their own futures. Here, we find an ambiguous social and political practice that does not withdraw into enclosed socialities but remains in open motion, always on the move. These environments and the lives within them remain in a state of incompletion, resisting any overarching narratives that consume vital, particular alliances into a "big picture" that obscures the tentative and textured realities of urban sociality (Simone and Pieterse 2017, 88–94).

These improvisational lives work outside and often in spite of the formal building practices within which they are entangled (Simone 2019, 99). Yet this does not have to be the case. Can building as a formal practice – while tied to codes, planning, governance, and existing power dynamics – support the vitality of such incoherence? It is perhaps a challenge, but not an insurmountable one.

Such ambiguous sociality is deeply tied to the senses of identity we bring to and alter through our actions within the city; an identity that changes – and can be changed – throughout our individual lives as we encounter both human and nonhuman bodies in the city (Amin 2008). An ambiguous, creative, and improvisational collective identity is a cultural attitude that enables, empowers, and expands our definitions of whose lives are grievable. It is grounded in the notion of "home" not as a single, static place where we are with "those like us," but rather a place of uncertain operations between strangers that are no longer strange (Simone 2019,

120). A specific cultural mode that is, to elaborate the 2020 World Urban Forum's argument, a key part of an innovative and sustainable future.

While we can see that ambiguous, interdependent senses of identity are not only possible but vital to future cities, the role of the built environment in enabling them remains opaque. Just how can architecture support these practices of living with ambiguity? To explore this, we must examine the links between social dynamics and the spaces that shape them. These urban spaces are sites for display, performance, and encounter, where our relationships with each other form and change as we engage in creative discourse (Toušek and Strnadová 2016; Sen 2009). They are common places for exchange, deliberation, contestation, and innovation – essential resources for creative action.

2 Composing Ambiguity: Communicative Structures for Urban Commons

Unlike common resources such as forests, fisheries and rivers, the great resources of the city are the intangible urban commons, produced through the relationships formed in the city (Hellman 2018; Martin 2016). At a macro-scale, it can even be argued that the city itself is a commons, a collective social resource on which all inhabitants depend and through which our vital networks take shape (Boonen et al. 2019; Foster and Iaione 2016). An urban common is "an unstable and malleable social relation" that is formed dynamically as people interact with their broader environments – which are both human and non-human (Harvey 2012, 125).

The practice of commoning, here defined as the ability to form "connective infrastructures," defines whether cities fail or succeed in the face of increasing insecurity (Amin 2016, 779; Hellman 2018, 46). The coming-together of seemingly unconnected lives allows for unexpected solutions, ways of thinking, and modes of adaptive action to arise. While the intangible urban commons are socio-economic, cultural, and political, they are inevitably tied to the built environment of the city. There is a twofold challenge to this practice of commoning: forming unfixed ensemble relationships relies on being able to see opportunity in an encounter with the strange, which in turn relies on space for collaborative action and on our existing conceptions of how we relate with one another. These challenges can be approached through the built environment in two broadly categorized ways: (1) through providing the disordering infrastructure that open social encounters require and (2) through structuring our sense of identity, forming a foundation for social dynamics.

The first approach stands in a long lineage of urban design, stretching back to the first cities and the social life that planned public spaces supported. Vitruvius wrote that the first architecture was built to sustain the deliberative assembly that arose from people coming together around a sacred fire (Vitruvius 1960, 38). Simplifying his parable, we can understand that social interactions require architecture of a certain size, organization, volume, amenities – a supportive infrastructure of function.

Building on Hannah Arendt's (2018) definitions of public commons as a discursive space of appearance and updating Jane Jacobs's (1992) work on urban street life, Sendra and Sennett (2020) reverse Vitruvius's story, exploring the potential for architecture to shape social assembly, putting the built environment before the sacred fire. They examine spatial infrastructures, compellingly arguing that an unstable, improvisational civil society can be designed for, providing examples of interventions and design processes that enable "fluid relationships of exchange" by "disordering" how people can use space (Sendra and Sennett 2020, 119). However, the design guidelines they explore are only half of the challenge. Spatial configuration in cities may indeed allow people to use the city in improvisational ways, but there is no guarantee that such use will take place. Allowing for engagement with strangers does not mean that will happen, and certainly does not mean that xenophobia will not perpetuate or even deepen through those interactions (Fincher et al. 2019).

This brings me to the second – and underexplored – aspect of commoning and architecture. Human dynamics in public spaces are deeply tied to the senses of identity that we bring to such spaces – identities that are continuously re-forming as a "pre-cognitive template for civic and political behaviour" (Amin 2008). This foundational template is often invisible to us, forming the "unknown knowns" that in turn structure the unacknowledged assumptions we make about each other (Bailey 2012). The way that architecture influences and structures these foundational assumptions is vital to understanding just how the built environment can enable people to form commons without adhering to the variety of prejudices we all hold. We need to better understand the operation of architecture at a pre-cognitive level if the practice of city-making is to have any hope in undermining the systemic injustices that mediate our ability to interact in common spaces. Thinking through architecture's position in relation with the urban milieu of communicative encounters can help to develop and guide such a destabilizing architecture.

Architecture is not representational but communicative; space structures our movements as we performatively re-construct our definitions of who we are in relation to others. These re-constructions happen throughout our lives as we link together and "edit" past experiences while moving through our environments, creating our frames of reference for reality. The fact that lived experience is fundamentally temporal means that our understanding of the present is contingent on the constructed narratives of our pasts that legitimate, sustain, and animate the facts of "what really happened" (Arendt 1970, 52; Saïd 2019, 255). Architecture organizes the experiential sequences that shape who we are today, providing an active ground for composing and re-constructing memory. Space is not a mute background to our lives, but rather an expressive agent, influencing how we act and how we remember our lives.

All spaces are to some extent expressive in that we create the stories of who we are through them. We do not live independently of our environments and so our stories are contingent on how the expressions of others, manifest in this case as architecture, intersect with and alter our existing frames of reference, which stay with us through memory (Deleuze and Guattari 1987, 316; Vesely 2004, 43–109).

The creation of meaning is a dynamic balancing act between what we bring to a space and what the space brings to us. Considering architecture as a communicative, expressive frame, we can understand that our environments contain messages that we interpret according to our existing systems of meaning-making. These existing systems exist for us as memory; our actions can appear to us as just or reasonable in relation to our stitched-together memories, which are shaped by our encounters with expressions of others and filtered by our own pre-existing attitudes.

If our existing frames of reference can limit the meaning we make from architectural experience, then can architecture ever provoke radical change? Our present sense of reality has the potential to be reinforced through our interpretations of expressive space. Yet, as I will examine in detail later, many of the existing frames of reference are conservative, but by no means fixed. This means that while most architecture today does not actively provoke change, this does not have to be the case: the experience of architecture can *diffract* our existing identities, potentially destabilizing the existing biases and prejudices we all hold.

In a world dominated by pervasive and often invisible racism, systemic violence and individualist apathy, the structure of these relational, diffracting frameworks – and not just the content – becomes important. Like all frames, architecture carries a message within its structure and uses the act of framing to implement and make real that underlying message (Butler 2009, 70). These messages are not always deliberate but are inherently tied into our personal perspectives, formed through our lived histories. These histories are, to paraphrase James Baldwin, controlling forces present in all we do (Baldwin 2018, 42–43).

This poses a serious challenge for designers hoping to affect social change. How do we know that even our most empathetic or emancipatory works are not a result of historical systems that actually serve to reinforce discrimination and exploitation? Recent postcolonial research has compellingly linked liberal movements and the development of modern democratic values, normatively conceived of as moving the world towards equality, to the opposite: an entrenchment and obfuscation of historic inequality and exploitation (Butler 2009, 131–35; Hudson 2019; Lowe 2015). Indeed, we cannot be sure that our stories and the frames that support them are not inadvertently tied to, if not in service of, systemic violence.

Yet there is a baseline intention that we can attend to. If there is always a chance that our framed messages reinforce systemic inequality and the enclosed realities it relies on, then perhaps we can aim to destabilize the frame itself? Simply put, the intentions behind any practice of framing should not be to find the right message, but to communicate a message in the right way. Like any creative work, architecture draws together the world beyond itself into a new composition (Deleuze and Guattari 1987, 312). The power of architecture to alter our lives lies in the specifics of this process of assembling (Grosz 2008, 13). Some architecture assembles with indifference; some with oppression; while yet others hardly assemble anything at all. In its mode of assemblage, architecture sets the terms within which we construct meaning from experience, influencing the living histories that ground our social dynamics. We need to find a structure, a way of framing, that does not circumscribe or predetermine our responses and ability to engage with each other. This mode of composing can help ensure that the diffraction of identities through architecture does not perpetuate the existing prejudices of the designer or the one experiencing the architecture.

In his work on media and morality, Roger Silverstone (2007) explores the rhetoric of evil that often accompanies mass violence such as war and genocide. This rhetoric is absolute, leaving no room for ambiguities, nuance, or the potential for critique (Silverstone 2007). By framing an event or group of people tied to an event in terms of evil, violent responses and actions against such people are presupposed as "good," limiting the possibility of unbiased assessments of injustices (Sen 2009, 123–40). The "domain of justifiability is pre-emptively circumscribed" by the form of communication (Butler 2009, 155). A framework of enclosed absolutes prescribes a response in equally absolute terms. In opposing such frames, an inconclusive, ambiguous framework may be able to communicate without presupposing and restricting meaning, thereby directing (rather, un-directing) the diffraction by which we create meaning from the experienced frame (Pérez-Gómez 2006, 213). These kinds of frames allow for the very terms of discussion to be criticized, opening the framed messages to new ways of being read, new justificatory schemas by which challenges and injustices can be apprehended (Butler 2020, 140; Butler 2009, 72). This lack of certainty keeps our critical faculties awake, allowing us to question the definitions of identity we construct through the experience of architecture (Butler 2020, 162; Arendt 1970, 8).

Our frameworks need to be ambiguous if they are to avoid reflecting and reinforcing normative systems through us. Instead, they may be able to diffract our unconscious prejudices, splitting them apart and rendering them in new, unstable ways. Given that the production of identity is an iterative, performative process, I argue that ambiguous architecture can destabilize our present identities and thereby hold "open the contingent and unpredictable forms that lives may take" (Butler 2020, 146; Butler 2009). To understand how an ambiguous structure can enable an improvisational identity, we now return to the discussion on architectural expression and memory.

3 Encountering Ambiguity: Body, Memory, Movement

While some buildings may mean generally the same thing to many people, the majority of architecture we encounter has either a specific significance to us (think of a childhood home) or little impact on who we are at all (think of that anonymous office block you passed by at some point). So, what then of architecture's supposed ability to structure identity and the actions undertaken through those identities? Given the infinite histories we each hold, the meaning we create from architecture will never be the same. However, it can structure a *kind* of identity, a mode of conceiving oneself in relation to others. As argued earlier in this chapter, improvisational identity is not a definition of "who I am" but a practice of enacting a worldview that treats strangers as potential accomplices. Moving towards this non-absolute way of living, we need to focus on architecture as a measure and scaffold of the rhythms of social life, working to both guide action and hold memory.

In its first operative definition, architecture is a malleable infrastructure that can be used to constrict, enforce and physically alter our ability to act, forcibly structuring actions that manifest the lived histories we bring to any social encounter (Mbembe 2019, 81; Simone 2019, 18). Architecture is the physical site for the performance of identity. As it guides that performance, it guides the way we claim and display public identity, affording free assembly and discourse, or prohibiting it. Protests, demonstrations, and riots happen between, through and against architecture; squares, alleys, boulevards, parks each afford different social performances. Our identities have a direct relationship with the spaces in which we can express ourselves: space, action and identity are intertwined.

The second operative definition of architecture is more complex and arguably more foundational to social dynamics. Beyond simply being a space for action, architecture also operates at a deeper level, informing the pre-cognitive foundations that influence how we act even in the least controlling spaces. The relationship between body, space and memory offers us a way to explore this operation. In so doing, we can begin to understand just how our urban environments could reframe our ability to form the crucial commons we need.

Our environments are experienced in time. Architecture is a spatial and temporal event, operating to inform our sense of identity as stories do: through unfolding spatial sequences that shift as we create meaning from continuous lived experience (Pérez-Gómez 2006, 131–33; Vesely 2004, 79). These orientational sequences are not a narrative in a conventional sense but rather function as a continuum, always working with what comes before, being shaped by it and in turn shaping it. Our sense of who we are comes from the shifting continuum of our lived memory (Vesely 2004, 101).

Memory, and therefore identity, is not fixed (McKim 2019, 107; Nora and Kritzman 1996; Young 1993). It seems a simple statement, but it is consistently ignored by universalist ontologies that strive for extra-temporal absolutes against which we must measure ourselves. As an alternative to such absolutes, the philosopher Gilles Deleuze and the psychoanalyst Félix Guattari have argued for an ontology of *becoming* that conceptualizes our sense of who we are as in-progress and multiple, transforming as we form new compositions and rhythms within the world (Deleuze and Guattari 1987, 232–309). Tied to the ever-changing lived experience of a world that is without beginning or end, this philosophical ontology is fundamentally experimental, dealing directly with the improvisational syntheses that can occur between seemingly disparate forces, rhythms, ways of living and thinking (Deleuze and Guattari 1987, 25). When considered in tandem with Deleuze and Guattari's thinking, we can begin to understand that both our memories and identities are in a perpetual state of unfixed becoming, re-forming in time as we experiment, engage, and are engaged with by the environments that we are embedded within (Pérez-Gómez 2011). Memory and identity change ceaselessly, "updating" every time we remember, incorporating our present experiences, altering their forms (Huyssen 1995; Virno 2015, 22–31).

This ongoing process of re-formation is inextricable from the environments in which it happens. Entangled with our lived environments, our identities are influenced by the way our choreographies through architectural space – which has use, history, message and intent inscribed in its frame – occupy a shifting place with our larger remembered movements through the city. Architecture has a particular affinity with memory, given that memory is a primarily spatial and sensory practice, rooted in the corporeal, situational world (Parker 1997; Vesely 2004, 99–100). Space is the site of memory. This means that architectural experience can be a major structuring event for how we connect seemingly disparate urban experiences; architecture ties our lived experiences of the city together, folding these memories into our story. The way that architecture connects with the city around it structures how we fold together our remembered experiences of the city around us into a sense of identity. Architecture is in this way a super-framework that impacts identity by bringing together other expressive forces in the city, filtering them, potentially replicating or transforming them.

We can take Peter Eisenman's well-known Memorial to the Murdered Jews of Europe in Berlin as an example of how the simple choreography of movement through the boundary of the architecture synthesizes normatively disparate events. This space, tasked with memorializing the Holocaust, blends into the street; you enter it subtly and unexpectedly, meaning that the experience of the architecture is found not as a rupture but as a linked part of the everyday experience of the street (Fig. 8.1). You slip, almost unknowingly, into the memorial (Fig. 8.2). However foreign and strange the subject matter or material experience may be, the territory

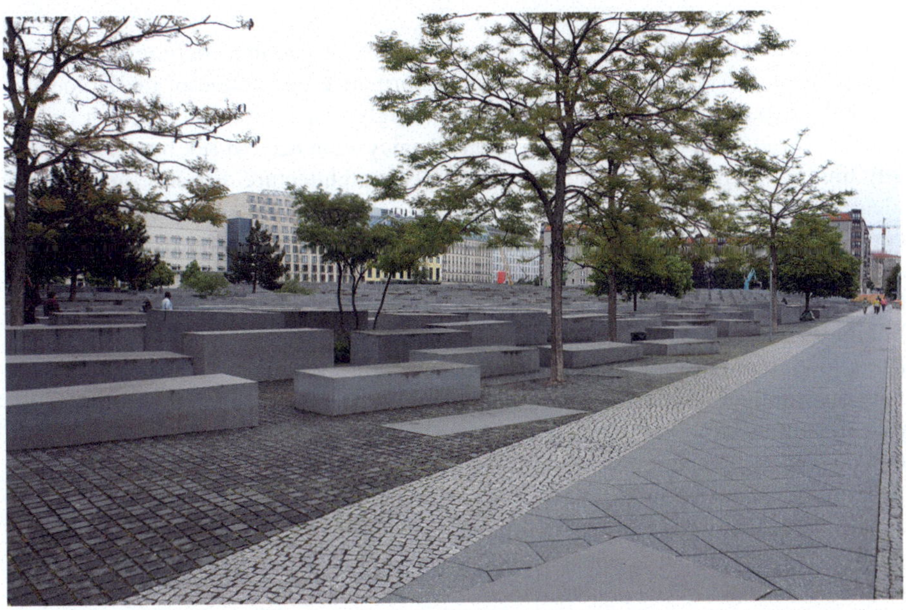

Fig. 8.1 Threshold of the Memorial to the Murdered Jews of Europe, Berlin 2019. (Photo: author)

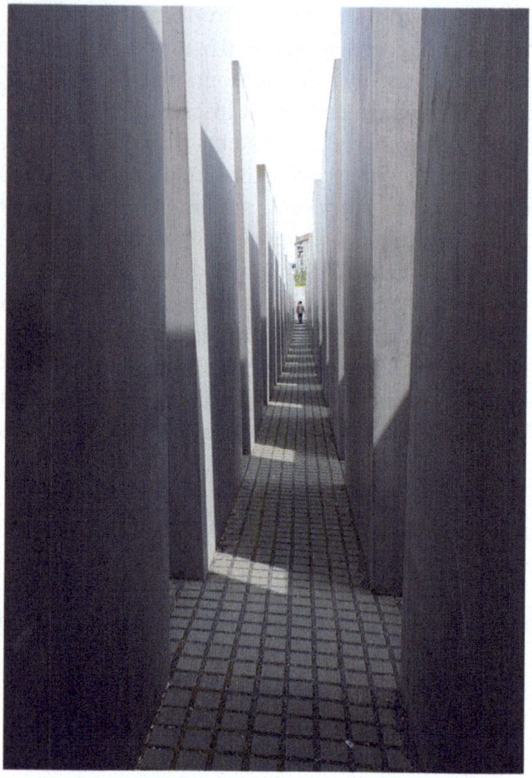

Fig. 8.2 Interior of the Memorial to the Murdered Jews of Europe, Berlin 2019. (Photo: author)

communicated through architecture is intentionally placed in a choreographed continuum with everyday events, reorienting a normative perception of the Holocaust as totally separate from "normal" life. Seemingly disparate events are not drawn together through the representation of a normalcy-violence hybrid, but by corporeally linking the two as a single experience. This architecture diffracts our existing identities back to us in relation to our existing conceptions, allowing for those conceptions to change in relation to architecture.

The transfiguration of whatever identities we bring to the architectural event relies on ambiguity – an encounter with the unknown in tandem with the known. By linking these seemingly disparate experiences, the larger architectural experience can allow seemingly foreign concepts, events, and identities to enter into concert with our familiar, normative rhythms, changing them. This encounter with ambiguity can "jolt us", forcing us to alter our existing frames of reference to account for the unexpected experience of familiar strangeness (Pérez-Gómez 2006, 100). Ambiguity cannot be incorporated into existing, absolutist frames of reference; it cannot be denied, rejected, or appropriated by existing norms given that it is both strange *and* familiar. It is uncanny, disturbing and destabilizing (Freud 1919). We need such strange spaces in our cities.

An encounter with ambiguity in architecture can happen both cognitively and, more importantly, affectively. Cognitive communication is the realm of the known. It is figurative, communicating through the forms of symbols, images, slogans, and signs (Deleuze 2002, 5–9). The cognitive experience of architecture is didactic: we are told what meaning to take away from the experience. Sometimes this is highly effective; cognitive messages are, after all, powerful modes of communication and can be sharpened tools for exposing injustices and critiquing normativity. The post-modernists played with messaging, using the language of architecture to encourage a messy vitality that supposedly balanced elitist modernist utopianism with populist historical signifiers (Venturi 1966, 22; Jencks 1987). Yet that conception of architectural communication relies on using culturally established routes of expression, a "universal grammar," to shape our sense of place, ignoring that the only grammar we can possibly make meaning from is based in an existing epistemology (Jencks 1987, 8). These routes of expression can presuppose and impose meaning, perpetuating what exists (De Raedt 2012). In a world characterized by entrenched exploitation and sustained inequality, we need to find ways of expressing our (potentially violent) messages beyond what is already established.

This means expressing the messages of the architecture affectively, through the figural realm of the diagram; forces, not forms (Deleuze 2002, 81–90). This kind of architectural expression relies on what the architectural theoretician Alberto Pérez-Gómez calls "poetic space": an unfixed, transformative "rhythmic experience" grasped as a "faint echo or evocation" (Pérez-Gómez 2006, 98–99). These kinds of spaces communicate non-didactically, as ephemeral events that are incomplete and non-prescriptive, offering a chance for us to become part of the event, and part of the story being made through the architecture (Pérez-Gómez 2006, 108; Pérez-Gómez 2011). This kind of architectural space is resonant and bodily, establishing spatio-temporal relationships that can be familiar but not re-cognizable (Grosz 2008). This uncanny familiarity forces a creative rethinking of what was once familiar and what was once strange by communicating through the thinking body, beyond the purely visual (Grosz 2008, 18; Vesely 2004, 174). By focusing on this ambiguous, bodily communicative forms of expression, architecture can limit the effects of our known, cognitive world that mediates – skews – our ability to make any extra-normative meaning from architecture (Pérez-Gómez 2011, 52; Vesely 2004, 56). This kind of expression works at a pre-cognitive level and allows for new experiences that reassemble and reorient our memories of the city in new ways. Ambiguous architecture can, to paraphrase Deleuze, reveal a Sahara in a scream (Deleuze 2002, 82). In so doing, this architecture reframes our lived histories and disturbs the foundational identities that shape our use and interactions with each other in space.

Ambiguity does not mean that such affective forms of communication have to be unclear or unspecific. They can and should be specific: vagueness, not taking a position, or staying silent are all abdications of responsibility and can result in empty architectures that fail to alter our lives, remaining ineffective accomplices in the continuation of a status quo. Expanding on the example that opened this chapter, the recently opened National Memorial for Peace and Justice in Alabama expresses a very specific message in non-prescriptive and destabilizing terms, opening that message to multiple interpretations and synthetic experiences while maintaining a level of clarity.

The message of the space is simple: thousands of recorded and unrecorded terror lynchings need to be remembered and seen in America's present systems of justice. The primary communicative elements for this memorial are the eight hundred steel figures that the visitor moves with, through, around, and below (Fig. 8.3). The entire experience of the memorialized event(s) is based on the figural resonance that we have with the metal "bodies," placing race-based violence within the familiar movement of our bodies through a crowd (Fig. 8.4). This experience is pre-cognitive; it does not tell us what explicit conclusions to draw and it does not use the violent dichotomy of white/black – a dichotomy based primarily on cognitive, visual differences – to communicate the event. Instead, we find normatively "abnormal" terror lynching within the familiar movements of our bodies with others that we perform regularly; no matter the apparent strangeness of terror lynching, it is now linked into our everyday lives, resulting in a rethinking of them to include the reality and possibility of this violence. The message is specific, but in being communicated through an ambiguous corporeal form, it is not caught in a racialized framework that keeps interpersonal violence within the one-dimensional confines of "white/black."

The stories that we make from architecture need this kind of ambiguity if they are to communicate beyond what exists and potentially destabilize normativity. Bounded by time, the experience of architecture can communicate pre-cognitively through our bodily relationship with architectural matter and space, imbued with

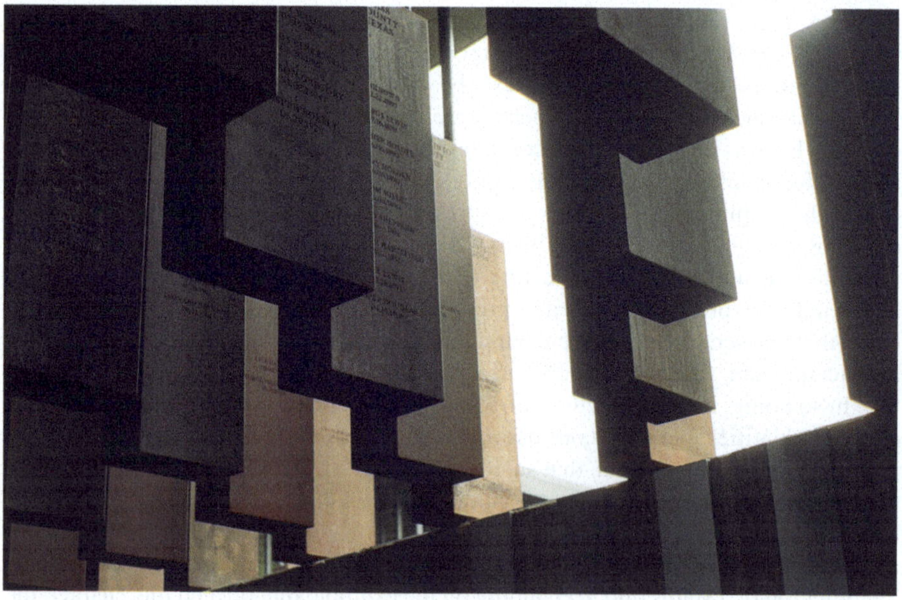

Fig. 8.3 Metal figures at the National Memorial for Peace and Justice, Montgomery 2019. (Photo: author)

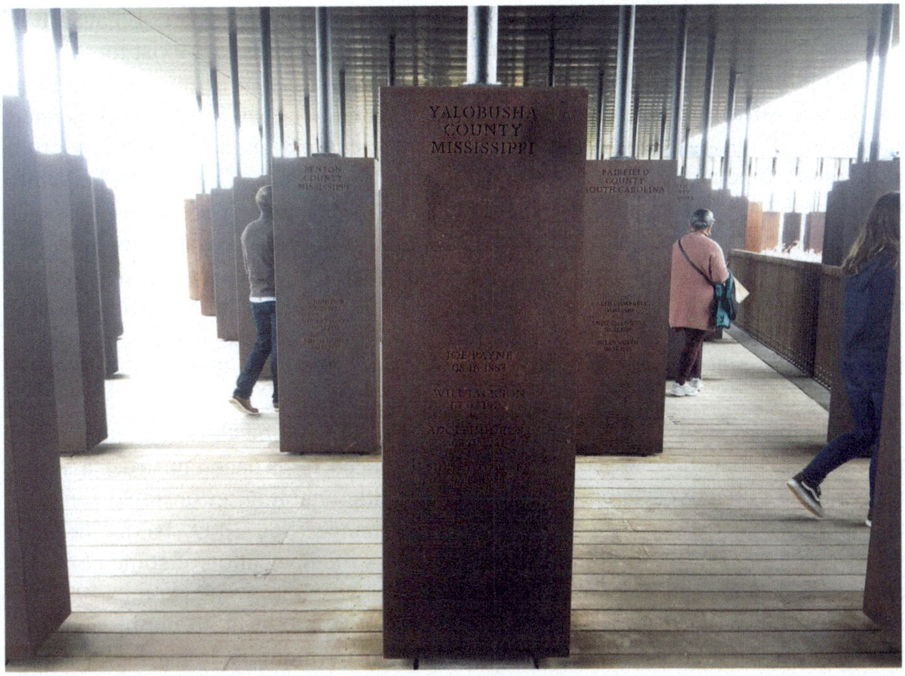

Fig. 8.4 Moving through an architectural crowd at the National Memorial for Peace and Justice, Montgomery 2019. (Photo: author)

meaning through its moving resonance with the familiar. This meaning is brought by us to the architecture from the rest of the city, transfigured through the experience of the architecture, which in turn un-structures our perceptions of that existing meaning, un-structuring our senses of identity. Ambiguous architecture opens us up to the potential that even the strangest among us are connected, or rather, connectable; it helps us to recognize the "strangers in ourselves" that may be a rampart against growing xenophobia and entrenched systemic oppression (Kristeva 1993, 27). It helps lay a foundation for new practices of resilience and improvisational adaptability on which our futures can be built.

4 Conclusion: Towards an Improvisational Architecture

This chapter does not provide a checklist for designers. Instead, it offers a concept to use when making cities. In light of mass urbanization amidst rapid population growth, a deteriorating climate, and growing social inequality, the focus of architecture needs to be on increasing the capabilities for people to form new collaboratives, new ways of thinking, and new tools for addressing the unprecedented challenges of the twenty-first century. Ambiguity as an architectural concept is part of that shift in how we conceive of city-making and the urban socialities that can exist there.

This philosophy of ambiguity can manifest in a variety of forms, ranging from the fertile infrastructural disorder explored in the work of Sendra and Sennett (2020), to affective, figural forms of architectural expression that can emerge from exploratory practices of thinking-making architecture (Pérez-Gómez 2011). While the examples of memorial architecture explored in this chapter are more message-oriented than other types of architecture, the lessons in communicative methods are no less applicable. Public institutions such as libraries, museums, city halls, and the supposedly common spaces that often emerge alongside them are primary sites where this philosophy can be used to create and understand architecture that destabilizes the potentially exclusory histories of who belongs to that public and who does not.

These spaces are key urban sites that deal directly with the tension at the heart of public space: that these spaces-for-commoning need to at once construct a seemingly stable collective vision of community, while enabling an unstable sense of community that fosters critical discourse and change (Sen 2009, 103). The construction of any sense of public identities requires the invocation of some common past, a mysterious and potentially violent "where we came from" (Young 1993). It is this expression of common memory and identity, present in all public architecture, that can limit or broaden our understanding of community, hindering or allowing us to take into account varying perspectives around justice and our global challenges (Huyssen 2011). By invoking community through ambiguity – a bodily, force-based pre-cognitive form of communication – specific histories and messages, which are inherently exclusory, can be opened so that any person can see themselves and others as part of that history. Or rather, part of the future of that history. Ambiguity as a communicative framework allows for the expression of a public to be at once specific and open, unable to appropriate and mobilize collective identity as a tool to enclose, a tool to control whose life can count as a life. This kind of framework is a non-didactic platform for the "enunciation of the exaltation required for collaborative practices" (McKim 2019, 139; Simone 2019, 22). Within the practice of making our built environments, it is a way to broaden the foundational habits of solidarity on which our futures must be built.

The concepts explored in this chapter place architecture in relation with the greater urban milieu, where our built environments can expand our ability to form improvisational relationships through the relational destabilization of normative identity. If we hope to live with ambiguity, our world-structuring environments cannot be absolute, and the creation of such environments cannot be conceived only within the cognitive realm of communication. Centered on destabilization as a structural framework, this philosophy directs architecture towards becoming a field concerned with the creation of situational environments that support contradiction, flux, and improvisational connectivity – a politics of transformative plasticity. (Unger 2004, xxxvii). These environments are vitally off-balance and ensure that our lives remain in motion.

In the expression of our urban environments and the commoning they must enable, this philosophy is particularly important. It is grounded in social relations that can and must be alterable, unstable, no matter how they are presently or historically defined. Even the most limiting histories, such as ones currently being written in ethnic, nationalist or racially supremacist terms, could be undermined in such a framework; ambiguity actively restructures illusions of absolute separation, moving towards a future based on radical interdependency. This future is not a fairytale, but an essential destination, a place that we need to move toward if we are to find the tools to address our unique and all-encompassing challenges today. Looking at the trajectory of over-consumption, exploitation, and oppression that we are on, it is clear that we remain ill-equipped to deal with these challenges. Given that these challenges are, and will continue to be, urban, our practices of city-making offer a way to intervene in that trajectory and alter the course of our collective futures. An architecture of ambiguity offers a mode of intervention; a way to ensure our commons cannot be enclosed, controlled, or destroyed any longer.

Acknowledgments This research was initially undertaken during my Masters in Architecture degree at the University of Waterloo, with funding from the Social Sciences and Humanities Research Council of Canada (SSHRC) under the supervision of Dereck Revington and Dr. Robert Jan Van Pelt. I also give thanks to Dr. Tracey Eve Winton and Jonathan Tyrrell who served on my examination committee and prompted the continuation of my research.

References

Amin, Ash. 2004. Regions Unbound: Towards a New Politics of Place. *Geografiska Annaler. Series B* 86: 33–44.

———. 2006. The Good City. *Urban Studies* 43: 1009–1023.

———. 2008. Collective Culture and Urban Public Space. *City* 12: 5–24.

———. 2016. On Urban Failure. *Social Research* 83: 777–798.

Amin, Ash, and Nigel Thrift. 2016. *Seeing Like a City*. Malden, MA: Polity Press.

Arendt, Hannah. 1970. *On Violence*. New York: Harcourt, Inc.

———. 1976. *The Origins of Totalitarianism*. New York: Houghton Mifflin Harcourt.

———. 2006. *Eichmann in Jerusalem: A Report on the Banality of Evil*. New York: Penguin Books.

———. 2018. *The Human Condition*. Chicago: University of Chicago Press.

Bailey, Robert. 2012. Unknown Knowns: Jenny Holzer's Redaction Paintings and the History of the War on Terror. *October* 142: 144–161.

Baldwin, James. 2018. *Dark Days*. Johannesburg: Penguin Books.

Boonen, Christiaan, Nicolas Brando, Samuel Cogolati, Rutger Hagen, Nils Vanstappen, and Jan Wouters. 2019. Governing as Commons or as Global Public Goods: Two Tales of Power. *International Journal of the Commons* 13: 553–577.

Butler, Judith. 2004. *Precarious Life: The Powers of Mourning and Violence*. New York: Verso.

———. 2009. *Frames of War: When is Life Grievable?* New York: Verso.

———. 2020. *The Force of Nonviolence: An Ethico-Political Bind*. New York: Verso.

Cuff, Dana, Anastasia Loukatiou-Sideris, Todd Presner, Maite Zubiaurre, and Jonathan Jae-an Crisman. 2020. *Urban Humanities: New Practices for Reimagining the City*. Boston: MIT Press.

De Raedt, Kim. 2012. Building the Rainbow Nation: A Critical Analysis of the Role of Architecture in Materializing a Post-Apartheid South African Identity. *Afrika Focus* 25: 7–27.

Deleuze, Gilles. 2002. *Francis Bacon: The Logic of Sensation*. Trans. Daniel W. Smith. Minneapolis: University of Minnesota Press.
Deleuze, Gilles, and Félix Guattari. 1987. *A Thousand Plateaus: Capitalism and Schizophrenia*. Trans. Brian Massumi. Minneapolis: University of Minnesota Press.
Fincher, Ruth, Kurt Iveson, and Helga Leitner. 2019. The Political Potential of Encounters: Being Together in Difference as Equals. In *Everyday Equalities: Making Multicultures in Settler Colonial Cities*. Minneapolis: University of Minnesota Press.
Foster, Sheila R., and Christian Iaione. 2016. The City as a Commons. *Yale Law & Policy Review* 34: 281–349.
Freud, Sigmund. 1919. *The Uncanny*. Trans. David McLintock. New York: Penguin Books.
Grosz, Elizabeth. 2008. *Chaos, Territory, Art: Deleuze and the Framing of the Earth*. New York: Columbia University Press.
Harvey, David. 2012. *Rebel Cities: From the Right to the City to the Urban Revolution*. New York: Verso.
Hellman, Jörgen. 2018. How to Prove You are Not a Squatter Appropriating Space and Marking Presence in Jakarta. In *Cities in Asia By and for the People*, ed. Yves Cabannes, Mike Douglass, and Rita Padawangi, 41–68. Amsterdam: University Press.
Hudson, Peter. 2019. The Reproduction of Racial Inequality in South Africa: The Colonial Unconscious and Democracy. In *Racism After Apartheid*, ed. Vishwas Satgar, 158–172. Johannesburg: Wits University Press.
Huyssen, Andreas. 1995. *Twilight Memories: Marking Time in a Culture of Amnesia*. New York: Routledge.
———. 2011. International Human Rights and the Politics of Memory: Limits and Challenges. *Criticism: A Quarterly for Literature and the Arts* 53: 607–624.
Jacobs, Jane. 1992. *The Death and Life of Great American Cities*. New York: Vintage Books.
Jencks, Charles. 1987. *The Language of Post-Modern Architecture*. London: Academy Editions.
Kristeva, Julia. 1993. *Nations Without Nationalism*. New York: Columbia University Press.
Lemkin, Raphael. 1944. *Axis Rule in Occupied Europe*. Concord, MA: Rumford Press.
Lowe, Lisa. 2015. *The Intimacies of Four Continents*. Durham, NC: Duke University Press.
Mabogunje, Akin L. 1990. Urban Planning and the Post-Colonial State in Africa: A Research Overview. *African Studies Review* 33: 121–203.
Martin, Reinhold. 2016. Public and Common(s). In *The Urban Apparatuses: Mediapolitics and the City*. Minneapolis: University of Minnesota Press.
Mbembe, Achille. 2019. *Necropolitics*. Translated by Steve Corcoran. Durham, NC: Duke University Press.
McKim, Joel. 2019. *Architecture, Media, and Memory: Facing Complexity in Post-9/11 New York*. London: Bloomsbury Visual Arts.
Nora, Pierre, and Lawrence D. Kritzman. 1996. *Realms of Memory: Rethinking the French Past*. Vol. 1. New York: Columbia University Press.
Parker, Rodney Douglas. 1997. The Architectonics of Memory: On Built Form and Built Thought. *Leonardo* 30: 147–152.
Pérez-Gómez, Alberto. 2006. *Built Upon Love: Architectural Longing After Ethics and Aesthetics*. Cambridge, MA: MIT Press.
———. 2011. Built upon Love: Towards Beauty and Justice in Architecture. *Mosaic* 44: 43–59.
Saïd, Edward. 2019. Permission to Narrate. In *The Selected Works of Edward Saïd 1966–2006*, ed. Moustafe Bayoumi and Andrew Rubin. New York: Vintage Books.
Saunders, Doug. 2010. *Arrival City: The Final Migration and Our Next World*. Toronto: Alfred A. Knopf.
Sen, Amartya. 1999. *Development as Freedom*. New York: Alfred A. Knopf.
———. 2006. *Identity and Violence: The Illusion of Destiny*. New York: W.W. Norton & Company, Inc.
———. 2009. *The Idea of Justice*. New York: Penguin Books.

Sendra, Pablo, and Richard Sennett. 2020. *Designing Disorder: Experiments and Disruptions in the City*. New York: Verso.
Silverstone, Roger. 2007. The Rhetoric of Evil. In *Media and Morality: On the Rise of the Mediapolis*. Malden, MA: Polity Press.
Simone, AbdouMaliq. 2019. *Improvised Lives: Rhythms of Endurance in an Urban South*. Medford, MA: Polity Press.
Simone, AbdouMaliq, and Edgar Pieterse. 2017. *New Urban Worlds: Inhabiting Dissonant Times*. Medford, MA: Polity Press.
Smith, John Charles. 2016. *Imperialism in the Twenty-First Century: Globalization, Super-Exploitation, and Capitalism's Final Crisis*. New York: Monthly Review Press.
Toušek, Laco, and Lenka Strnadová. 2016. What Is "Out of Place"? Intolerance in Public Space. In *Can Contemporary Cities Be Places of Tolerance?* ed. Stefano Moroni and David Weberman. Budapest: Central European University Press.
Unger, Roberto Mangabeira. 2004. *False Necessity: Anti-necessitarian Social Theory in the Service of Radical Democracy*. New York: Verso.
UN-Habitat. 2020. Cities of Opportunities: Connecting Culture and Innovation. In *Abu Dhabi Declared Actions*.
Venturi, Robert. 1966. *Complexity and Contradiction in Architecture. The Museum of Modern Art Papers on Architecture*. New York: The Museum of Modern Art.
Vesely, Dalibor. 2004. *Architecture in the Age of Divided Representation: The Question of Creativity in the Shadow of Production*. Cambridge, MA: MIT Press.
Virno, Paolo. 2015. *Déjà Vu and the End of History*. Trans. David Broder. Verso: New York.
Vitruvius. 1960. *The Ten Books on Architecture*. Trans. Morris Hicky Morgan. New York: Dover Publications Inc.
Waller, James. 2007. *Becoming Evil: How Ordinary People Commit Genocide and Mass Killing*. Oxford: University Press.
Watson, Vanessa. 2016. Shifting Approaches to Planning Theory: Global North and South. *Urban Planning* 1: 32–41.
Young, James E. 1993. *The Texture of Memory: Holocaust Memorials and Meaning*. New Haven: Yale University Press.

Part III
Design, the Commons and Ecology

Chapter 9
Intriguing Human-Waste Commons: Praxis of Anticipation in Urban Agroecological Transitions

Markus Wernli

Abstract In recent years, citizen designers have been working with urban communities on the ecological reuse of human waste. In this commoning effort, practitioners reclaim body-expelled resources for exploring the metabolically enabled household as a networked site of radical, co-productive transitions that harnesses nutrients and boosts local value chains. The commoning of human excrement is understood in the context of agroecological urbanization that seeks to empower urban dwellers to become contributing actors in the food-energy nexus by making the city more food-enabled for storing and proliferating feeds, fertilizer, and food. By introducing three cases of human-waste commons in Brussels, Hong Kong, and Berlin, this study approaches commoning design as a process grounded in the praxis of anticipation. In this way of life, consistent with the anticipatory nature of living systems, the transformative potential in people, their waste, and social arrangements stem from the dynamic continuum of mutual purpose, trust, and vigilance. Collective desire, resolutions, and statuses are a result of direct involvement, context, and relationships. The three examples show how citizen designers draw energy from anticipating regenerative, life-giving value chains around human waste that give momentum to overcome the given thresholds with perseverance and resourcefulness.

Keywords Value chain design · Ecological sanitation · Food pedagogies · Collectivized resourcefulness · Metabolizing infrastructure

Agroecological commoning places urban soil care centrally within urbanization and community development to bring regenerative, biophysical relations into decision-making processes, and move beyond narrow functionalist approaches (Schneider and McMichael 2010). In recent years, some urban designers have been actively developing a public commons through the agroecological use of human excrement. What initially may sound suspect is the sincere attempt of building pragmatic

M. Wernli (✉)
School of Design, The Hong Kong Polytechnic University, Hong Kong, SAR, China
email: mswernli@gmail.com

G. Bruyns, S. Kousoulas (eds.), *Design Commons*, Design Research Foundations, https://doi.org/10.1007/978-3-030-95057-6_9

working alliances from human-waste commons that reconcile, whereas the ecological disconnections of human-nature and producer-consumer become the central concern in food system transformations. Human-waste commons with the inserted hyphen between 'human' and 'waste' is a reminder that producing "muck" (Worster 2017) is an inextricable part of being human. The question is how to go about remediating this human wastefulness. To meet the vital need of recovering humanly released nutrients that otherwise charge and overburden the environment, human-waste commoners work hard and engage in grassroots experimentation to demonstrate the feasibility of closing material circulations originating in food stocks and expelled from one's body.[1] Even at their small scale, human-waste commons are radically potent and political because they are about resource access, regulation, and sovereignty.

Increasingly, public sewage filtration plants extract fecal nutrients for agrarian use through very costly, tax-subsidized retrofitted technologies (Etter et al. 2015). Similarly, pharmaceutical companies are privatizing the biome of human waste for gastroenterological treatment with the blessing of industry-friendly regulations (Bollier 2019). Human-waste commoners must go through the trouble of acquiring tried-and-tested sanitary remediation skills to regain control over what is their own: bodily excreted nutrients and biomes. Excremental commons thus circumvent the enclosures of market order and enforced hygiene regimes to safeguard their free access and affordability.

Before sedentary settlement and agriculture, foraging nomads already understood that when the woods, tundra, or desert reached a surfeit of their bodily wastes, it was time to resettle, escape the polluting waste products, to find new grounds and food sources. While the metabolic exchanges of material and energy in organisms are the "natural chemistry of staying alive" (Worster 2017), the waste products expelled contain inherently pathogens that pose health threats for people and the environment when space and precaution are lacking. Urbanization complicated the conservation and transformation of metabolic pollutants into nutritious wealth. As city populations swell alongside their metabolic demands on eating and excreting, conventional sanitation infrastructure focused on waste elimination that breaks material circulations of water/nutrients/carbon, thus depleting instead of replenishing the biophysical foundation of life (Waltner-Toews 2013). Until recently, sanitation research was focused on pathogen removal and avoidance of human health risks.

In contrast, applied research that emphasizes agricultural reuse aspects have been side-lined (cf. Carr et al. 2004). With a multitude of contextual factors influencing the reuse options for human waste, one-size-fits-all solutions are not viable. In response, there is a call for reconsidering sanitation through the discipline of agroecology (Weckenbrock and Alabaster 2015) that offers a range of value-based

[1] Human waste is part of a massive "global translocation of feeds." The nutrients, water and energy extracted from an ecosystem on one side of the world are transported as packaged crops or food across the world, then consumed and eventually deposited as excrement into ecosystems on the other side of the world. While these effluent nutrients lead to toxic manure lakes, suffocating water bodies, and potent greenhouse gas emissions, petrochemical fertilizers applied to soils do not sufficiently replenish them in the long run (Waltner-Toews 2013, 120).

methods for explicitly addressing the social, cultural, and environmental integration of human waste.

1 Radical Agroecological Transitions and Design Commons

Agroecology conjoins participatory and culturally sensitive progression toward non-extractive, resource-conserving access to food, energy, and self-efficacy that define it simultaneously as a scientific, practical, and political tool (Tornaghi and Hoekstra 2017). It is rooted in non-hierarchical social learning frameworks that shape the way humanity's biophysical foundation is replicated and socialized across communities and generations. Inside the vision of agroecological transitions, human waste becomes a matter of combined hygienization, irrigation, and fertilization strategy that links environmental and health protection with food and energy production while developing entrepreneurial opportunities (Mateo-Sagasta et al. 2013).

Reintegrating human waste and commoning of urban resources is not just about replenishing soil ecologies. In part, it permits communities to regain control over their social reproduction. It implies that composting, soil care and food cultivation are recognized as essential, biopolitical relations that substantiate people, society, and economies (Gidwani and Ramamurthy 2018; Katz 2001). The agroecological tradition originates from South American peasants' constant struggle for regaining food sovereignty over dominant forces of colonized or 'civilized' life (Altieri and Toledo 2011) that entails soil degradation, food deserts, land speculation, exhausted populations, and squandered resources. The interlinking of southern agroecological methodology with the northern concept of commoning in food systems can spur collective practices where natural resources and human behaviors can be co-constructed, co-valued, and co-regulated in locally appropriate arrangements that overcome incremental reformist attitudes in consumerist societies (Ferrando et al. 2020).

Agroecological transitions require radical pathways for shifting the current logics behind short-term growth and risk aversion toward long-term "thrivability" (Russell 2013) in urban metabolic and social ecologies. Agrarians established the Foodsystem Transformation Taxonomy (Holt-Giménez and Shattuck 2011) to understand how agroecological transitions can retain their integrity and change potential. It differentiates reformist/aid-oriented, progressive/empowerment-seeking, and radical/redistribution-enabling approaches. Radical/redistributive here means not only creating isolated alternatives or supporting agroecological farmers but also systematically dismantling the self-perpetuating processes and disempowering arrangements of current urbanization (Tornaghi and Dehaene 2020).

Agroecological transitions understood as commoning of urban nutrient systems propose a pragmatic praxis of knowing, doing, eating, eliminating, and being beyond the current conditions of the market and the state. Reproductive social practices intertwine the shared domains of domesticity, associations, and environment, thus moving beyond dualistic conceptions of subjectivities like public/private (Sohn et al. 2015) or consumer/producer. Current design discourse points to a monistic individuality where selfhood is continuous, adaptable, and coterminous (Braidotti and Vermeulen 2014). This insight stems from the realization that the capitalist order is flexible and unbreakable. Radical change does not materialize from awaiting a distant, otherworldly future but from the interior reorientation of subjectivity (Bendell 2018). If commoning is indeed an action concept for social learning or pragmatic working alliances (Linebaugh 2008, p. 279) – rather than "abstract, compulsive repair" (Berlant 2016) – it will involve "transformational infrastructures of attention and aversion" within the messy ambivalence of an ecologically situated, social life (Braidotti and Vermeulen 2014).

Parting from a fading era based on specific state and market definitions of what it entails to be a citizen, consumer, or participant, commons activists are demanding a new vocabulary that can yield social paradigms of caring, exchanging, and collectivizing (Bollier and Helfrich 2019). Participatory design researchers are also arguing that such concepts need to better account for sharing practices that may challenge or stretch the notion of the commons if they want to promote pluralistic, intercultural perspectives in design (Botero et al. 2020). Moreover, design commons are challenged to better account for shared frailty and bodily relations in a world of exchanged resources and mutual interdependence of material cycles (Savazoni and Andrade 2019; Escobar 2018a).

Design commons that seriously aim at empowering communities to become agents in circulations of the food/energy nexus need to devise new strategies that are not resigned to merely imagining utopian agroecological futures. Instead, they involve stakeholders in a rich constellation of wholesome practices, pervasive and engaging enough to disrupt prevalent arrangements and bring about change (Dehaene et al. 2016). If local, home-cooked food is to replace industrial readymade meals, if recovered nutrients instead of finite petrochemicals are to nourish the land, if adept producers are to emerge from inept consumers, then making and taking *time* is imperative. Thus, social arrangements need to be in place, and situated knowledge reproduced that makes dedicating time *worthwhile*. Design research describes this transition process as the "infrastructuring of everyday life" (Karasti 2014; Marttila et al. 2014). If engaging with human waste is more than an activist stance, it demands a healthy dose of humility, acumen, and social support. In response, this chapter seeks to understand better the dynamics underlying community formation and sustainment in alternative resource systems. It also seeks to contribute to heuristics and terminology in design commons that account for the visceral materiality and temporality evident to agroecological transitions.

In assessing their transformational potential, this research considers human-waste commons that (i) revere the ecological use-value of soils and organic residues for land and food cultivation; (ii) partner with natural forces, microbes, insects, and plants to render visible the ecological separation in urban life; (iii) link citizen-led collaborative arrangements with resource sovereignty; and (iv) challenge previous decisions and path dependencies that have been "hardwired in the 'food-disenabling' city" (Tornaghi 2017) to enact inventive social arrangements, cultural practices and public infrastructures that endorse and reward agroecological resourcefulness and food growing.

This research looks at human-waste commoning examples in three cities: Brussels, Hong Kong and Berlin. They were selected because of their socio-cultural diversity and focal complementarity as they represent varied cultural-political-economic contexts and contrasting population densities. Deliberately these examples are all situated in the global North and run counter to the prevailing assumption that predicate efforts in alternative human-waste management are exclusive to informal and indigenous settlements of the global South.

In order to respect local and traditional knowledge, this research adopts the participatory action research method (Heron and Reason 2001), constituted from social platforms that include placemaking events, urban living labs and social enterprises specific to each example. Each platform convenes actors from distinct domains for staging pre-design studies, workshop series, agroecological practices and in-depth conversations that shape strategies and active implementation plans. Consequently, this chapter synthesizes exchanges with facilitators and analyzes content from participants' responses, project presentations, and media coverage, areas of similarity, contrast, and mutual learning of these projects. Grounded in day-to-day experimentation within the fluid, contradictory social arrangements, and with disparate expertise engaged, the examples productively navigate the tension between what-is-now and what-is-possible-tomorrow, for testing of place-based alternatives (Siltanen et al. 2015; Harcourt and Escobar 2002). By acknowledging resourceful interdependence as a 'distributed problem' to be addressed within everyday life, the examples avoid playing citizens against regulatory authorities by situating their efficacy-seeking practices squarely inside sanctioned communities and places.

The human-waste commons provides a practical context for mobilizing diverse actors on different scales. Beyond articulating debates around urban nutrient sinks, localized food systems, co-evolutionary health, and land access, it outlines possible changes in economic values, spatial relations, and collective improvisation. As described in the following sections, the examples illustrate constructive disruptions to the deep-seated unsustainability of current urbanism. Abandoning normative positions and conceiving urban resourcefulness as a way of life and praxis, the examples seek personal and social accountability for the regeneration of the bio-physical foundation, knowledge, and skillset required for long-term thriving (Wernli 2020).

2 Examples of Human-Waste Commons in Brussels, Hong Kong and Berlin

Through thoughtful content review and practitioner exchange, this section illustrates the assets, ambitions, collective practices and value systems impelling the human-waste commons as opposed to the compelling determinism of industrial waste infrastructure. Each example in their local setting explores alternative social configurations toward radical agroecological trajectories (Tornaghi 2017). In the latter sections, I deduce the primary transformative dynamics across the examples and discusses how they relate to debate and furthering the design of the commons.

Brussels shows an agroecology movement that is fragmented alongside language and regional divisions (Fig. 9.1). Recent local food and climate-sensitive regulations support urban community growers and market gardeners who cultivate interstitial areas in the dense city. This policy context also spurs young entrepreneurship in the agricultural reintegration of by-products from the brewery and food industry. The action-research here gravitates around a cohort of architects, artists, and citizen-designers that go by the name of Collective Disaster (Amaya 2016). Initially, the group formed in response to a call in 2014 by the Belgian Ministry of Environment to revitalize a derelict downtown park. In the ensuing collaboration with neighbors and authorities, over the summer of 2015 Collective Disaster realized a community-run, ecological public toilet facility (Fig. 9.1). The resourceful place-making activation became known as *L'Usine du Trésor Noir* ('The Temple of Holy Shit'). To overcome resentment and gain social support, the collective involved neighbors in planning, building, and operating facilities that incorporated spacious, urine-separating toilets and designed a nutrients-processing system for renewably-powered hot tubs.

The collective conceived the pyramid-like structure to make the fermentation stages from waste to soil into an enjoyable, social experience. The toilet units were operated over six months and integrated on top of an elevated platform with arena-like stair access. On the back side, the toilets could be exited on two slides while the front stairs served as a public stage. The basement below the toilet structure housed the sealed collection barrels that separated urine from solids to process them onsite with microbially activated charcoal dust (biochar). Using Terra Preta Sanitation, the combined fermentation/compost process eliminates pathogens, stores nutrients, and upgrades human waste (alongside woody residues) into veritable fertilizer within one year (Andreev et al. 2015; Schmidt 2014).[2] From the collection chamber, narrow-gauge tracks allowed the easy transportation of full collection barrels to the adjoining composting site. The excess heat in the metabolic process was harnessed for operating hot tubs installed above the compost bays. The park-enlivening public toilet has garnered several ecological awards and international acclaim (Karga 2014; Sollazzo et al. 2014). Ever since, Collective Disaster combines the resurrection of organic wastes with the resourcing of novel social constellations through outdoor structures and urban interventions that are materially performative. The collective brings together soil experts, authorities, and local communities to reconceive, at least temporarily, physical and operational infrastructures (for composting, water provision, greenery, and mobility) that cross the divide between pro-environmental resource conservation and social capital at peri-urban scale. The

[2] Compared to conventional human-waste composting that requires up to five years for pathogen-removal in temperate climates, Terra Preta Sanitation is considered a speedy bioremediation process (Andreev et al. 2015).

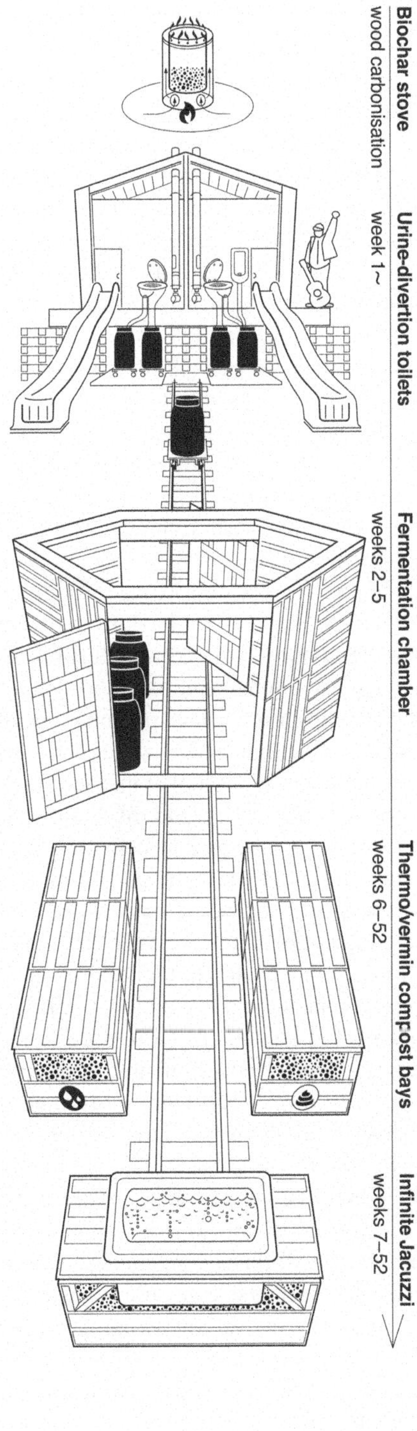

Fig. 9.1 Linking public eco-toilet and renewable hot tubs with place-making in a derelict park. The ecological public toilet was architecturally and socially arranged to make the metabolic stages into a collective experience. (Illustration: author)

challenge here in commoning human waste is to create strategies and synergies for public tolerance and land access that cut across long-standing inertia and revulsion against life-supporting, microbial partnerships. Pandemic fears can trigger unsubstantiated, adverse reactions towards everything microbial since it is unseen, unknown, and inevitably associated with disease.

Hong Kong presents a more dissonant context, where activists and practitioners are resisting rampant, speculative development, concentrated land ownership and food-skill loss by carving out niches for agroecological thinking and acting (Fig. 9.2). Hong Kong's highly fertile delta region was until the 1970s home to substantial rice and fish cultivation that appreciated human manure inside organized commodity markets to sustain yields (Xue 2005; Shiming 2002). Even today, smallholder farmers in southwestern China employ human waste for private use (Leung 2020). The university-endorsed action-research responded to mounting food safety and environmental health concerns, and that evolved around The Zero Organic Waste Home initiative instigated by the author for engaging the community in a reskilling and practical exploration of agricultural opportunities. A citizen-design initiative coming out of this urban living lab is the promotion of collective urine upcycling towards cultivation of indoor plants as a means to enable urban residents deprived of balconies and land access (Fig. 9.2). In the ensuing work alliance, 22 households of diverse socio-cultural backgrounds were invited into an 'urban ecology adventure' for fermenting their urine to fertilize a substrate to grow edible plants. Thereby they created a simple material relationship between their bodies and the environment. Each fermenting urine specimen became part of an annotated self-examination passage (Meiselman and MacFie 1996) that involved medical dipstick testers, diet monitoring, and plant development tracking. Participants consolidated this into an intricate 'mutual thrivability' journal. The citizen-design research spearheaded a closed-loop resource system that was untested and required participants to overcome technical and affective ambiguities jointly. The exercise sparked curiosity and a unifying purpose, strengthening social engagement, inventiveness, and environmental connections for over three months. In the context where continued destruction of arable land and leisure-fixated use of green areas are curtailing land access for composters, farmers, and foragers, this human-waste commons shaped and sharpened an ethic of (soil) care for activating practices of personal nutrient sovereignty. The agroecological experimentation reframed human waste as a responsibility-triggering agent. Here more-than-human affinities run counter to visions of the urban as an inevitable nutrient sink, reimagining instead the functioning of agroecological households in closing and embracing their material circulations.

Berlin appears in many respects, a contrasting example (Fig. 9.3). Here the legacy of the centralized, state-owned economy of former East Germany and the fall of the Berlin Wall has contributed to a socio-natural landscape with many undeveloped urban allotments. Thus, many residents have access to food growing, neighborhood composting and urban foraging, in parks or other land. Concurrently, Berlin's surrounding agrarian region (historically known as Central Europe's Bread Belt) suffered in recent decades from acute soil degradation, droughts and water shortages that are spurring experimentation in agroecological urban sanitation systems. Within

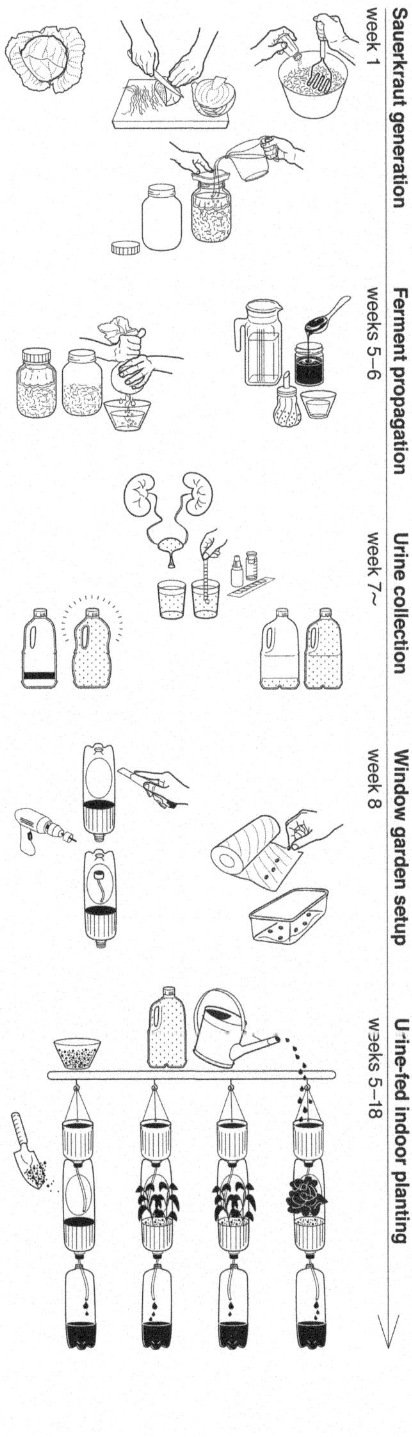

Fig. 9.2 Linking urine upcycling and indoor planting with food pedagogy. Fermentation allowed for making urine into a viable medium for collective food pedagogy and self-discovery. (Illustration: author and Sarah Daher)

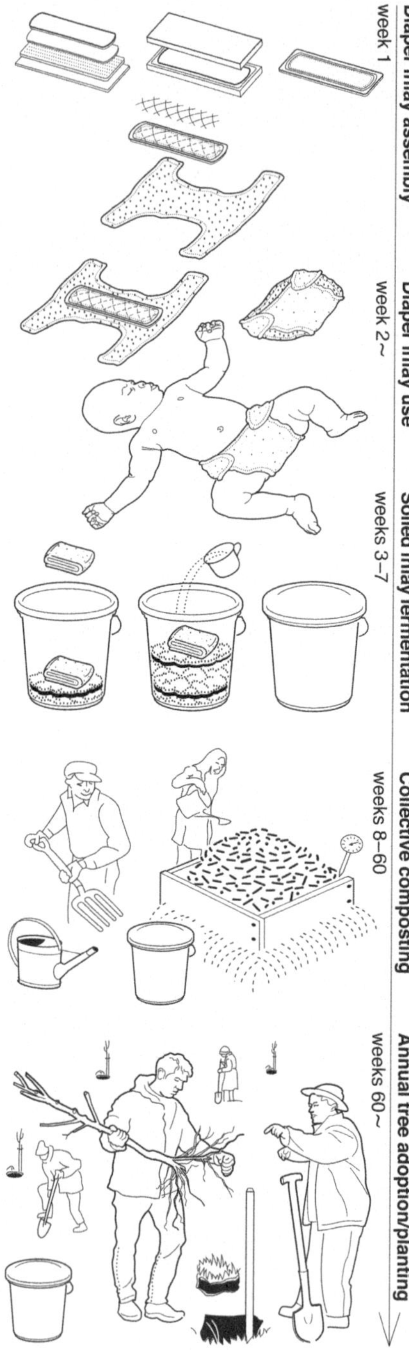

Fig. 9.3 Linking biodegradable diapers, collective composting and tree cultivation with social enterprise. The social enterprise builds on the revenue of fruit tree adoptions that support the manufacture of custom diaper inlays and the metabolic processing. (Illustration: author)

this context, an aspirational, artist-led human-waste reuse program has evolved over the last decade into a communal start-up that pioneers the eco-friendly transformation of baby nappies into fertile soil and fruit orchards (Fig. 9.3). Building on revenue from tree adoptions, rather than sale of diapers, the DYCLE social enterprise entails custom production of biodegradable diaper inlays and communal composting into Terra Preta black soils (akin to the Brussels example) for cultivating heirloom fruit trees. Participating young families meet weekly at a central processing point for exchanging soiled inlays with fresh ones (Debatty and Matsuzaka 2019; Reynaert 2016). Beyond biodegradable product, DYCLE creates a complete value chain that enables the local community, soil care, food forests and employment. Each step in this value chain was a matter of iterative testing over six years before the agroecological diaper cycle was fully established and ready to launch. In this instance, commoning is about inviting people into the uncertainties of agroecological circulations by offering them multi-level discovery of reaffirming, environmental interactions. Catering to curiosity by emphasizing novelty overcame initial aversion towards human waste and stirred latent aspirations for enacting more fulfilling, ecologically contributing lifestyles. Besides, the mindset of generations that grow up with diaper-fertilized orchards helps manifest the interdependence of humans with the earth's shared metabolism. The enterprise helps normalize the reintegration of human waste in agroecological urbanism.

3 Findings: Praxis of Anticipation as an Operational Mode for Design Commons

The three examples of agroecological human waste reuse illustrate practices and struggles that disrupt spatial conditions, cultural validation, economic assumption, and planning processes in the context of Western urbanism. Acting on the level of resourceful toilets, composting cycles and regenerative afforestation, these examples illustrate important principles for enlisting and guiding community development, which can also be relevant for design commons more generally.

The practitioners of the human waste reuse examples find themselves in a dynamic of anticipation that encapsulates three areas, including (i) advance awareness and care, (ii) deconstructing existing worldviews, and (iii) novel programmatic affirmation. Advance care for the human-waste commoner starts with a concern for the suitability of her metabolic waste products for reuse in growing food since modern diets and lifestyles currently compromise ecological use-value. Deconstructing worldviews is inevitable when advance care is put into action, and the human-waste commoner realizes how to shift the public away from short-term, purification-fixated agrarian and sanitation regimes that rely on synthetic inputs and impede long-term soil, human and ecological health; a conundrum that questions the very logics of 'being human' in this world. Programmatic affirmation comes from the mutual realization that the current resource-squandering social arrangements can be confronted while rebuilding resourceful communities, and novel infrastructures can be a delightful prospect.

4 Advance Care: Embodying More-Than-Human Solidarities

All three examples above go a long way to challenge human selfishness embedded in current social and urban arrangements. The primary concern for human-waste reuse practitioners is always how the nature and quality of their excrement influence the plant life sprouting out of it. When humans recognize they are creatures of the soil alongside metabolizing lactobacilli, mycorrhizae, and earthworms, it stimulates more-than-human alliances and raises possible solidarities (Heynen et al. 2005). When the person opts to eat healthier food to produce better quality excrement-to-be-fertilizer, the person then exists to the extent of anticipation, not just participation and reaction. Anticipation from Latin *antecapere* denotes 'understanding (of harm) beforehand', what Mihai Nadin (2005) encapsulates as *anticipo ergo sum*: I anticipate, therefore, I exist. Thus, living here is contingent on the "sixth sense" (Nadin 2005) of taking care of matters ahead of time to forestall detriment or decay. In contrast to prediction, which tends to fixate positions and probabilities, anticipation is an ongoing scanning and prefigurating of possibilities that acknowledges uncontrollable dynamics in the biological world. Through anticipation, the actor responds to and integrates with complexity.

In Hong Kong, urine-recycling participants came together to become directly involved in monitoring human and plant health in a self-diagnosing, pedagogical arrangement. In Berlin, diaper-composting families maintain awareness of the inextricable connection between plant prosperity, soil health, and human flourishing with support from soil and tree experts. In Brussels, the functional requirements and underutilized potential of composting energy and its excess heat dictated a social and architectural re-structuring of the public toilet. As a social form, advance care suspends careless consumption and other degenerating practices. In all three cities, human-waste commoners recover nutrients through fermentation and mulching, regenerating growing media and committing to native species wealth. These practices disrupt productionist rationales that have depreciated advance care ecologies and social reproduction as 'unproductive' (Puig de la Bellacasa 2017, 177). The revival of communal composting, fermentation cohorts and heirloom plantations, as well as personal accounts of traditional resourcefulness practices with human waste – most prominently in the Hong Kong example – also contribute to reshaping eating and excreting behavior in the context of culturally entrenched, socially divergent, politically charged and fragile ecosystems.

The hosting of a public composting toilet in the neighborhood, cultivation of urine ferment in the bathroom, and processing diapers into viable humus for orchards rely on complex socio-biological timings and precarious grassroots coordination. It highlights the centrality of householding routines in urban toilets, joint compost yards and peri-urban food forests as essential sites of social reproduction and livelihood enablement. Anticipatory commoning thereby moves beyond the institutional workplace and shifts focus to the enlivening opportunities inherent in a collective reorganization of domestic spaces (Hester 2017). If metabolic transformations become the baseline for how society approaches modes of production and destruction, then it lends urgency to increasing the permeability between those two spheres. Placed within the demands and responses of living compost and fermenting cultures, humans reveal themselves as individuals of a species working together with other species; as knowledge-holders drawing from experience or intuition; as problem-causers burdening ecosystems, and as solution-holders grounded in trust and collaboration (Fletcher and Tham 2019, 40).

This opens up temporal spaces of attention, connection, and production. It means that underutilized resources in people and community gain value when shifting from market to affect economies (Sohn et al. 2015). It is the adventuring between self and system. In a privatized, time-wasting world, 'in-betweening' efforts demand and necessarily raise issues about replenishing and the dangers of exhaustion (Hester 2017). The public eco-toilet commons in Brussels focused on participatory planning and implementation, yet efforts ceased at the end of the project, and the black soil generated remained unused. Similarly, with the urine reuse experiment ending in Hong Kong, only a few participants remained motivated to continue. In contrast, the diaper recycling venture in Berlin incorporated entrepreneurial opportunities and the persistent neediness of the human infant that sustain the human-waste commons over the long arc of generational time.

Human-waste commons as a co-evolutionary process of cohabitating species contrasts with another approach to reorganizing the social. When predictive computing and "anticipatory algorithms" extract data deemed "useful" from other sources deemed "useless," then such reality-formation has become an asymmetric economic process (Reed 2017). Therefore, machine intelligence is "sociomorphic" (Pasquinelli 2016) since "sharing economies" or "economies of suggested content" boost a sense of affiliation with like-minded people. Yet these algorithms are devised for and restricted to monetization of notions of friendship, hospitality, domesticity and volunteering (Han 2014). Thus, perhaps the process of commoning human waste can be appreciated as a place to counteract the digital encroachment and enclosure of human living, since such resource systems require and uphold human qualities of individuality, kinesis, anticipation, and deliberation.

5 Deconstructing Worldviews: Resisting the Logics of Substitution

Currently, global economies are founded on monetary exchange where the validation of wages for labor and the cost of goods is fixed through mechanisms of markets and policy that often do not account for ecological damage and human equity. Instead, these logics substitute land speculation, unseasonably produced food imported from far afield, and extravagant waste regimes causing nutrients loss, transportation pollution and hunger. All three examples of human-waste commoning prefigure and test practices that break free from such pervasive and harmful wastefulness.

In Brussels and Berlin, a quest for more pastoral use of land and establishing alternative livelihoods as citizens designing 'grower-eater-digester-fertilizer' systems have inspired human-waste reintegrating practices that align with the local food movement based on soil-proliferating validation. In Berlin, the diaper-fertilized apple orchards expand to protect the green belt from further real estate development. In Hong Kong, lack of land prompted urine-recyclers to explore suitable rooftops, balconies, and windowsills as sites of planting experiments. In all three examples, human-waste commoners realized and demonstrated how agroecological use of excess nutrients is ignored in the speculative real estate market driven merely by rents and mortgages. Agroecological commons denote the foundational dependence of humans on soil for provisioning of food and resources as a vital concern and political contention (Tornaghi and Dehaene 2020, 605). Thus, it opposes submitting to soil-less types of urban agriculture, indoor warehouse growing, vertical farming, or aquaponics. By substituting the natural ecologies of soil with costly, energy-intensive and wasteful infrastructures that rely on synthetic nutrient formulae and extractive mining, they cement the dominant, financially speculative market logics rather than providing protection of land and resources (Trejo-Téllez and Gómez-Merino 2012).

This contention became particularly prevalent in Hong Kong, where due to land-deprivation, the citizen-research initially tried to apply the upcycled urine to a water-based growing method. As participants had to learn the hard way, this soil-less system lacked the necessary microbiological and respiratory functions to convert the human-derived nutrients into legitimate plant fertilizer. The urine commoners eventually replaced hydroponic water with coir, a fiber by-product of coconut shells, and thus able to mimic the resource-metabolizing, transformational powers of the soil.

Currently, the unhealthy substitution of quick profits for soil depletion and maximized returns of a privileged few for equitable work conditions seem to prevail. For deconstructing (and overriding) these imbalances design commons is tasked with leading transitions (agroecological and otherwise) toward alternative value proposition (Escobar 2018b; Lane 2011). It seeks to socially enact practices that anticipate and stipulate non-extractive forms of urbanism. Feminist philosopher Rosi Braidotti indicates how navigating, let alone manipulating, the self-perpetuating, hegemonic conventions of capitalist urbanism requires guidance through "fight[ing] inertia with creativity, negativity with affirmation (Braidotti and Vermeulen 2014, 188)." All three human-waste commons seek renewal across life forms as a radical answer to the general abandonment of human nutrients, brownfields and self-efficacy. Since human waste tends to be unfit for field crops, designers needed to propose and demonstrate alternative, advantageous applications such as jacuzzi heating, interspecies health monitoring, or intergenerational afforestation. In anticipation of such regenerative purposes, dry-toilet patrons, urine-recyclers, and diaper-composters became part of a living constellation and stratified sensemaking where subjects are multiple and ever-becoming, therefore always in flux. Braidotti (Braidotti and Vermeulen 2014) also emphasizes how only such processual, multiple personalities will be able to pervert and overturn the current substitution logics.

Philosopher John Dewey (1938, 69) already acknowledged how moving from inertia to affirmation is the result of anticipation coupled with joint desire; the discernment of consequences blending with a purpose that then "gives direction to what otherwise is blind." In the three examples, the pairing of the direct experience of human-waste reuse with the reflection thereof through feedback sessions, a journaling routine, and teamwork builds subjectivities based on negotiating and mediating the thresholds of shared living. Consequently, the ethical orientation of the subject is shaped by the influence of the power of the subject's actions on the environment rather than external impositions (Dolphijn and van der Tuin 2012, 35–36). This fluctuation of deliberate and repressive power potentials increases the subject's capacity to enter into multiple, nuanced relational "intimacies and associations that make life sticky and interesting for it," as cultural anthropologist Lauren Berlant (2016) notes.

In the social learning proposed by the three examples, citizen-designers opened up wicked problems with multiple others searching for viable responses instead of trying to solve them in preconceived and homogenized ways (Carolan 2017, 173). In this "researchful" power potential, the facilitators or instigators ("experts") learn about the challenges from and alongside participants, household members, and neighbors ("students"). Bringing urban dwellers into action-research and exposing scientists to the messiness of the field is key to unbinding scientific development, generating yet unexploited knowledge, and tapping into previously unimagined relational agency (Sciannamblo et al. 2018). Thus, commoning practices become the substantiation of differentiated networks and practices where attending to shared metabolism is the membership in what food sociologist Melanie DuPuis (2015) calls "the world of eaters." The commoners in the three examples exchanged the logics of substitution (alongside unhealthy inertia) with the "digestive paradigm" that recognizes the vitality of materials, the importance of nature's nonhuman actors, and the need for contingent socio-natural alliances to redevelop potentials, reservoirs, buffers, and resources. In human-waste commons, subjects sanitize their excrement by way of positive, microbiotic colonization. This anticipatory process of paying attention to spatial and temporal conditions propagates a succession of beneficial, fermenting cultures that eventually out-compete the malevolent and untrustworthy kind.

6 Programmatic Affirmation: Infrastructures Enabling Resourceful Communities

Although the quest for resources and food sovereignty has made agroecological transitions a topic in design discourse, the community of practice upon which they depend requires investing in solidarities, collaborative arrangements, and infrastructures beyond the level of the single household, urban living lab, and farmstead. Commoning efforts that envision the centrality of resource cycling and food growing in the urban environment face a staggering absence of suitable infrastructure, which is necessary to establish resourcefulness (MacKinnon and Derickson 2013). The ill-equipped agroecological city expels superfluous, yet barely recoverable resources. Strategically investing in permanent improvements for leading plentiful and resourceful urban lives needs to be viewed in opposition to the utterly selective, ecologically-blind drive for ever-expanding mobility or IT infrastructure. If the city is considered as a common good and a layered outcome of accrued achievements and improvements (Stavrides 2016), then constructive opposition puts agroecological practices squarely into collective processes aimed at providing resources for food growers and urban dwellers as soil stewards over time.

The commons examples in Hong Kong, Brussels, and Berlin formulated a community among strangers who work together in response to a basic need (bowel movement), a threat (wastefulness), a desire (neighborly relations), or pure curiosity (novelty). Rosi Braidotti calls this a "programmatic affirmation" where people come together to realize what they are missing and, in turn, anticipate who they want to become: "We need to borrow the energy from the future to overturn the conditions of the present" (Braidotti and Vermeulen 2014, 188). The forward-invested spatial organization ranging from compost-friendly architecture and indoor urine reserves to edible greenery, call at once on different responsibilities as a householder, resourceful worker, or landscape provider. In such transformational potential, the commoners rehearsed different implications of constituting *we's* from conjoint action and desirability. As Dewey (1938) notes, these anticipatory collectives raise the issue of consequences over intentions, whereas responding to a common cause or threat becomes a matter of pragmatically solving problems rather apportioning blame. Commoning design then is about endorsing a social movement of assembling (novel solutions) and dissolving (inadequate fixations). It acknowledges the complex dynamics of attraction and aversion among actors within contexts while facilitating a progressive blurring of roles that allows for collective desire to emerge – including toilet-led urban revitalization, urine-based plant/human flourishing, or intergenerational agri-forestry.

This emerging collective desire for a different world can only be sustained longer-term if strangers can be enlisted to collaborate in anticipation of mutual benefit and trust (Felstead et al. 2019). Similarly, Eleanor Ostrom (1990, 88) gauges the success of novel infrastructure creation in anticipatory communities that "share a past, and expect to share a future." Long-term agroecological commoning means imagining an adequately equipped, food-enabling urban landscape that proliferates socio-natural resources and bio-cultural diversity. Since commoning is a situated condition, passing on generic best practices poses challenges (Botero et al. 2020). Therefore, food-enabling visions aim beyond scarcity and private property at community-based and community-led agroecological "resource hubs" (Tornaghi and Dehaene 2020) that are spatially pervasive, seasonally adaptive, and widely socialized.

Nonetheless, attempting to make urban social reproduction more agroecological entails a way of being in the world that is comfortable with visceral materiality and exposure to risk. Lauren Berlant, referring to Paolo Virno, notes: "the ordinary of the contemporary commons [is] a dispossessedness in its awkward, convoluted, observational, comic, noisy, and diversely manifest vulnerability" (Berlant 2016, 408; Virno 2004). Coping with unpredictable environmental conditions, tolerating exuberant, urine-fermenting cultures in the bathroom, adjusting to changing affinities and demands of a diverse working alliance, the infrastructuring human-waste commoners attest how moving in concert with social beings, human or not, is difficult, inconvenient and demanding (Wernli 2020). Therefore, attempting to shift from the normative infrastructures of the state and the markets into affective infrastructures of ordinary life necessitates close consideration for the social dynamics of attraction and aversion (Berlant 2016). Anticipatory infrastructuring thus is simultaneously about ethical and technical enablement that reaches beyond activist or utopian aspirations to the quotidian routines of everyday life as experienced by frugal neighbors with common needs and a wasteful commons.

In all three example cities, the agroecological infrastructure for human-waste reuse is patchy, informal, and confined to a legal grey zone. In Hong Kong, the reuse of organic waste is left to grassroots and commercial initiatives, since dominant development and hygiene regimes impede agroecological ambitions. The Berlin example is part of a larger agroecological movement for reintegrating organic waste in urban greening efforts, where with institutional support, public eco-toilets and neighborhood-managed composting operations are on the rise. Similarly, in Brussels, nutrient resourcing is part of recently launched negotiations between socially diverse urban farmers and authorities to identify better and address their respective needs. In the global South, the compounding urgency of soil depletion, the wealth gap, and nutrient pollution are already spurring the establishment of adequately-scaled and community-owned agroecological infrastructure (Koop 2020; Gianella-Estrems et al. 2015). In Rosario, Argentina, this entails (i) facilities for gathering, storing, and transporting nutrients whereas organic clippings, industrial by-products and food waste are composted and distributed to farmers in and around town; (ii) holding capacity for rainwater harvesting in response to climate change; (iii) shared infrastructure for food preservation and processing, including access to markets; and (iv) programs and spaces for seed exchanges backed by a public seed bank. In this multidirectional context, the self-organizing basic needs of food producers and consumers have yielded social infrastructures and complementary economies that mutually support each other.

Regardless of the differences in the local context and developmental stages at which the actors are building up resourcefulness, the praxis of anticipation and its three-way transformation dynamics can be seen as a strategy in agroecological transitions and beyond. Table 9.1 correlates the key concepts mentioned in this chapter upon which the anticipatory praxis builds.

Table 9.1 Operational matrix of anticipatory praxis in commons design

	Domain	Epistemology	Heuristics	Ontology	Source
		Knowing	*Doing*	*Being*	
i	Commons	Resource system	Community formation	Social code	Ostrom, Linebaugh
ii	**Praxis of anticipation**	Advance care	Deconstructing worldviews	Programmatic affirmation	Braidotti, Virno
iii	Subjectivity configuration	Social reproduction	Conjoining trust, desire and action	Assembling and dissolving movement	Dewey
iv	Agroecological transformations	Energy-food nexus (science)	Metabolic co-proliferation (practice)	Participative diagnosis and planning (politics)	Tornaghi, Dehaene
v	Publicness of difference	Domesticity	Socio-technical infrastructure	Co-programming	Sohn, Stavros, Bruyns
vi	Temporality	Commitment	Consequences over intention	Desirability for (dis) continuum	Carolan, Dewey
vii	Roles	Producer/ consumer, knowledge-holder	Value creator	Problem-causer/ solution-holder	Tham, Fletcher

Correlation of the key concepts that anticipatory praxis in design commons builds upon

7 Conclusion: Praxis of Anticipation as Operative Mode in Commoning

Building on the insights from agroecological human-waste ventures in Brussels, Hong Kong and Berlin, this chapter evaluated the critical dynamics behind resourceful transitions and its implications for design commons. The radical integration of the commons within resourcefulness and social reproduction implies a paradigmatic shift in subjectification processes, bio-economic value creation and programmatic facilitation – identified as praxis of anticipation. This operative term for commoning is distilled from the actual nature of human nutrients, peri-urban resourcefulness, food pedagogies, and compost-friendly infrastructure. Practitioners of anticipation confront and engage the contradictions implicit in social reproduction and its care ecologies, including domesticity, hospitality, resourcefulness and community. These shared domains and principles of commoning become values on which the exhaustion of people and the environment can be reversed. Anticipation in design commons challenges the linear and causal determinist narratives of temporary values. Anticipation offers open-ended trajectories that allow for composition and decomposition of oblique relationships, vague aspirations, and negotiated alliances across time and space. Anticipation, through its manifold meaning of prospecting, taking care ahead of time, forestalling harm, and enthusiasm is intricately linked to dynamic renewal processes of social lives. Anticipation thus is a proposition to admit radical pedagogies of unlearning, re-learning, and aspirational ambivalence for honing not just technical but also affective ecological infrastructures that enhance the prospects of a world worth inhabiting. Thus, the expression of the human waste commons is a praxis of anticipation that makes it possible to draw energy from engagement with something, or someone, yet to be grasped. It is a driving force for overturning the entrenched negating present and stepping onto the threshold of affirmative transformations. Anticipation as the *force* of nature is a tool to rework human nature, return humans to nature, and restore human waste to humus.

Acknowledgements Heartfelt thanks to Nathan Felde, Britta Boyer, and Sarah Daher for their valuable feedback and advice. This work is supported by a seed grant from the Design Trust in Hong Kong and an internationalization grant from Dutch Creative Industries in Rotterdam.

References

Altieri, Miguel A., and Victor Manuel Toledo. 2011. The Agroecological Revolution in Latin America: Rescuing Nature, Ensuring Food Sovereignty and Empowering Peasants. *Journal of Peasant Studies* 38 (3): 587–612. https://doi.org/10.1080/03066150.2011.582947.

Amaya, Sasha. 2016. An Interview with Collective Disaster. *MVT Journal for Art x Architecture x Landscape*, October 2016. http://www.mvt-journal.com/collective-disaster. Accessed 15 Sep 2020.

Andreev, Nadejda, Mariska Ronteltap, Piet Lens, and Boris Boincean. 2015. *Improving the Efficiency of Resource Oriented Sanitation: A Critical Review on Terra Preta Sanitation*. UNESCO-IHE Institute for Water Education. The Netherlands: Delft.

Bendell, Jem. 2018. Deep Adaptation: A Map for Navigating Climate Tragedy. *IFLAS Occasional Paper* 2: 36. www.iflas.info. Accessed 15 Sep 2020.

Berlant, Lauren. 2016. The Commons: Infrastructures for Troubling Times. *Environment and Planning D: Society and Space* 34 (3): 393–419. https://doi.org/10.1177/0263775816645989.

Bollier, David. 2019. Human Waste: The Latest Enclosure of the Commons? *P2P Foundation*, March 2019. https://blog.p2pfoundation.net/human-waste-the-latest-enclosure-of-the-commons/2019/03/11. Accessed 15 Sep 2020.

Bollier, David, and Silke Helfrich. 2019. *Free, Fair and Alive: The Insurgent Power of the Commons*. Gabriola Island: New Society.

Botero, Andrea, Giacomo Poderi, Joanna Saad-Sulonen, and Frederick Van Amstel. 2020. Commoning Design and Designing Commons. In *PDC 2020: Proceedings of the 16th Participatory Design Conference*. Manizales: Association for Computing Machinery (ACM).

Braidotti, Rosi, and Timothy Vermeulen. 2014. Borrowed Energy. *Frieze* Issue 165, August 2014. https://www.frieze.com/article/borrowed-energy. Accessed 15 Sep 2020.

Carolan, Michael. 2017. *Society and the Environment: Pragmatic Solutions to Ecological Issues*. 2nd ed. Boulder: Westview Press.

Carr, Richard, Ursula Blumenthal, and Duncan Mara. 2004. Health Guidelines for the Use of Wastewater in Agriculture and Aquaculture. In *Wastewater Use in Irrigated Agriculture: Confronting the Livelihood and Environmental Realities*, ed. Christopher Scott, Naser I. Faruqui, and Liqa Raschid-Sally, 41–58. Wallingford: International Water Management Institute.

Debatty, Régine, and Ayumi Matsuzaka. 2019. Turning Human Waste into Beer and Fruit Trees. *We-Make-Money-Not-Art*, August 2019. https://we-make-money-not-art.com/turning-human-waste-into-beer-and-fruit-trees. Accessed 15 Sep 2020.

Dehaene, Michael, Chiara Tornaghi, and Colin Sage. 2016. Mending the Metabolic Rift: Placing the "Urban" in Urban Agriculture. In *Urban Agriculture Europe*, ed. Frank Lohrberg, Lilli Lička, Lionella Scazzosi, and Alex Timpe. Berlin: Jovis.

Dewey, John. 1938. *Experience and Education*. New York: Kappa Delta Pi.

Dolphijn, Rick, and Iris van der Tuin. 2012. *New Materialism: Interviews & Cartographies*. Ann Arbor: Open Humanities Press.

DuPuis, Melanie E. 2015. *Dangerous Digestion: The Politics of American Dietary Advice*. Oakland: University of California Press.

Escobar, Arturo. 2018a. *Designs for the Pluriverse: Radical Interdependence, Autonomy and the Making of Worlds*. Durham/London: Duke University Press.

———. 2018b. Patterns of Commoning: Commons in the Pluriverse. *P2P Foundation*, June 2018. https://blog.p2pfoundation.net/patterns-of-commoning-commons-in-the-pluriverse/2018/06/08. Accessed 15 Sep 2020.

Etter, Bastian, Kai Udert, and Teddy Gounden. 2015. *Project Report: Valorisation of Urine Nutrients – Promoting Sanitation and Nutrient Recovery through Urine Separation*. Dübendorf: Swiss Federal Institute of Aquatic Science & Technology.

Felstead, Aimee, Kevin Thwaites, and James Simpson. 2019. A Conceptual Framework for Urban Commoning in Shared Residential Landscapes in the UK. *Sustainability* 11 (21). https://doi.org/10.3390/su11216119.

Ferrando, Tomaso, Priscilla Claeys, Dagmar Diesner, and Jose Luis Vivero-Pol. 2020. Commons and Commoning for a Just Agroecological Transition: The Importance of Decolonising and Decommodifying Our Food Systems. In *Resourcing an Agroecological Urbanism: Political, Transformational and Territorial Dimensions*, ed. Chiara Tornaghi and Michiel Dehaene. London: Routledge.

Fletcher, Kate, and Mathilda Tham. 2019. *Earth Logic: Fashion, Action, Research, Plan*. London: The J.J. Charitable Trust.

Gianella-Estrems, Teresa, Teobaldo Pinzás, and Antonio Latucca. 2015. Urban Agroecology: A Tool for Social Transformation. *Farming Matters*, June 2015.

Gidwani, Vinay, and Priti Ramamurthy. 2018. Agrarian Questions of Labor in Urban India: Middle Migrants, Translocal Householding and the Intersectional Politics of Social Reproduction. *Journal of Peasant Studies* 45 (5–6): 994–1017. https://doi.org/10.1080/03066150.2018.1503172.

Han, Byung-Chul. 2014. Why Revolution is Impossible: On the Seductive Power of Neoliberalism. *Süddeutsche Zeitung*, September 2, 2014.

Harcourt, Wendy, and Arturo Escobar. 2002. Women and the Politics of Place. *Development* 45 (1): 7–14. https://doi.org/10.1057/palgrave.development.1110308.

Heron, John, and Peter Reason. 2001. The Practice of Cooperative Inquiry: Research "With" Rather Than "On" People. In *Handbook of Action Research: Participative Inquiry and Practice*, 179–188. London: Sage.

Hester, Helen. 2017. Promethean Labors and Domestic Realism. *E-Flux Architecture: Artificial Labor*, 2017. https://www.e-flux.com/architecture/artificial-labor/140680/promethean-labors-and-domestic-realism/. Accessed 15 Sep 2020.

Heynen, Nik, Maria Kaika, and Erik Swyngedouw. 2005. *In the Nature of Cities: Urban Political Ecology and the Politics of Urban Metabolism*. https://doi.org/10.4324/9780203027523.

Holt-Giménez, Eric, and Annie Shattuck. 2011. Food Crises, Food Regimes and Food Movements: Rumblings of Reform or Tides of Transformation? *Journal of Peasant Studies* 38 (1): 109–144. https://doi.org/10.1080/03066150.2010.538578.

Karasti, Helena. 2014. Infrastructuring in Participatory Design. *ACM International Conference Proceeding Series* 1: 141–150. https://doi.org/10.1145/2661435.2661450.

Karga, Valentina. 2014. Collective Disaster Helps Community with Sustainable Temple of Holy Shit. Designboom. 2014. http://www.designboom.com/art/collective-disaster-temple-of-holy-shit-factory-of-the-black-gold-brussels-11-06-2014/. Accessed 15 Sep 2020.

Katz, Cindi. 2001. Vagabond Capitalism and the Necessity of Social Reproduction. *Antipode* 33 (4): 709–728. https://doi.org/10.1111/1467-8330.00207.

Koop, Fermin. 2020. Depleted Soils Drive Argentina to Sustainable Farming. *Diálogo Chino: China, Latin America and the Environment*, May 2020. https://dialogochino.net/en/agriculture/35279-depleted-soils-drive-argentina-to-agroecology/. Accessed 15 Sep 2020.

Lane, Ruth. 2011. The Waste Commons in an Emerging Resource Recovery Waste Regime: Contesting Property and Value in Melbourne's Hard Rubbish Collections. *Geographical Research* 49 (4): 395–407. https://doi.org/10.1111/j.1745-5871.2011.00704.x.

Leung, Daren. 2020. Strange Encounter with Poo: East-West Roots of Organic Farming. Medium: Foodprint. 2020. https://medium.com/foodnote/a-strange-encounter-with-poo-east-west-roots-of-organic-farming. Accessed 15 Sep 2020.

Linebaugh, Peter. 2008. *The Magna Carta Manifesto: Liberties and Commons for All*. Berkeley: University of California Press.

MacKinnon, Danny, and Kate Driscoll Derickson. 2013. From Resilience to Resourcefulness: A Critique of Resilience Policy and Activism. *Progress in Human Geography* 37 (2): 253–270. https://doi.org/10.1177/0309132512454775.

Marttila, Sanna, Andrea Botero, and Joanna Saad-Sulonen. 2014. Towards Commons Design in Participatory Design. *ACM International Conference Proceeding Series* 2: 9–12. https://doi.org/10.1145/2662155.2662187.

Mateo-Sagasta, Javier, Kate Medlicott, Mamzoor Qadir, Liqa Raschid-Sally, Pay Drechsel, and Jens Liebe. 2013. Safe Use of Wastewater in Agriculture. In *Proceedings of the UN-Water Project*, ed. Jens Liebe and Reza Ardakanian. Bonn: UN-Water Decade Programme on Capacity Development.

Meiselman, Herbert L., and Haliday J. MacFie, eds. 1996. *Food choice, Acceptance and Consumption*. London: Chapman & Hall.

Nadin, Mihai. 2005. Anticipatory Systems. *Ubiquity* January: 1. https://doi.org/10.1145/1052128.1046683.

Ostrom, Elinor. 1990. *Governing the Commons: The Evolution of Institutions for Collective Action*. Cambridge: University Press.

Pasquinelli, Matteo. 2016. Abnormal Encephalization in the Age of Machine Learning. *E-Flux*, September 2016. https://www.academia.edu/28335333/Abnormal_Encephalization_in_the_Age_of_Machine_Learning. Accessed: 15 Sep 2020.

Puig de la Bellacasa, Maria. 2017. *Matters of Care: Speculative Ethics in More Than Human Worlds*. Minneapolis: University of Michigan Press.

Reed, Patricia. 2017. Xenophily and Computational Denaturalization. *E-Flux Architecture: Artificial Labor*, March 2017. http://www.e-flux.com/architecture/artificial-labor/140674/xenophily-and-computational-denaturalization/. Accessed 15 Sep 2020.

Reynaert, Maaike. 2016. From Soiled Nappies to Just Plain Soil. *RESET*, February 2016. https://en.reset.org/blog/soiled-nappies-just-plain-soil-02012016. Accessed 15 Sep 2020.

Russell, Jean M. 2013. *Thrivability: Breaking Through to a World That Works*. Charmouth: Triarchy Press.

Savazoni, Rogrigo, and Oswald Andrade. 2019. The Crossroads of the Commons: Citizen Laboratories in Transit. In *The XVII International Association for the Study of the Commons*. Lima, Portugal.

Schmidt, Hans-Peter. 2014. Terra Preta – model of a cultural technique. *The Biochar Journal*. July 2014. https://www.biochar-journal.org/en/ct/4. Accessed: 4 Feb 2022.

Schneider, Mindi, and Philip McMichael. 2010. Deepening, and Repairing, the Metabolic Rift. *Journal of Peasant Studies* 37 (3): 461–484. https://doi.org/10.1080/03066150.2010.494371.

Sciannamblo, Mariacristina, Peter Lyle, and Maurizio Teli. 2018. Fostering Commonfare: Entanglements Between Participatory Design and Feminism. In *DRS2018: Catalyst*, vol. 2. Limerick: Design Research Society. https://doi.org/10.21606/drs.2018.557.

Shiming, Luo. 2002. The Utilization of Human Excreta in Chinese Agriculture and the Challenge Ahead. In Proceedings of the First International Conference on Ecological Sanitation 2001(37–41): Nanning. EcoSanRes, Stockholm Environment Institute. http://www.ecosanres.org/Nanning_Conf_Proceedings.htm.

Siltanen, Janet, Fran Klodawsky, and Caroline Andrew. 2015. "This Is How I Want to Live My Life": An Experiment in Prefigurative Feminist Organizing for a More Equitable and Inclusive city. *Antipode* 47 (1): 260–279. https://doi.org/10.1111/anti.12092.

Sohn, Heidi, Stavros Kousoulas, and Gerhard Bruyns. 2015. Commoning as Differentiated Publicness. *Footprint* 16: 1–8. https://doi.org/10.7480/footprint.9.1.895.

Sollazzo, Andrea, Louisa Vermoere, Valentina Karga, and Pieterjan Grandy. 2014. *Usine Du Trésor Noir: User's Manual*. Brussels: Collective Disaster.

Stavrides, Stavros. 2016. *Common Space: The City as Commons*. London: Zed Books. https://doi.org/10.1007/1-4020-0613-6_3100.

Tornaghi, Chiara. 2017. Urban Agriculture in the Food-disabling City: (Re)Defining Urban Food Justice, Reimagining a Politics of Empowerment. *Antipode* 49 (3): 781–801. https://doi.org/10.1111/anti.12291.

Tornaghi, Chiara, and Michiel Dehaene. 2020. The Prefigurative Power of Urban Political Agroecology: Rethinking the Urbanisms of Agroecological Transitions for Food System Transformation. *Agroecology and Sustainable Food Systems* 44 (5): 594–610. https://doi.org/10.1080/21683565.2019.1680593.

Tornaghi, Chiara, and Femke Hoekstra. 2017. Urban Agroecology. *Urban Agriculture Magazine* 53 (9).

Trejo-Téllez, Libia I., and Fernando C. Gómez-Merino. 2012. Nutrient Solutions for Hydroponic Systems. In *Hydroponics: A Standard Methodology for Plant Biological Researches*, 1–23. Montecillo, Mexico: Texcoco. https://doi.org/10.5772/37578.

Virno, Paulo. 2004. *A Grammar of the Multitude: For an Analysis of Contemporary Forms of Life*. Cambridge: MIT Press.

Waltner-Toews, David. 2013. *The Origin of Feces: What Excrement Tells Us About Evolution, Ecology, and a Sustainable Society*. Toronto: ECW Press.

Weckenbrock, Philipp, and Graham Alabaster. 2015. Designing Sustainable Wastewater Reuse Systems: Towards an Agroecology of Wastewater Irrigation. In *Governing the Nexus: Water, Soil and Waste Resources Considering Global Change*, 153–188. Cham: Springer. https://doi.org/10.1007/978-3-319-05747-7.

Wernli, Markus. 2020. Anthroponix: Collective Urine Upcycling to Grow Plants and Material Responsibility. In *Culture, Community & Climate: Conversations and Emergent Praxis*, ed. Richard Povall, 68–89. Kingsbridge, Devon: art.earth imprint.

Worster, Donald. 2017. The Good Muck: Toward an Excremental History of China. *Rachel Carson Center Perspectives: Transformations in Environment and Society* 5: 1–54.

Xue, Yong. 2005. "Treasure Nightsoil as If It Were Gold": Economic and Ecological Links Between Urban and Rural Areas in Late Imperial Jiangnan. *Late Imperial China* 26 (1): 41–71. https://doi.org/10.1353/late.2005.0009.

Chapter 10
The Secondary Use Group: Unlocking Waste as a Common Pool of Resources in the 1970s

Piero Medici

Abstract Today, the evident need for more efficient conservation, management and redistribution of natural and human-made common resources have inspired thinkers, researchers, and designers to redefine the organization of our societies. For example, Silke Helfrich and David Bollier argue that the common-pool resources (CPR) defined by Elinor Ostrom require new "practices of commoning" that reconsider the conventional discourses of market economy and state intervention. Several contemporary architectural firms have introduced innovative design strategies concerning the collective collection and reuse of local materials, the commons and the circular economy.

However, already after the oil crisis in the early 1970s, practices like the Secondary Reuse Group (SUG) engaged with circular reuse of materials but did not correlate to discourses concerning the commons. This essay analyzes SUG's projects during the 1970s using a lens calibrated on the contemporary debate of the commons, to unveil and highlight some relevant aspects of their work. This lens will refer to Michel Bauwens and Tom Avermaete who differentiate between material commons, that is, human-made and -handled reserves of materials from our environments and cities; immaterial commons, knowledge and craft skills existing in a particular place; and commoning processes, social practices of mutual collaboration. The first goal of this research is to describe the work of SUG concerning its material and immaterial commons. The second goal is to inform the contemporary debate regarding waste and materials as a CPR to be unlocked by architects and users through commoning processes of materials reuse.

Keywords Commons · Common pool resources · Commoning process · Materials reuse · Circular economy · Architectural approach · Secondary use group · Martin Pawley

---

P. Medici (✉)
Faculty of Architecture, Delft University of Technology, Delft, The Netherlands
e-mail: P.Medici@tudelft.nl

G. Bruyns, S. Kousoulas (eds.), *Design Commons*, Design Research Foundations, https://doi.org/10.1007/978-3-030-95057-6_10

Today, the continuous growth of the world population, the increasing wealth inequality and the depletion of the resources have inspired thinkers, researchers, and designers to redefine the organization of our societies, economic systems and design processes. The need for more efficient conservation, management and redistribution of natural and human-made common resources is evident. For example, the concept of common-pool resources (CPR) described in *Governing the Commons* by Nobel Prize winning political economist Elinor Ostrom (1990), considers our everyday assets beyond the prevailing discourses of market economy and state intervention (Avermaete 2018). In recent years, sociologist, economist and activist Silke Helfrich and activist, writer, and policy strategist David Bollier argued in *Die Welt der Commons* (Bollier and Helfrich 2015) that the CPR requires new "practices of commoning" that challenge our conventional understanding of economy, politics and culture. The British economist Kate Raworth referred to the commons and the circular economy when creating "doughnut economics" (Raworth 2017), Calisto Friant referred to the commons when illustrating the "circular society" (Calisto Friant et al. 2020).

Several contemporary architectural firms are outlining innovative strategies considering the commons, the doughnut economy, the economy for the common goods (Felber 2015) and the circular economy. These practices have introduced architectural approaches concerning the collective collection and reuse of local materials, also for different functions from which they were originally designed. However, practices regarding circular reuse of materials already appeared in the late 1960s and after the oil crisis in the early 1970s (Pawley 1975a). One representative example was the Secondary Use Group (SUG) operating during the 1970s (Kan and Poteliakhoff 1978), when mainstream architecture, academics and intellectuals largely ignored experimental ecologic design (Ingersoll 2012) (Bonnemaison and Macy 2003; Medici 2017). Unfortunately, the design and research of SUG and other similar practices were easily cast aside during the 1980s when the oil price decreased again (Borasi and Zardini 2007). During the 1970s, practices like SUG proposed alternative economic models such as industrial production of materials easy to reuse, but they did not correlate to notions and discourses regarding the commons. It is quite understandable since in 1968, Garrett Hardin wrote that the common resources are tragically exploited when open to all and not regulated (Hardin 1968).

Reading the work of SUG with contemporary knowledge of the commons can unveil and highlight some relevant aspects of their work, which did not emerge sufficiently in the 1970s and the 1980s. For example, several years after the 1980s, Helfrich stated that a CPR becomes a commons only when communities use and sustain it (Helfrich 2011). Belgian political theorist Michel Bauwens further elaborates on the CPR, differentiating "material" commons as reserves of materials, and "immaterial" commons as knowledge and skills (Bauwens 2017; Bollier and Helfrich 2015). Belgian architectural scholar Tom Avermaete distinguishes among "res communis" as materials commons, "lex communis" as common codes and norms, and "praxis communis" as collaborative practices (Avermaete 2018).

In this essay I analyze the projects of SUG during the 1970s using a lens calibrated on the contemporary debate of the commons. Through this lens, I examine some crucial aspects of the projects, like the CPR composed of waste such as used industrial products. In this case waste represents material commons and res communis, converted into new objects, buildings, and spaces in the city. Architects and users then unlocked the CPR through immaterial commons, lex communis and praxis communis. For example, SUG defined a strategy for the collection and reuse of local second-hand materials related to the distance from the building site; they embraced do-it-yourself construction with technologies such as solar panels, rainwater collectors and greenhouses integrated into their buildings. SUG designed and crafted external cladding from used food cans, and floors and walls built from beer crates as a structure and shredded paper as insulation (Pawley 1975c). With the collaboration of some students, SUG tested structural beams composed of cut cans, and they reused sections of pipes as shelves for a shop (Kan and Poteliakhoff 1978).

The first goal of this research is to illustrate the work of SUG concerning its immaterial commons as design methods, referring to the British architect and researcher Martin Pawley, who taught, researched and published about reuse in the 1960s and 1970s. The second goal is to inform the contemporary debate regarding waste and materials as a CPR to be unlocked by architects and users through lex communis and praxis communis of reusing materials also for a different function than they were originally designed for.

This research has employed two methods to achieve the research goals: literature review explained in Section One and Two and case study in sections from Three to Eight. With the title *The 1970s as a Forge of New Concepts Concerning Common Resources*, Section One gives an overview of the literature throughout the 1970s regarding the commons, the bottom-up participative, and cooperative processes. Section Two, entitled *Contemporary Discourses of Urban Commons*, explains the concepts of CPR, lex communis, praxis communis, and other literature concerning the commons developed after the 1970s. The first two sections define the main theoretical elements used to calibrate the lens adopted in this essay to read the SUG works. Section Three illustrates the work of Martin Pawley, which is the primary reference of the architects composing the SUG. Section Four to Eight then analyse SUG's works through a series of lenses calibrated by the literature view, such as collaborative practices, the CPR, praxis and lexis communis.

1 The 1970s as a Forge of New Concepts Concerning Common Resources

At the end of the 1960s and the beginning of the 1970s, prior to the 1973 global oil crisis, different literature introduced new economic and social approaches regarding natural and human-made resources. On the one hand, discourses emphasizing direct participation and citizen inclusion often through mechanisms of bottom-up governance were described in a number of influential publications: *Limits to Growth* by Donella Meadows and others (Meadows et al. 1972), *The Entropy Law and the*

Economic Process by Nicolas Georgescu Roegen (1971), *The Closing Circle* by Barry Commoner (1971), *Post-Scarcity Anarchism* by Murray Bookchin (1971), *Small is Beautiful* by Ernst Friedrich Schumacher (1973) and *Tools for Conviviality* by Ivan Illich (1973). On the other hand, texts such as the renowned *The Tragedy of the Commons* by Garrett Hardin (1968) and *The Population Bomb* by Paul R. Ehrlich (1968) advocated population control and resource efficiency strategies from the top down.

In this context, literature concerning architecture and design reacted through texts like *Design for the Real World* by Victor Papanek (1971), Charles Jencks and Nathan Silver's *Adhocism: The Case for Improvisation* (1972), *The Alternative: Communal Life in America* by William Hedgepeth (1970) and Peter Cook's *Experimental Architecture* (1970), to mention some. In 1971 and 1973, *Architectural Design* magazine devoted two covers to "garbage housing", also the subject of English critic Martin Pawley's book *Garbage Housing* (1975a). In the book Pawley published the research carried out with his students at the Architectural Association (AA) in London and the School of Architecture at Cornell University (Ponte 2006). The magazines also described the work of SUG, mentioned in Pawley's book as well. They were a group of British architects, former students of Pawley at AA and mainly active in London, who reused second-hand materials to build housing prototypes and interiors.

Pawley and the group described their work, mainly focusing on technical aspects regarding reused materials and construction details, contributing to solving the housing crisis. They did not center the discourse on sharing the common resources through design and cooperative processes, which is quite understandable since the commons had recently been described as a tragedy by Hardin. Nor did they highlight the importance of the collective and cooperative aspects of their work relating, for instance, to concepts described in *Tools for Conviviality* such as autonomous and creative intercourses among persons and their environment (Illich 1973). Against this background, in this essay I will read the work of SUG and Pawley through the knowledge developed about the commons after the 1970s. I will draw on some of the sources mentioned above including the titles advocating bottom-up participative and inclusive processes and more recent literature concentrated on the commons in urban territories.

2 Contemporary Discourses of Urban Commons

Concerning the concept CPR (Ostrom 1990), which this essay takes as a reference, Ostrom maintains that these common-pool resources can be found in the environment and consist, in the first place, of a "resource system" such as groundwater basins, grazing areas, fishing grounds, irrigation canals and bridges. Secondly, they are composed by "resource units" which entail the "water withdrawn from a groundwater basin or an irrigation canal, fish harvested from a fishing ground, the fodder consumed by animals from a grazing area" (Ostrom 1990, 30). Ostrom claims that the common-pool resource system and unit maintain a reciprocal relationship that needs to be governed and regulated (Avermaete 2018).

Silke Helfrich states that a CPR becomes a commons only when it is turned into a commons by its users; when communities use and sustain it (2011). Michel Bauwens further elaborates on the CPR to be found in urban territories. He differentiates between inherited commons which he links to resources such as earth, water, and forests; material commons such as human-made and human-handled reserves of materials from environments, and immaterial commons such as the intellectual and cultural knowledge and craft skills existing in a particular place (Bauwens 2017; Bollier and Helfrich 2015). From this perspective, cities can be looked upon as a stock of materials that is constantly used and reused, and where waste materials can be considered both a stock of material resources and a potential CPR, prevalent with contemporary professionals engaging with reuse and sustainability (Avermaete 2018).

In recent years, Tom Avermaete has focused on the architecture of the city as "one of the main tangible forms in which the commons exist in society" (2018, 33). Avermaete explores how conceptions of the commons have been part of the development of the architecture of the city, distinguishing among "res communis" as CPR stocked in the city (2018); "lex communis" as common codes and conventions and norms of art posed by architecture (Avermaete 2018); and "praxis communis" as social practices of mutual collaboration, support, negotiation, communication and experimentation that are needed to create systems to manage CPR and to engage with common codes and conventions (Avermaete 2018). Concerning the process of commoning, David Bollier argues that in a commons, care work (2020), also defined affective labor by geographer Neera Singh (2013), has primary importance. However, market systems like capitalism consider labor as mainly motivated by monetary reward and usually ignore the care economy characterized by social conviviality that is essential for a stable, sane, rewarding life (Avermaete 2018). In this essay, while analyzing the work of SUG, I will refer to the Bauwens material and immaterial commons, definitions that are related to Avermaete's res communis and lex communis. I will also refer to the concept of the commoning process, which relates to praxis communis. The goal will be to read the work of SUG in particular and the concept of waste and design as a commons in general, using a lens calibrated towards contemporary discourses about the commons.

3 Martin Pawley, Waste Materials Reused to Build Houses

Martin Pawley was the founding editor of the London-based Architectural Association (AA) School's weekly newssheet *Ghost Dance Times* (1975b), and author of several articles and books including *Garbage Housing* (Pawley 1975a). In his publications, he does not directly address the topic of the waste as a common resource to share and to use through cooperative, bottom-up processes, similar to the processes of commoning. He is critical towards linear processes of production, consumption and disposal of materials and in favor of introducing recycling and reuse strategies to improve the efficiency of industrial cycles. However, he does not significantly promote practices of cooperative, bottom-up initiatives regarding reuse and collection of waste materials. Even if not deliberately, Pawley highlights the potential of waste materials as a CPR, and he describes first steps of possible immaterial commons and commoning processes.

In the *Architectural Design* issues of 1971 and 1973 (Pawley 1971; Pawley 1973), he discusses ideas that would later form part of his book *Garbage Housing* published two years later. In this book, he calls for better use of waste and second-hand materials, for affordable housing "built from the detritus of a society of conspicuous consumption" (Pawley 1975a). Pawley focuses on the shortage of houses in the UK, the expensive and lengthy processes of construction, especially compared to industrial consumption products such as bottles, televisions and cars. He describes examples of material reuse in the work of his students and in different practices, also briefly mentioning the work of SUG. He refers to some examples including the WOBO (World Bottle) project by the Heineken corporation, where beer bottles designed in the shape of bricks are reused as a construction material for housing (Pawley 1975a). Pawley tends to concentrate on topics such as cost, materials and industrial production; social and behavioral issues only form a tangential part of his work. He maintains that technology at the time was a linear process, which consumes raw materials to generate products which then become waste. The recycling of waste materials was unusual, expensive and time-consuming, with waste from industrial production causing problems like chemical, visual and noise pollution, and workers striking due to unhealthy working conditions. Pawley thus signals the need for converting industrial waste material into secondary use, to treat it as a new raw material for other processes, including housing (Pawley 1971).

In the last pages of the book, he mentions the student work of Shiu-Kay Kan, one of the future members of SUG. In 1973, at the Central London Polytechnic (PCL), Kan built some metal structures assembled from spot-welded catering cans. The structures could sustain the minimum dwelling unit designed by Kan at the school, in the following year. Kan's four by three-and-a-half meter enclosure used a raised floor of beer crates with welded-can wall units and Masonite cladding. The design incorporated aluminum solar heating panels developed at PCL (Pawley 1975a). The minimum dwelling unit, an open space with kitchen, bathroom, bed and desk, could be suitable as a temporary dwelling for students. Pawley then proposes a garbage housing project for a few hundred students, self-built, self-designed and self-administered by the students with some support from manufacturers and the approval by the local authorities The goal of the project is to contribute to further developing the secondary use of materials discipline and to extend its implementation scale. Pawley identifies a student housing crisis in most of the British universities, both in terms of shortage of available housing and of cost. He explains that students, a social group still outside the spiral of consumption, would build their own dwellings. His suggestion is to locate the garbage housing on the large areas of the abandoned, unmaintained, unused urban waste land available for a minimum of two years during the cycle of redevelopment. Pawley considers the unused land, which could be leased by the owners, the local authorities, and the labor force of students willing to build their own housing, a winning combination. In addition, he considers short-life garbage housing settlements rotating around redevelopment sites on two-yearly cycles with the expertise from the students of architecture, design and engineering, as a first step to implement secondary use on a bigger scale (Pawley 1975a).

Pawley does not directly refer to notions about the commons nor cooperative, collective and communal practices. However, he claims that the second life of materials is indeed a resource essentially not used despite its great potential. Although not emphasized by Pawley, *Garbage Housing* illustrates several elements related to the notion of immaterial commons, such as the skills and the knowledge needed to use waste as a resource, and the DIY skills of the students cutting, welding and connecting parts of cans and beer crates. Pawley's idea about the garbage dwellings for students has many features of a commoning process. The waste materials as a CPR would be used by the community of students to build their own houses. The originally unused urban area would then host a community of student dwellings with the potential of accommodating more functions. The students would share knowledge, craft skills, codes and norms posed by architecture regarding the collection of the materials and the construction of the houses as immaterial commons. Furthermore, the minimum dwelling unit would be provided with solar panels developed at PCL. Students and researchers of the community would also be able to share knowledge about these technologies regarding the possible partial autonomy from the energy grid. The described social practices of mutual collaboration, support, negotiation, communication and experimentation between students, researchers, local authorities and manufacturers would be needed to create a system to manage waste as a CPR and to engage with craft skills, architectural codes and norms (Avermaete 2018).

4 A First Attempt to Reuse Urban Waste

In 1975 Pawley wrote again in *Architectural Design* about a small house in London designed and built by architects Shul-Kay Kan (no longer a student) and Michael Poteliakhoff (Pawley 1975c). Pawley this time does not describe an academic or a corporate research project but a real built one, even if a prototype (Figs. 10.1 and 10.2). From the description of the project and the design process, some more details emerge concerning the notion of the commons. Pawley explains that the project was born in 1973, designed originally by Kan during a period at the Architectural Association School (AA) in London, where Pawley was working. The spot-welding techniques of the tin cans were developed with the assistance of the technical staff in the PCL technical laboratories. He explains that both the designers had been students of architecture at PCL. Together, they presented the house structure as part of their final year portfolio, and together with Andy Robson they belonged to the practice called SUG (Pawley 1975c). Their intention was to explore to what extent it was

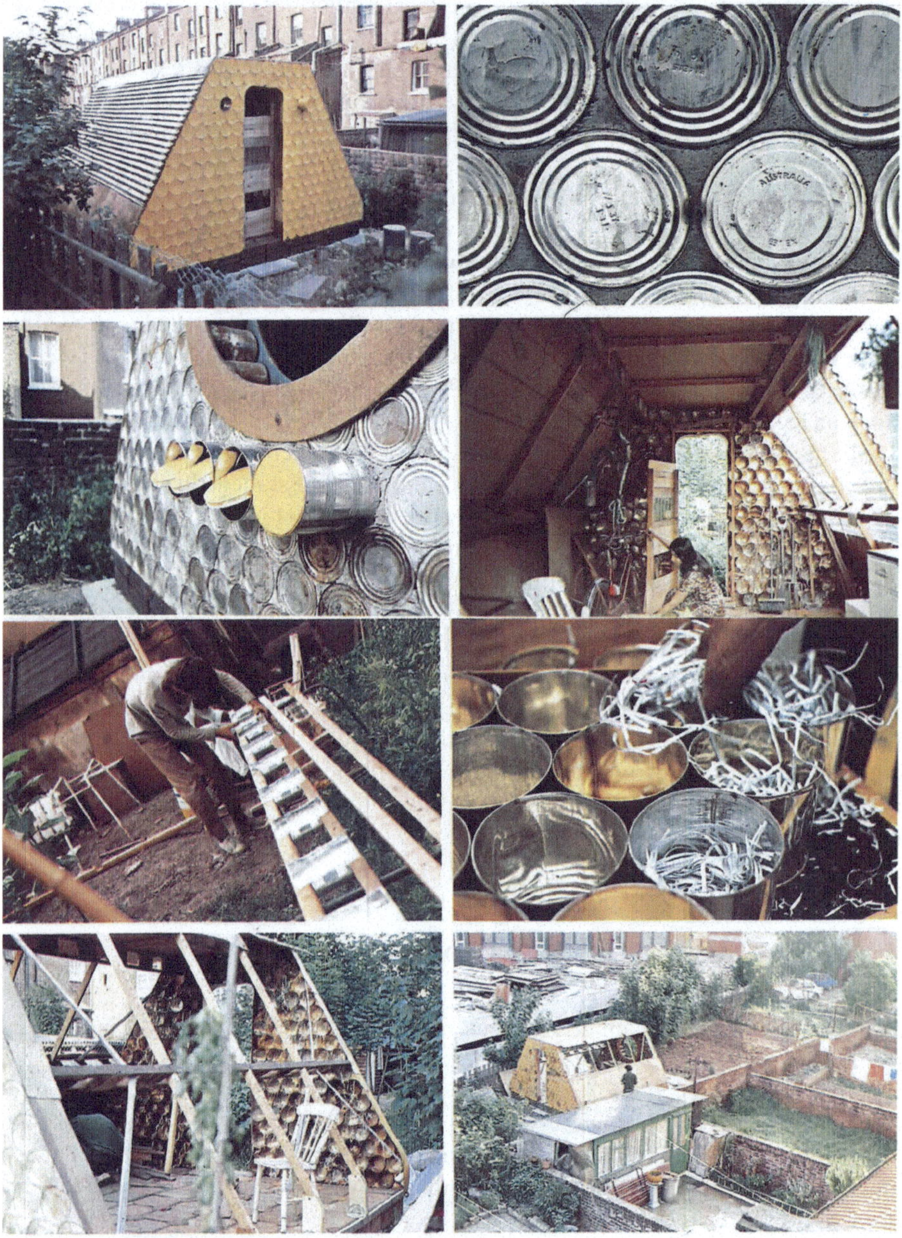

Fig. 10.1 Small house in London, SUG, details from the inside and the outside of the reused tin-can wall, tin-can structural beam, reused tin cans filled with shredded paper, and location of the house in the backyard. (Source: Kan and Poteliakhoff 1975)

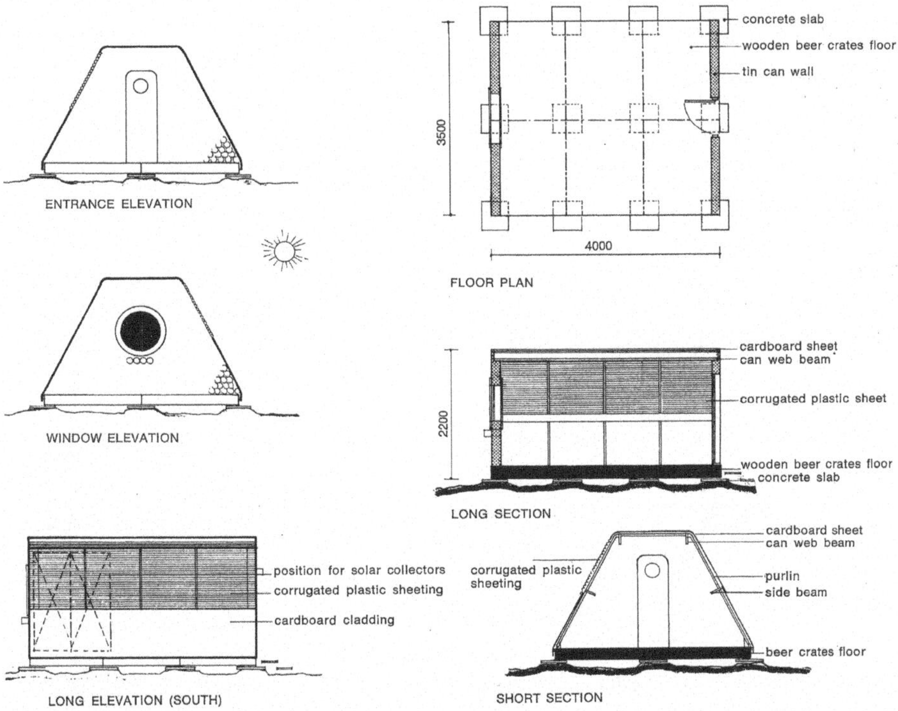

Fig. 10.2 Small house in London, SUG, elevations, plan, section. (Source: Kan and Poteliakhoff 1975)

possible to make practical use of cardboard, paper, wood, metal or plastic waste, which in any city are disposed daily in enormous quantities (Kan and Poteliakhoff 1975).

In this article, Pawley not only describes in detail how the house is constructed but also mentions where and how the materials are collected. He notes that the two end walls of the small house are made of used food cans collected from locations such as restaurants, bars and canteens (Fig. 10.3). The insulation is provided by shredded paper from office refuse, placed inside the cans. The floor is composed of beer crates and it rests on pieces of damaged cement slabs retrieved from building sites. The beer crates are donated by the British company Watneys and are connected with can lids that are nailed over four corner junctions (Pawley 1975c). The outer shell is constructed from a corrugated plastic sheet, which is damaged and therefore low in cost. The interior cladding is realized with packing case cardboard (Pawley 1975c). The cans – welded together and given a lightweight cement infill – are collected from local businesses, starting with a Chinese restaurant nearby. After persuading some chefs and businesses owners, the tin cans came in at the rate of 200 per week.

The harvesting of used materials is a relevant aspect of their work in the context of commons. Pawley mentions several locations where the cans are collected, such

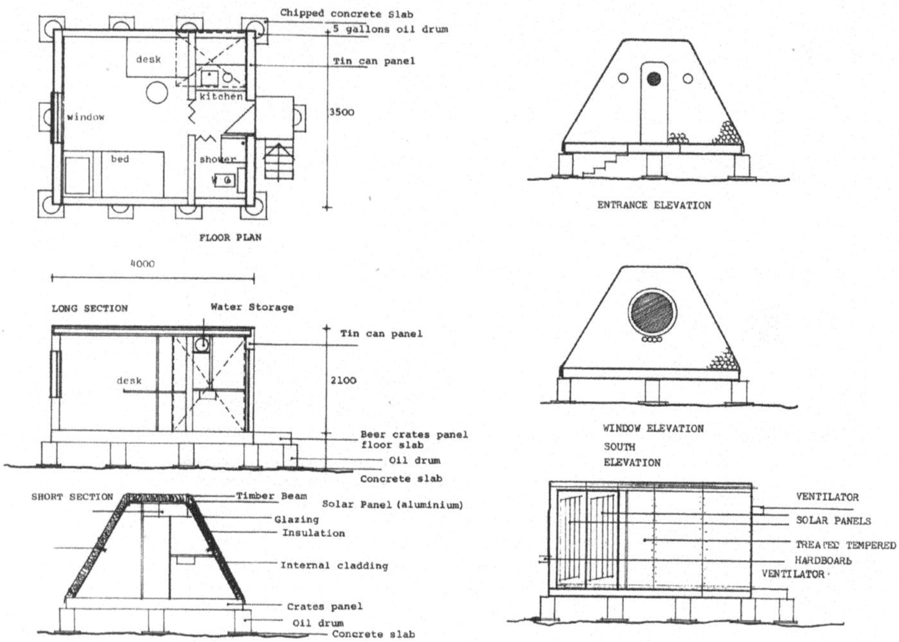

Fig. 10.3 Small house in London, SUG, elevations, plan, section. (Source: Pawley 1975a)

as restaurants, offices and businesses; and also describes the group's negotiations with some providers of materials; other materials were received as a gift. Waste, in this case, can be considered as taking a step towards becoming a CPR, since more community actors are taking part in the process. SUG collect the waste to reuse in the project, saving the businesses the work of disposing of it. Furthermore, the knowledge and craft skills acquired by SUG in operating with waste materials – in terms of negotiations, communication, mediation, explanation of the innovative practice to businesses, collection, design and manufacturing – can be regarded as immaterial commons.

In the same article, Pawley adds an explanation about the design and the construction process and the different tasks of the two architects. Kan designed a prefabricated wall section system for the two end walls of the house, storing the spot-welded sections ready for site assembly. Poteliakhoff designed a composite lattice beam using flattened cans as spacers between thin timber planks. The walls built by Kan were assembled and connected through the beams designed by Poteliakhoff. The roof consisted partly of tri-wall cardboard, partly of Masonite, partly of damaged and therefore cheap corrugated plastic sheeting. The latter was lapped on the side in order to mask its damage. The roof was seam-sealed with Sylglas and waterproofed with yacht varnish. Pawley says that the house "possessed … evident strength and promising durability" (1975c, 699). The house was only 14 m^2, the minimum dwelling unit, and it was built on a private back garden in

Holloway, London. The article also relates some particulars about the design process, specifically two false starts during the assembly phase: the first selected building site had to be changed because the owners complained. In the second chosen site, the Crumbles Adventure Playground in Islington, London, some youngsters set fire to the project during the construction phase. The project was completed at the third site in Holloway, London. Pawley specifies that the transportation, due to the site changes, increased the total cost up to approximately five British pounds for square meter, which was still very affordable compared to the standard housing of the time (Pawley 1975c).

In this case, we see a missed opportunity of commoning process. Due to a lack of mutual collaboration, support, negotiation and communication, the local actors not only did not cooperate in the collection of materials, construction and use of the house but also complained about the project, and they even burned down the first house. Some other notions also emerge from the Pawley's article. In the construction of this house, several materials are reused for a different function than they were initially designed for, and sometimes transformed before being reused. They are reused both for cladding and for structural purposes. In the latter case, they usually need a higher degree of manufacturing. The tin cans composing the walls, for instance, are welded together and linked with a light concrete infill. Most of the materials used are reclaimed. Only some materials are new, such as the concrete slabs, or the wooden part of the tin-can beam. The reused materials are cut, drilled, flattened and reused as walls, floors and structure. For instance, the tin cans are cut, painted and welded together; at first glance, they might appear familiar. Nevertheless, only close observation of the tin-can wall and construction details makes it possible to detect that the objects that compose the wall are well-known. This constitutes an innovative aesthetic value: the experience of slowly rediscovering a familiar object reused unexpectedly. This aesthetic feature can be related to the concept of commons aesthetics discussed by urbanist Adam Greenfield and urban sociology researcher John Bingham-Hall referring to contemporary urban spaces usually not governed by either the market or the state, characterized by the use of low-cost, reclaimed materials and DIY construction (Greenfield 2018; Bingham-Hall 2016). From this perspective, the house in Holloway could be considered an innovative precursor of and reference for a contemporary commons aesthetics found for example in urban projects such as R-Urban in Paris and Institut for (X) in Aarhus (Greenfield 2018).

From an architectural perspective related to the adaptable use of space, the trapezoidal walls supported by the tin-can beam allows for the realization of flexible open space. It can be used as a shelter or as a greenhouse because of the translucent southern façade. Part of it can also be opened as a window. In effect, the translucent plastic sheets on the south side permit the access of natural light. Even if in this case the house does not incorporate a greenhouse for food production, the house itself has a similar structure and some features of a greenhouse. It is in effect a semi-transparent unique space with the southern wall inclined towards the sun. Because of the trapezoidal shape of the house, one of its two long façades can support solar panels. The possible location of solar panels is drawn with a dashed line on the southern elevation. In this case, the solar collectors would obscure only one third of the southern façade, leaving the remaining two thirds of the façade open to natural light access and a view towards the outside.

Compared to the design of the house developed at the university two years earlier, some minor differences emerge. The houses are very similar in terms of dimensions, structure and shape. In the real house SUG did not build the bathroom and the kitchen. The design shows the solar panels on the southern façade completely covered by treated tempered hardboard. In the real shelter, half of the surface of the southern façade is translucent with corrugated plastic sheeting. Judging from the pictures, the building seems to have been used, at least temporarily, as a greenhouse with some plants inside, and as a tool shed. The similarity of the shelter to a greenhouse and the explicit indication of the location of the solar collectors suggest that even if the house does not have energy and food autonomy as a declared goal, it represents a design concept in favor of at least partially disconnecting from national grid services. From this perspective is worth citing a thinker about the commons like David Bollier, who maintains that "rather than look[ing] to state authorities as guarantors or administrators of their interests, commoners generally prefer to seek direct sovereignty and control over spheres of life that matter to them: their cities, neighborhoods, food, water, land, information, infrastructure" (Bollier 2020).

5 The Operation of Harvesting Materials as Immaterial Commons

Three years later, *Architectural Design* further illustrated the work of SUG. The magazine, which on the first page welcomes unsolicited manuscripts and other material for publication, does not specify the author of the article. It seems safe to assume that someone from SUG at least contributed to the writing. The article, which explains how waste was managed in the UK at the time, gives an insight into the SUG strategy of material collection and describes four of the group's projects: a two-person dwelling, a further development of the structural beam, a demountable greenhouse and an installation inside a shop (Kan and Poteliakhoff 1978). From these works, illustrated in detail in the article, several aspects regarding the commons can be outlined.

The article emphasizes that at the time in the UK, the amount of waste materials reused by the construction sector was meagre. Buyers of recycled waste were discouraged because though the materials were often damaged or of low quality, they were still quite expensive. Moreover, at the time costs were rising for new standard building materials too. The article observes that in a time of crisis, with high unemployment, new jobs could be generated by a newly organized re-use industry, which could also satisfy the increasing demand for urban renewal. The article goes further, proposing a strategy that entails the design of objects and packaging with a reuse afterlife in building construction planned beforehand. An example of this is replacing one year's production of bricks with recycled bottles designed with a specific shape in order to be reused as bricks (Pawley 1975a). However, the article recognizes that the implementation of such objects with second use planned beforehand would encounter some difficulties. First, because they would be more expensive to produce compared to standard bottles. Secondly, they would need to be incorporated in traditional construction processes and to be weather-proof. The presence of many kinds of standard garbage would require a strategy to make better use of it.

The article maintains that SUG assesses the feasibility of reuse depending on the distance between the source of waste materials and the actual place of use. According to their study, if the distance is greater than 20 kilometers, it was unlikely to be cost-effective in the UK at the time. This assessment is probably related to what is intended by the so-called economic radius (Kan and Poteliakhoff 1978) in the diagram below (Fig. 10.4). The image is a map realized by SUG showing the distance of the materials to be collected around the city (1), processed in the workshop (2) and assembled at the building site (3). The building site is represented with a sketch of the SUG two-person dwelling project. The diagram can be used as a tool to collect materials and a map with which to estimate and trace where the materials are located and if it is worth reclaiming them in terms of time and costs. Distances up to 50 kilometers are indicated on the diagram.

Knowledge of the collection, storage and remanufacturing of materials to be ready for use can be defined as an immaterial commons, in the sense of locally specific knowledge and craft skills. In this case, too, the article attends to technical details related to distances and logistical costs. Unfortunately, not much is explained about the even more interesting actions and possible commoning processes such as the operation of collecting the materials through cooperation or negotiations with the materials providers. Nevertheless, the map designed by SUG traces an innovative representation of the building site and logistics and manufacture of materials. The source of waste materials, the workshop where materials are manufactured and

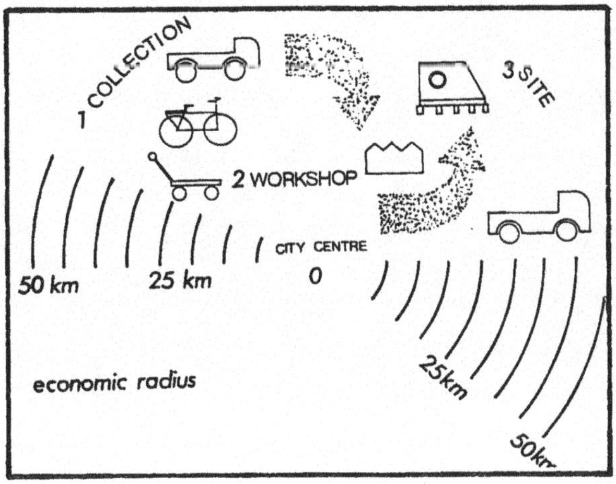

Fig. 10.4 Diagram representing the reused materials process and the distances from the city center, SUG. (Source: Kan and Poteliakhoff 1978)

stored, and the building site are all within a maximum 50-kilometer radius, but generally closer to 20 kilometers. The collection of materials is again at the center of the architectural design. The common resource of waste is represented in its potential and its importance related to the context. The map shows the different phases (1, 2, 3) where possible commoning processes and new jobs, could materialize. SUG, relying on their own knowledge, but also by communicating with local businesses and actors, has drawn an additional layer on the map of part of the city, indicating the location of a valuable common resource. Furthermore, they have transformed the design and construction process from linear to circular, sourcing locally available materials, designing, storing the building components, assembling on site and eventually disassembling again.

6 Two-Person Dwelling and a Greenhouse

In 1977, architects Shiu-Kay Kan, Mike Merhemitch, Michael Poteliakhoff and Andy Hobson from an expanded Secondary Use Group, designed a two-person dwelling (Kan 1978; Kan and Poteliakhoff 1978). The structure consultant was Jime Tyne (Kan 1978) (Figs. 10.5, 10.6, 10.7, 10.8, and 10.9). This project was derived from an experiment where a structure was erected composed of external walls of the same type as their first house built in 1975. A further development of the

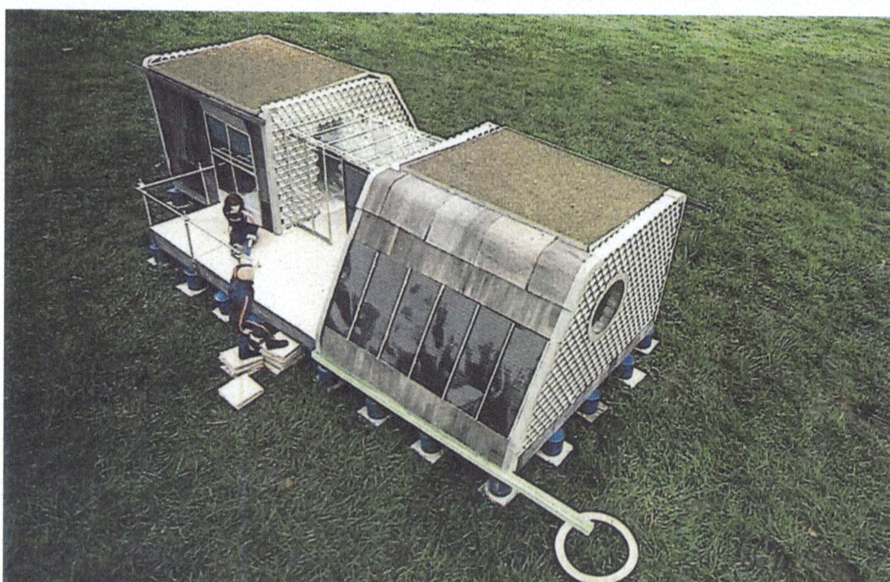

Fig. 10.5 Two-person dwelling, SUG, south-east side, pending wall with solar collectors. (Source: Kan 1978)

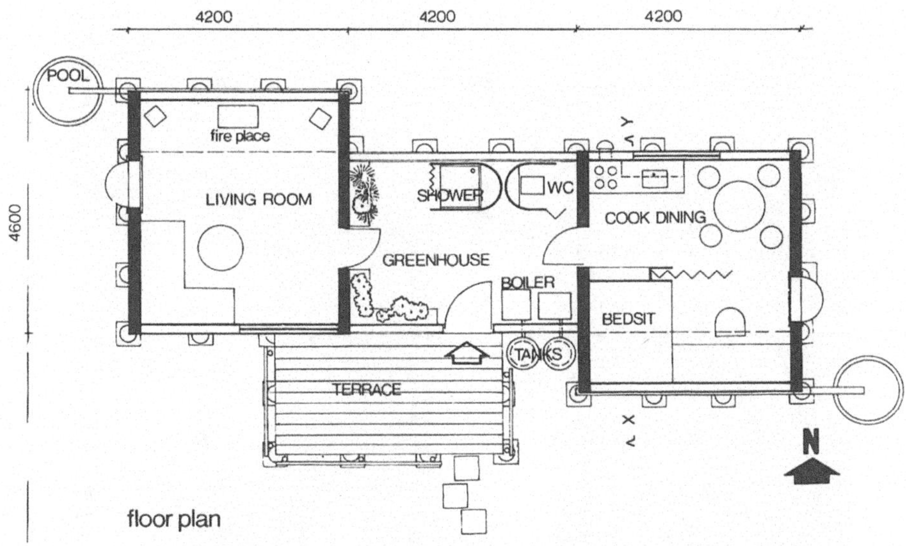

Fig. 10.6 Two-person dwelling, SUG; ground floor plan; the greenhouse in the middle is also the entrance, containing shower, toilet and boiler. (Source: Kan and Poteliakhoff 1978)

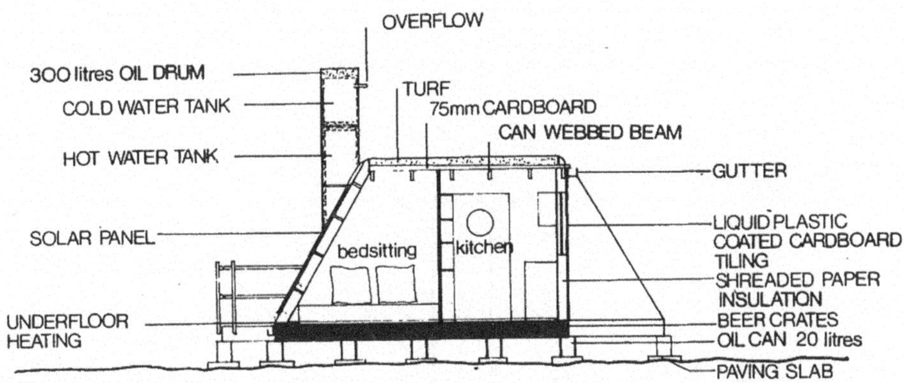

Fig. 10.7 Two-person dwelling, SUG, section along the eastern volume. (Source: Kan and Poteliakhoff 1978)

structural beam applied to the first house was used in this second house as well: the four-meter tin-can beam was tested at PCL (Kan and Poteliakhoff 1978). The two-person dwelling architecturally consists of two identical units, one for living and the other for sleeping. The units are alternatively positioned and their dimensions are 4 × 4.2 × 2.3 m. A glazed area acting as an acoustic barrier that accommodates a toilet, a shower and a greenhouse connects the units. The southern pitch roof has an inclination of 60 degrees and it has a solar collector installed. The other vertical wall with a double-glazed window allows the sunlight to penetrate into the living area.

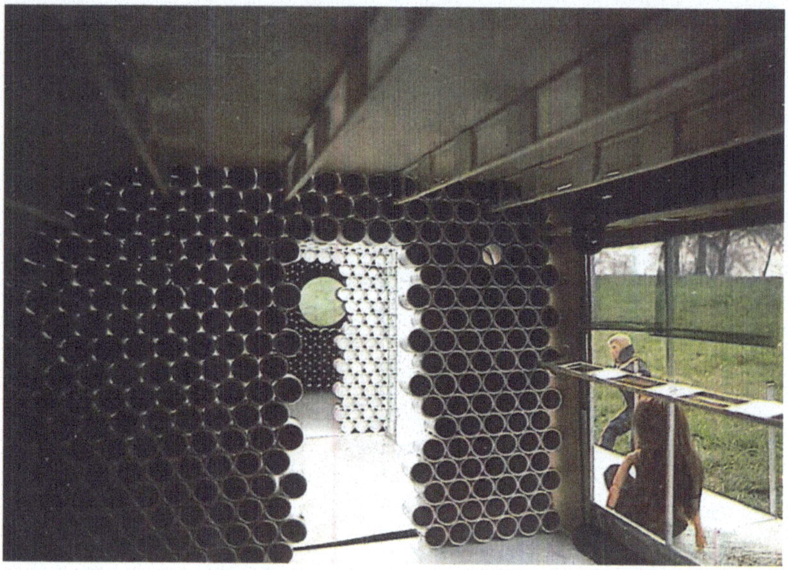

Fig. 10.8 Two-person dwelling, SUG, view of the living room area, tin-can wall and beams. (Source: Kan 1978)

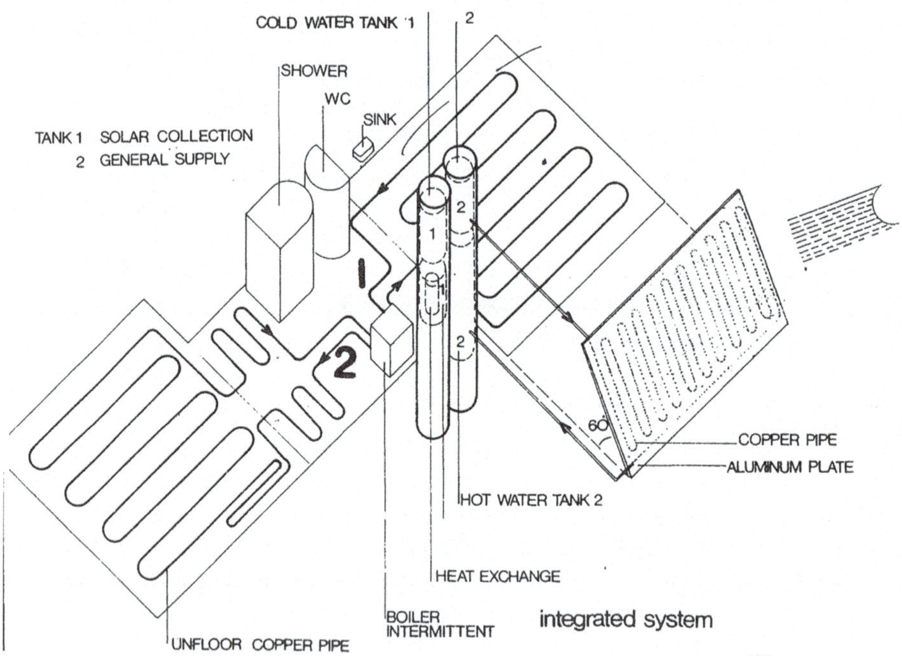

Fig. 10.9 Two-person dwelling, SUG, Integrated system. (Source: Kan 1978)

The article describes the simple construction of the building and how the components prefabricated in portable sections could be transported and assembled on site by two workers. Most of the building materials are recycled. The floor panels are made from wooden beer crates jointed together. They are lifted up by 20-liter oil drums and paving slabs. The tin can webbed beam is four meters long and rests on a spot-welded wall at 600 mm intervals. The external cladding is a plastic-coated cardboard sandwich panel while the rooftop is 100 mm thick turf, drained towards the vertical side of the wall. The rainwater is channeled into little pools and the water is stored for gardening during dry periods. SUG predicts that if the walls are painted and looked after yearly, the building could last for 15 years. The total cost of the materials and the transportation on the site are around 500 British pounds. This amount includes a solar central heating system, but excludes labor (Kan 1978).

The architectural composition of the house consists of two volumes connected by a greenhouse. The greenhouse also serves as the entrance and contains the shower, the toilet and some technical installations like the boiler. Apart from the greenhouse area, the living room has a window facing south and the bedroom has a window facing west. The solar collectors for water heating are embedded in the southern wall of the building volume containing the bedroom and the kitchen. This example represents the paradox of catching the solar light with solar collectors, which block the view towards the outside and the access of natural light from the south. In this case the compromise was to place the solar collectors only on one of the two building volumes. This house with greenhouse, solar panels, and rainwater collection moves towards an alternative economic and social model. The use of waste materials and the social practices and behavior of the inhabitants allow the house to become the tool to unlock rainwater and solar energy as a common resource to produce energy and grow food. The building could, for example, fit perfectly in the commoning process of student housing on urban waste land described by Pawley in *Garbage Housing*.

The article does not extensively describe the detailed role of the various actors involved in the processes of collection, storage and manufacturing of the various components. Since the house was still a prototype, not much can be said about the client, stakeholders and the users of the house. However, social practices of mutual collaboration between the members of SUG, the materials providers, PCL and the structure consultant (even if it is not clear if they were voluntarily helping or hired) could all be considered part of commoning processes. Also, in this case, waste can be considered a CPR since it is used and sustained by the small community composed of SUG, the materials providers and the few other actors involved in building the prototype. As an example, the prototype could become part of student housing, built by the inhabitants as in the project described by Pawley in Garbage Housing; in this case, the social practices of mutual collaboration, negotiations, communications and experimentation would not only focus on waste materials but also energy and food.

SUG also built a demountable greenhouse (Fig. 10.10). Shiu Kay Kan, Michael Potehakhoff, Mick Merhemitch, Andy Hobson and Melanie Sainsbury participated in the project. The tin cans were once again the main reused material but this time

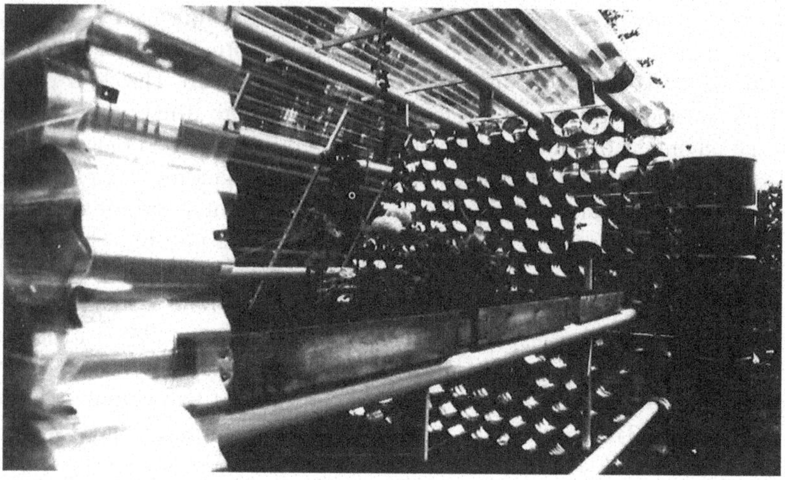

Fig. 10.10 Demountable greenhouse, SUG. (Source: Kan and Poteliakhoff 1978)

the construction method was slightly different. The group explains that the cans opened at both ends, could be linked by 'spike' clips which could be easily purchased at most auto parts suppliers. The greenhouse was built using this method for the cross walls while the beams were made from cardboard tubes with a tensioning system made of wire. The article explains that the structure was clad entirely with reused corrugated plastic sheeting while discarded commercial bread trays were reused as plant boxes and the cans on the walls were used for tool storage. The greenhouse was successfully dismounted on two occasions for transportation and the project was also presented twice on BBC television (Kan and Poteliakhoff 1978). Analyzing the pictures, the greenhouse has a similar shape to one of the units of the two-person dwelling. This greenhouse shows the interest of the group in autonomy intended as self-reliance in terms of energy and food. In the TV broadcast, the group tries to spread to a broad public their immaterial commons like their intellectual and cultural knowledge, craft skills, and collaborative DIY practice.

7 Waste as a Resource for Interior Design

SUG applied the same method of second-hand materials reuse to interior design. The article in *Architectural Design* also describes a Christmas décor realized for a shop in Milan by Shiu Kay Kan, Mick Merhemitch and Andy Hobson (Fig. 10.11) (Kan and Poteliakhoff 1978). The client required a highly decorative quality for the display shelves. Moreover, the design had to be flexible in order to suit a large variety of clothes and jewelry. The group could rely on an Italian can manufacturer who generously collaborated on realizing the installation. The Italian company double

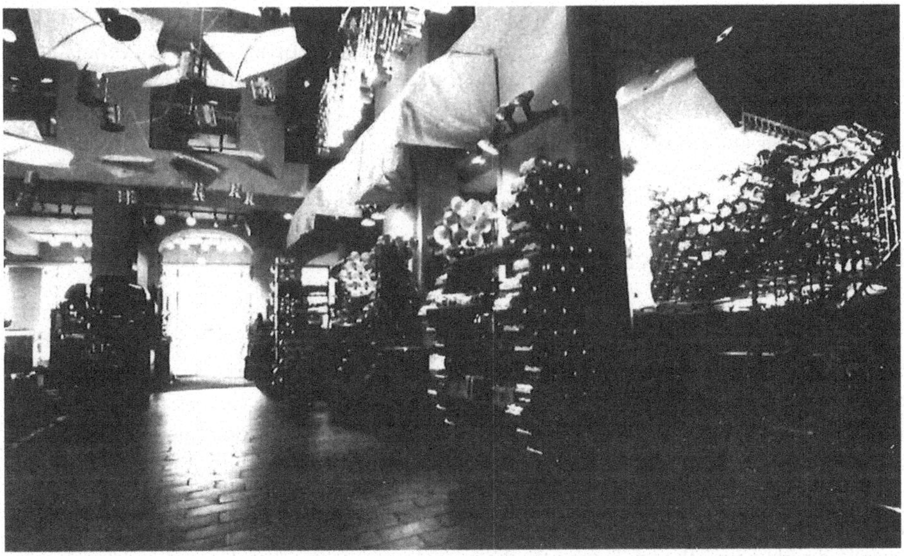

Fig. 10.11 SUG installation in a shop in Milan. (Source: Kan and Poteliakhoff 1978)

end-sealed 2000 rejected catering cans, giving the required decorative quality to the raw material. The majority of mounting details were either entirely designed or at least finalized on site. Specially developed universal plastic clips connected the display units that required re-arrangement and regular dismounting. Certain stands were built with all spot-welded joints in order to withstand the heavy use of a busy shop. The article explains that the contract was completed in seven days, with a degree of cooperation by client and suppliers unknown in England, a detail probably not highlighted by chance. In fact, cooperation between parties could be very useful to initiate a new design method such as the reuse of reclaimed materials structured on a national scale. The essay concludes with another curious detail, describing that on the introduction day, some demonstrators attacked the shop with paint, and on the following days even with firebombs, luckily missing the group's work (Kan and Poteliakhoff 1978).

This interior design project was the first of SUG's built and commissioned by a client. In this example, waste represents a CPR sustained and used by a community, although mainly for commercial reasons. Still, the degree of collaboration between the different parties is remarkable and can be considered related to a commoning process. The Italian manufacturer collaborated in the project, generously sealing the cans, probably without being paid. Plastic clips were specially developed, many mounting details and re-arrangements were realized on site with a high degree of cooperation not only by the suppliers but also by the client. The cooperation between practices was fundamental for the success of the project, which experimented with new methods of supply and construction.

It is ironic and remarkable that on one hand local actors strongly cooperated to unlock waste materials as a CPR through a commoning process; on the other hand, some other local actors, violent demonstrators during the Italian historical period named the years of lead, attacked and tried to destroy what they probably considered a regular shop built with standard industrial materials.

8 The Tin-Can Beam as a Waste Backbone

The Tin-Can Beam project by Michael Poteliakhoff from SUG is published in the same issue of *Architectural Design* (Kan and Poteliakhoff 1978). The beam is a spin-off from the project initiated by Kan at PCL in 1975 (Pawley 1975c). The beam is a compromise, because it combines reclaimed tin cans with new timber. The article describes the beam's technical details, comparing it with a new one. In the prototype, the new timber is a section of 50 mm x 25 mm, while the ten reclaimed tin cans are 175 mm tall and have a 150 mm diameter each. The construction process is divided into three phases: the base and any remains of the ends of the tin can are removed; the metal is snipped, flattened and folded to an oval profile to fit the timber; tin cans are placed between timber and pattern and nailed down with 25 mm flat-head nails. The Tin-Can Beam prototype, of which the alignment could be adjusted to give a pre-determined curvature, was tested with the support of the PCL engineering department for deflection over a clear span of 3.58 m. The beam was compared with a standard new solid joist, with negative results on several criteria. The beam deflection was excessive for many uses, it could not be used in standard fire-resistant timber floor constructions due to its steel content, and it required two hours of extra labor compared to the standard one. However, the cost of the Tin-Can Beam, almost two British pounds, was an advantage compared to more than four British pounds for a standard beam. Furthermore, it was composed of a smaller quantity of timber, so it was environmentally friendly and convenient for do-it-yourself construction. The article concludes with a very interesting detail. It explains that the design has not been patented due to the group's clearly stated goal that secondary use discipline and knowledge must be spread freely to increase the opportunities for re-using raw materials (Kan and Poteliakhoff 1978).

SUG's care work as defined by Bollier is visible even in this structural element. The beam first developed at PCL and further tested and transformed by Poteliakhoff is one of the backbones of several SUG projects. The last lines of this article state the importance to the group of diffusing the knowledge of their care work, craft skills and social practices. The decisions not to patent the design and to broadcast their work on television are ways to promote the development of waste as a CPR. The group tries to scale-up the commoning process as a social practice of potential mutual collaboration, support, communication and experimentation between more potential involved actors in the future. This project shows some limits of reusing materials on a big scale. In general, at the time, new materials were more efficient and affordable. On the other hand, the beam is still functioning very efficiently.

Moreover, the fact that it is possible to recognize the SUG style even from a structural component due to the tin cans is remarkable in terms of design and aesthetics. The user would recognize something familiar in the beam, and discover that it is composed of everyday use objects only after observing it in detail. In this case, too, it is possible to refer to a commons aesthetics.

9 Conclusions

Architects often represent the mainstream trend of economic profit as a primary goal for project developments in the built environment. However, architects like SUG operate differently, moving away from a strict client-architect relationship to more collaborative practices and commoning processes in which they take several roles not only as designers but also as researchers, technologists, DIY-builders, facilitators, developers, and initiators (Havik and Pllumbi 2020).

In the 1970s SUG adopted the same design method with reused tin cans for a house, a greenhouse and an interior design project. The generative and repetitive feature of the reused components created a specific design style belonging to the group, recognizable whether the cans were used for a house, a greenhouse, a structural beam or a piece of furniture. This feature can relate to the concept of commons aesthetics discussed by Greenfield and Bingham-Hall. SUG, with their research and development about recycled materials for construction, attempted to propose a new architectural approach. The building elements designed with reclaimed materials must perform to building specifications such as being weather proof and structurally tested. The author of the 1978 *AD* article defines SUG's architectural approach as maintaining a social and historic perspective to the design, as well as being environmentally friendly. *AD* emphasized the historic perspective, probably because SUG reclaimed local materials collected from no more than a couple of dozen kilometers from the building site. *AD* pointed at the social perspective because during a time of economic and social crisis, SUG's approach opened up new practices of cooperation and job opportunities related to commerce, collection, manufacture and assemblage of reclaimed materials. The architects of SUG, instead of purchasing the materials, collected the used materials from different sites and stored them in different locations before the application to new buildings. They were in charge of the entire reuse process, with innovative collection methods, construction of building components with reused materials. In SUG and Pawley's projects, like the construction processes of cans walls and beams, urban waste materials were reused as material commons, through craft skills and norms of art representing immaterial commons. Commoning processes as social practices of mutual collaboration, support, negotiation, communication and experimentation among architects, researchers, authorities, users, manufacturers, clients were needed as a system to manage the CPR.

The work of SUG and Pawley seems to be characterized by the concept of care work described by David Bollier or affective labor as developed by Neera Singh.

Singh reminds us that while in market systems labor is motivated by monetary reward, it seems that processes of commoning are instead characterized by affective labor (Singh 2013). David Bollier adds that rather than looking to the state as the main guarantor of their interests, commoners prefer to directly control resources and cities that matter to their life. From this perspective, Pawley and SUG can be considered commoners who see waste materials not only as an unused resource, but also as a symptom of an unhealthy consumption society. They claim that a change of culture is needed, which would accept the concept of secondary use and garbage housing. In the last lines of his book, Pawley notes that the act of building with any type of garbage "encapsulates too much truth about the nature and weakness of our society to be ignored" (Pawley 1975a, 114). With this essay I hope to open up future discussions about the commons concerning architectural thinking and practices for logics and processes that go beyond the state and the market, too often considered the main driver of the development of the city.

References

Avermaete, Tom. 2018. Constructing the Commons: Towards Another Architectural Theory of the City? In *An Atlas of Commoning: Places of Collective Production*, ed. Anh-Linh Ngo, Mirko Gatti, Stefan Gruber, Christian Hiller, and Max Kaldenhoff, vol. 232, 32–43.

Bauwens, Michel. 2017. The History and Evolution of the Commons. Commons Transition. https://commonstransition.org/history-evolution-commons/. Accessed 3 Mar 2020.

Bingham-Hall, John. 2016. Future of Cities: Commoning and Collective Approaches to Urban Space. In *Theatrum Mundi, LSE Cities, London School of Economics and Political Science*. London: Government Office for Science.

Bollier, David. 2020. Commoning as a Transformative Social Paradigm. TheNextSystem.org. https://thenextsystem.org/commoning-as-a-transformative-social-paradigm. Accessed 8 Oct 2020.

Bollier, David, and Silke Helfrich. 2015. *Die Welt der Commons: Muster gemeinsamen Handelns*. Bielefeld: Transcript.

Bonnemaison, Sarah, and Christine Macy. 2003. *Architecture and Nature: Creating the American Landscape*. London: Routledge.

Bookchin, Murray. 1971. *Post-Scarcity Anarchism*. Montreal and Buffalo: Black Rose Books.

Borasi, Giovanna, and Mirko Zardini. 2007. *Sorry, Out of Gas: Architecture's Response to the 1973 oil crisis*. Montreal: Canadian Centre for Architecture.

Calisto Friant, Martin, Walter J.V. Vermeulen, and Roberta Salomone. 2020. A Typology of Circular Economy Discourses: Navigating the Diverse Visions of a Contested Paradigm. *Resources, Conservation and Recycling* 161: 1–19. https://doi.org/10.1016/j.resconrec.2020.104917.

Commoner, Barry. 1971. *The Closing Circle: Nature, Man & Technology*. New York: Bantam Books.

Cook, Peter. 1970. *Experimental Architecture*. London: Universe Books.

Ehrlich, Paul R. 1968. *The Population Bomb*. New York: Buccaneer Books.

Felber, Christian. 2015. *Change Everything: Creating an Economy for the Common Good*. London: Zed Books.

Georgescu Roegen, Nicolas. 1971. *The Entropy Law and the Economic Process*. Cambridge, MA: Harvard University Press.

Greenfield, Adam. 2018. On the Received Aesthetic of the Urban Commons, and Transcending It. Adam Greenfield's *Speedbird* (blog). https://speedbird.wordpress.com/2018/04/28/on-the-received-aesthetic-of-the-urban-commons/

Hardin, Garrett. 1968. The Tragedy of the Commons. *Science* 162: 1243–1248.
Havik, Klaske, and Dorina Pllumbi. 2020. Urban Commoning and Architectural Situated Knowledge: The Architects' Role in the Transformation of the NDSM Ship Wharf, Amsterdam. In *Architecture and Culture*, 1–20. London: Routledge. https://doi.org/10.1080/20507828.2020.1766305.
Hedgepeth, William. 1970. *The Alternative: Communal Life in New America*. New York: Macmillan.
Helfrich, Silke. 2011. The Commons: Marginalized but Rediscovered, Year One of the Global Commons Movement, *CommonsBlog*. https://commons.blog/2011/01/29/the-commons-year-one-of-the-global-commons-movement/. Accessed 1 Mar 2020.
Illich, Ivan. 1973. *Tools for Conviviality*. New York: Harper & Row.
Ingersoll, Richard. 2012. The Ecology Question and Architecture. In *The SAGE Handbook of Architectural Theory*, ed. C. Greig Crysler, Stephen Cairns, and Hilde Heynen, 573–589. London: SAGE. https://doi.org/10.4135/9781446201756.
Jencks, Charles, and Nathan Silver. 1972. *Adhocism: The Case for Improvisation*. Cambridge, MA: MIT Press.
Kan, Shui-Kay. 1978. Secondary Use Experiments. *Domus* 578: 16.
Kan, Shui-Kay, and Michael Poteliakhoff. 1975. A Londra con rifiuti (Re-used refuse). *Domus* 553: 30–31.
———. 1978. Tin Can Tech. Re-use of Wastes in Industrialised Societies. *Architectural Design* 48 (2): 192–193.
Meadows, Donella H., Dennis L. Meadows, Jorgen Randers, and William Behrens. 1972. *The Limits to Growth: A Report for the Club of Rome's Project on the Predicament of Mankind*. London: Universe Books.
Medici, Piero. 2017. The Trombe Wall During the 1970s: Technological Device or Architectural Space? Critical Inquiry on the Trombe Wall in Europe and the Role of Architectural Magazines. *SPOOL* 5: 45–60. https://doi.org/10.7480/spool.2018.1.1938.
Ostrom, Elinor. 1990. *Governing the Commons: The Evolution of Institutions for Collective Action*. Cambridge: University Press.
Papanek, Victor. 1971. *Design for the Real World*. Chicago: Academy Chicago.
Pawley, Martin. 1971. Garbage Housing. *Architectural Design* 41 (2): 86–94.
———. 1973. Garbage Housing. *Architectural Design* 42 (12): 764–776.
———. 1975a. *Garbage Housing*. 1st ed. Manchester: Architectural Press.
———. 1975b. Stop This Nonsense Now. *Ghost Dance Times* 5: 1.
———. 1975c. Garbage Housing in London. *Architectural Design* 45 (11): 699.
Ponte, Alessandra. 2006. Garbage Art and Garbage Housing. *The Log* 8: 99–111.
Raworth, Kate. 2017. *Doughnut Economics: Seven Ways to Think Like a 21st-Century Economist*. New York: Random House.
Schumacher, Ernst Friedrich. 1973. *Small is Beautiful: Economics as If People Mattered*. New York: Harper & Row.
Singh, N.M. 2013. The Affective Labor of Growing Forests and the Becoming of Environmental Subjects: Rethinking Environmentality in Odisha, India. *Geoforum* 47: 189–198.

Chapter 11
Reclaiming the Habitat: Food, Fire and Affordance in Designing and Living the Urban

Liana Psarologaki and Stamatis Zografos

Abstract This essay is a critical and radical proposition to reclaim habitats of life, investigating ecological and pedagogical models of critical and spatial design practices. We examine these in relation to commons which we associate to urbanism and life. We do so via politics of food, fire, and affordance, which we explore as agents shaping the assemblage of sapiens life through design, and reposition as cues for methods of thinking, learning, designing, and constructing (conditions of life). The vilification of fire (when associated to fear) and food (when associated to desire) has a tremendous effect on how we think of and practice life, and presents ramifications to how we design, make, and consume objects and environments. We must affirm such phenomena as part of advanced capitalist networks and societies, and reconsider commoning and its affinity with social capital in practice. We attempt in response to propose ecologies and pedagogies of food and fire (as methods for urban commoning) in urban design practice, promoting larval affordances between bodies, buildings, and the commons as urban conflict. We approach conflict as affordance and the collective shaping agent allowing public sharing to form through dynamic and non-stagnant networks and patterns. As with urban commoning practices, these propositions may diverge from dominant institutions and can open up to opportunities that allow us to rethink life and live differently. The proposal will therefore offer a space for theorization, interrogating contemporary citizenship in learning to design for and by such experiences and events, and highlight pathologies of urban living.

Keywords Food · Fire · Urban commons · Affordance · Disorder

L. Psarologaki (✉)
School of (EAST) Engineering, Arts, Science & Technology, University of Suffolk, Suffolk, UK
e-mail: L.Psarologaki@uos.ac.uk

S. Zografos
Architectural History and Theory, UCL Bartlett School of Architecture, London, UK
e-mail: stamatios.zografos@ucl.ac.uk

G. Bruyns, S. Kousoulas (eds.), *Design Commons*, Design Research Foundations, https://doi.org/10.1007/978-3-030-95057-6_11

This transdisciplinary essay is a propositional whisper about reclaiming habitats of life in relation to commons, via politics of food, fire, and affordance. We explore these as design methods and agents shaping the assemblage of sapiens life. It is an attempt to instigate a theoretical and practical repositioning of food, fire, and affordance as cues for thinking, learning, designing, and constructing (conditions of life). Our starting point is the work of anthropologist Jean-Pierre Vernant and the affirmation of cooking as humans' primary self-reference of technical ability (Prometheus myth) (1989, 39). In the last 30,000 years, fire and food have presented a complex genealogy. We (humans) have progressively shown affinity to associate them less with conviviality and more with danger sometimes in controversial ways. At the same time, we imperatively recognise that fire and food should be seen as perishable and therefore retained for collective life maintenance and preservation of the wider ecosystem(s). Their demonisation has a tremendous effect on how we think of and practice life, through designing and consuming objects and spaces, yet we can only affirm that they are by-products of advanced capitalist networks and societies where "meals are socially wounded and emphatically enshrined in architectural space" (Jones 2007, 32).

This is pertinent to theorising and practicing methods to do with how people live and shape their living with their (natural and built) inhabited environments in formal and informal ways from systems of thinking and emergent philosophies to everyday language and relationships and from strategies and typologies to practices and praxes. Craig Verzone and Christina Woods examine the complex relationship between food and city in their book *Food Urbanism: Typologies, Case Studies, Practices* and note that "to contain the sprawl, the inward growth and densification of cities carry the risk of loss of quality of life and space… As a result, urban dwellers became disconnected from one of their most basic needs: food" (Verzone and Woods 2021, 9). We can straightforwardly extend this note to people's relationship with fire and its controversial existence (or absence) within urban environments (safe when regulated, lethal when emergent uninvited). It is very important that beyond strategies of infrastructure and governance we start discussing and practicing food and fire as urban commons. We can do so following the "recipes" in *The Urban Commons Cookbook: Strategies and Insights for Creating and Maintaining Urban Commons* where Mary Dellenbaugh-Losse, Nils-Eyk Zimmermann, and Nicole de Vries define commoning as the practice "fostering the deepening of social relations, the forging of connections between people who are not tied by tradition or family, and more interpersonal trust… a solidarity practice [which] produces 'social glue' between individuals" (2020, 108). The authors furthermore assign the term "bonding" (2020, 152) to the notion of social glue in opposition to Robert Putnam's "bridging of social capital" (2000, 22). This is very relevant to our topic because it creates a new dimension of solidarity related to commoning that emerges within groups themselves therefore strengthening the importance of and accountability for sharing from the person to community, which then can result to societal change from the community to the welfare state and a rediscovery of the commons overall (Dellenbaugh-Losse et al. 2020, 108).

As such, it is becoming apparent that we should aim to reconsider what we value as common and how to practice commoning in urban living. We must turn to the forgotten, the imperative and the misunderstood or vilified to reconsider how we learn to live or live to learn. We can also approach and attempt affordance in the urban commoning as the collective shaping agent allowing public sharing to form through dynamic and non-stagnant networks and patterns that go beyond governance and infrastructural techniques, such as conflict and disorder. As with urban commoning practices, these propositions may diverge from dominant institutions, and theorising them is important to "hint at the possibility of a future different society" (Stavrides 2016, 176–77). They are at the edge – a peri-urban zone of contested ground; a "productive periphery" (Parham 2015, 197). For instance, and most pertinently, while food urbanism expands and becomes progressively a design approach for architects and planners as well as clear agenda for environmental professionals, artists, and activists, this remains largely focused on agriculture and the sharing of resources, with only a few examples of approaching food beyond its nutritional and economic value. Fire is also rarely associated to the practice of food urbanism and commoning even when the kitchen becomes an apparatus for practicing the commons. This is a reciprocal omission or diversion as many of the strategies and governance that regards fire ceremoniously puts a limit to its assertion of humanity through cooking and eating for the sake of life safety and security. Therefore, let this proposal of affording food and fire in as commons offer a space for theorisation, interrogating contemporary citizenship in learning to design for and by such experiences in the built environment, and the practices of urban commoning.

1 Ecosystemic Emergency

In the book *Designing Disorder* Urbanist Pablo Sendra with Richard Sennett make an honest statement about who we are as citizens living within environments by returning to Hegel's and Constant's civil society: we crave acceptance and recognition which is less a cliché and more of a "deeper and darker issue" (2020, 15). In the essay "Space Caviar" Hunter Doyle and Sofia Pia Belenky highlight that "gossip is a sign of ... currency" (2019, 4). We must not forget that gossip is situated as well as global (just like contemporary life) and takes place in the same country that we all live in, namely capitalism, as Bong Joon-ho, the director of seminal film *Parasite* (2019), noted in a recent interview (which became a meme on Reddit). In this essay, we will engage in non-insidious deep (and dark) gossip about humanity, to then take (in)action and offer propositions for change towards a more hopeful turn for the future.

The Institute for the Future (IFTF) – a DELL research initiative – report predicts that 85 percent of the jobs we are likely to practice in the year 2030 have not yet been invented (2017, 14). This is both ominous and encouraging. It also means that by modes of life and resilience we – humanity – are at best a generation behind in recognizing our indicators of societal progression: working, designing, and training for the future. We are humbly witnessing a unique moment in time facing the next extinction (this would be the sixth in the genealogy of the earth), immunological emergency (check your mask while thinking COVID-19), biodiversity dangers (think bees and bats), and rapidly manifesting climate change as well as a post-millennium future that we had innocently promised ourselves and which never came. What we offer to the future is a neurologically ill generation (Han 2015, 1) of young adults and built environments so rigid that they are unable to democratically absorb (and protect us from) the tremors and therefore change dynamically. As eugenics blink at us flirtatiously from Silicon Valley during a Zoom video call, our cities prove unable to form part of any ecosystem because they inertly fail to "allow organic growth" (2020, 60) and unwittingly rely on relationships of self-recognition and processes of eating their own tail. This is a question of erudition, cultural deficit, and humanity – a question of design in the broader sense. It is current and ancient, contemporary, and out of zeitgeist. It is also however a question of practical tactics and of human practices more broadly.

"Only after re-designing our common sense can we exit the cultural deficit we find ourselves in society" (2019, 52) Eric Lower writes, while in *Examined Life* (2008) Judith Butler in conversation with Sunauna Taylor stresses the need for "rethinking the human as site for interdependency." In response, Doyle and Belenky pose an "invitation to continue imagining ways in which we can challenge the normalizing and segregating tendencies of space upon bodies and communities through their occupation" (2019, 25). This ecosystemic (and perhaps humanitarian) emergency calls out for something more than awareness and the public eye. Collective awareness is not enough when we need to prompt collective responsibility and action (or inaction) for the re-assembly of humanity against its extinction. It is a matter of realizing what we are missing and what we can possibly and usefully reclaim and how, without polemically demonizing what we have achieved and advanced in. This is both an ecological and a pedagogical emergency call against cultural starvation and please note… this is not a drill.

2 Theory as Reciprocal Reclaimer for Practice

We have accepted here that the crisis we face is both current and ancient and that is why we are discussing a humanitarian emergency – which implies an emergency about humanity as species within an ecosystem (not coping so well) and an emergency that we can turn to design humanities to address and overcome. This is becoming pertinent as the latest generation of humanity (as the wider domain of systematic thought and logos that defines the human) approaches practical matters in a constitutional way: thought is *for* practice and stems *from* practice; an approach that is widely known as post-theory (and to some extent linked to meta-philosophy). Although we – authors – note specific issues with the term post-theory (and specifically the prefix) we cannot turn our backs on the truthfulness and applicability of Don Ihde's observation in *Postphenomonology*: *Essays in the Postmodern Context*

(1993) that we live in the "post-" era. In the context of this essay, we will use the principle that this post-era is prescribed by technological innovation. This era is fundamentally capitalistic (and therefore post-human) as Mark Fisher notes: "a strange hybrid of the ultra-modern and the archaic" (2009, 6). Deleuze and Guattari mention in *Anti-Oedipus* that "capitalism is profoundly illiterate" 2000, 240). For us, technology is not a matter of prescription but an event of immersion. We base this on the axiomatic statement that immersion is a profoundly distinctive human condition that involves logos manifesting either as contemplation and therefore mediated concentration or ecstasy – beyond intelligence that can be machinic, and beyond affect that can be animalistic. The animal (or machine) is not capable of contemplative immersion because it is "forced to divide its attention between various activities" and "process background events" (Han 2015, 12–13).

Surgeon and transhumanist Jean-Michel Besnier debates with philosopher Laurent Alexandre in the book *Do Robots Make Love? From AI to Immortality* (2018) about the trajectory of our species and makes a statement that summarizes our past, present, and perhaps not so luminous future. They say that "the indices of our species and by which our species alone can claim to have a history" (2018, 21) are technology as tools (practicum) and language as words (*theoria*), by referring to Plato in *Protagoras* (2018) and Aristotle in the *Nichomachean Ethics* (Roochnik 2009, 69). Tools themselves by default make our bodies "qua human" (Stiegler 1998, 152) and we must now question respectively what the words themselves can do in a tools-only-driven learning world. It is the value of technology and language assigned to learnt capacities that have so far determined income and capital and not the "differences in mental ability" (Alexandre and Besnier 2018, 103–4). This means that we have lost half of what defines our species as *anthropotis* (humanity). We lost the words to be able to keep carrying the tools. We are helpless Prometheans holding a fire we were so preoccupied with sustaining that we do not know how to extinguish it. We have already and long ago been destined to become cyborgs: post-human ahistorical creatures moving towards another extinction in the dawn of the post-Anthropocene.

Alexandre and Besnier reflect (and often quote) architectural historian Beatriz Colomina who, together with architect, Colomina and Wigley (2019) wrote the pivotal book *Are We Human?* presenting humanity in constitutional and ontological analogy with the genealogy of design and vice versa. To be human means to have the capacity to design. This does not equate to the praxis of inventing tools, according to Colomina and Wigley (2019, 51) or using the tools for their assigned purpose. It is fundamentally and in principle ensuring that the tools made can be assigned to a plethora of uses and applications and to foresee a trajectory of evolution for the tools themselves – and that is thought; "thinking ahead" (Colomina and Wigley 2019, 33). Now, it has become apparent that *theoria* has a role in our investigation and the proposed action or inaction in question (perhaps also part of the solution). We will at this point start assigning this role to theory and thought (logos, words, and language) which were never singular in the turbulent (or immensely tedious) philosophical treatise of design practices so far in the long history of human and built things. For this we will be turning to Nigel Thrift and his "Steps to an Ecology of Place" in *Human Geography Today* (1999, 295–322) and the interrogation of the so-called building perspective by Tim Ingold that "the worlds are made before they are lived in" (1995, 66). Following Thrift, we will be using theory as a practical reminder that moves us forward (Thrift 1999, 304) to discover "different possible ways in which we might relate ourselves to our surroundings" (Shotter 1995, 14) and evolve our outward-facing, not discursive but disclosive, capacities (Spinosa et al. 1997).

3 Post-Human Habitations

It is evident that Baudelaire's *flâneur* who "curses the hours he must spend indoors" (Reed 1996, 8) is long gone, hidden in his man-cave with a noise cancelling headset and a monstrous graphics card. This room is the crematorium of psychogeography as we know it so far. With the flaneur's ashes, thought morphs new fields for studying and designing for life such as future anthropology. Pronouns here are significant, as we note (with Reed) that the flâneur-turned-domestic pet (the modern cosmopolitan citizen) becomes a spectacle and craves to be witnessed from outside rather than establishing a sphere of domesticity (Reed 1996, 8). The latter of course has been lurking in the background since the first act of design: the use of hands to keep and carry stuff, later to be replaced by a plethora of objects, vehicles, and wearable devices. Donna Haraway writes beautifully about the same act in her introduction to feminist speculative fiction writer Ursula Le Guin's (2019) *The Carrier Bag of Theory of Fiction*. Le Guin highlights that the hero (and owner of the definition) in the storyline of civilization has never been the gatherer but the warrior. For human civilization, to be fully human means to make a weapon and kill with it (as primary method for life and supposedly prolonging the presence of the species in the ecosystem). Not doing so, says Le Guin, meant to be a woman – therefore a defective human or not human at all (2019, 31–32). We are proposing to embrace theory as a carrier bag and reclaim our X chromosome of a carrier, keeper, becoming primary containers ourselves.

There is another pathology of the contemporary late-capitalist world that is pertinent here and which leads to a breakthrough. Our generation – the same generation that is neurologically suffering and culturally ahistorical – measures value by achievement. We seek the levelling up and we (think we) unlock it by reconfirming our recognition by audience. This event has been moved from the city to the home and has become primarily domestic. This home however is the 'Home' page/button of our situated, pinned-on-a-map projection; reterritorialized, connected, "qua human," and "qua built." The modern accomplishment asserted by the right to inhabit the urban and natural landscape has been extinguished, chewed up and digested in a command centre of one's "Home," which is still a temple for rituals (Tweeting) and mundane routine (scrolling) that leads to the most human characteristics now: experiential poverty (*Erlebnisarmut*) (Doehlemann 1991) and boredom (Svendsen 2005).

Nonetheless, this story is not by any means new. It is deeply rooted in the habits and habitations of domestic interiors that have become in one way or another (semi-) public; again, a condition that is both ancient and current – out of time. This is deeply architectural at the same time, concerning life by virtue of communal ontology and conviviality – a matter of commons. Architect Uwe Schröder in his poetic collection of essays *The Spatiality of the City* notes that 'in evolutionary terms, the system of architecture is a constant which can be traced back to prehistory," developed from "a space with a fireplace sufficed for all life functions in a community" (2018, 13). Constructing the commons into architectural conditions of life will always lead "one hand to continually new places, and on the other to continually similar ones" (2018, 49). The most notable spatial turn, however, took place when the bed became the centre of the individual world, and when food and work found a common place in the most notoriously labour-less element of domestic life. Since the invention of the concept for the labour-saving apparatus in the post-war household (primarily the kitchen), the bed has been assigned more roles than any other furnishing of the home. Colomina once more stresses that the bed has become "the ultimate prosthetic and a whole new industry... to facilitate work while lying down... and of course eating, drinking, sleeping, making love, activities that seem to be turned into work itself" (2015, 121). From Hugh Heffner to the contemporary COVID-19 locked-down worker-from-home, humans have progressively and perhaps insentiently lost a zone of contemplation; the grounds for commoning.

Concomitantly, the phasing out of the fireplace from the domestic sphere is becoming progressively more so evident. This comes with the loss of intimate, meaningful, and inspiring contact with the element of fire. This is a ubiquitous symptom of late-capitalist habitation; a symptom lurking suppressed and compromised, awaiting occasional flare-ups in festive events of bonfires, and burning effigies. Fire once predominated over the mind of the child and the poet, and its flames were used conceptually to theorize consciousness, reveries, and dreams. Hence Gaston Bachelard's assertion that the "fire confined to the fireplace was no doubt for man the first object of reverie, the symbol of repose, the invitation to repose" (1968, 14). Once, fire dominated architectural spaces too. In the primitive hut and the early house, the hearth was placed centrally, and its flames provided for feeding and warming up (energy of maintenance). The transformative capacity of fire on materials was soon discovered with ovens placed directly on top of flames, melted steel, baked clay, and formed glass. Fire now fuelled the process of construction (energy) offering a wide range of new building materials. Commenting precisely on this close relationship between buildings and fire, architect Louis Fernández-Galiano playfully comments that "in the hearth, fire dwells in the building; in the oven, fire builds the building" (2000, 22). The fire of the hearth later proliferated in architectural space and moved to recesses in external walls to create what we know as the fireplace. With the advent of modernism and the emergence of central heating systems, fire still burnt in houses but was now dislocated, hidden away either in basement boiler rooms or external to the building. Heat travelled silently through an elaborate network of pipes concealed within walls and floors so that "we can still sense the heat radiated through them, but the lack of direct visual contact leads to an abstract, decentralized appreciation" (Zografos 2019, 104). In late-capitalist habitations, which, as Naomi Klein firmly spells out, should no longer follow the logic of a "New Deal painted green, or a Marshall Plan with solar panels" (2019, 39) but instead promote a global, blanket Green New Deal, the combustion of materials for energy production ends. Boiler rooms, fireplaces and chimneys, all spaces once designated to accommodate processes of combustion, now become redundant and the "memory of fire is projected outwards and is replaced by external processes. The heat produced by the sun and the earth replaces combustion, and architectural form is defined by an absence of the latter" (Zografos 2019, 107). The fireplace today is reduced to a mere indication of social status with the occasional guest appearance in scenes that call upon collective memory and popular fantasy that is almost cinematic: the romantic play of lovers sipping red wine with flames in the background and children's overly enthusiastic anticipation of gifts and sweets in stockings brought by Father Christmas.

4 Disorders of Affordance

Our aim here is to discuss the neurasthenic and reflectively impotent citizen of today who is sedentary, hyperallergic, and consumed by immobility, (Colomina and Wigley 2019, 114) in a state of regression. Most importantly, we will attempt to see how this quasi-human can reclaim a lost commonality as common sense, common ground, and common life of significance by design of urban life. We have accepted that we are indeed attempting to educate an ill generation (if not an ill species) and that what was once called neurasthenia is nothing but the collective symptoms of intellectual drain in the age of late capitalism enforced and aggravated by nomophobia – the fear of losing one's mobile device– and another constant: the law of least effort as the economy of effort towards efficiency. The latter has transformed into asthenia and *aboulia* (Rabinbach 1990, 176) which are conditions of life lacking regulated expenditure of physical and mental vigour respectively and have now come to manifest as "the nervosity of our species as a kind of malady for development… [a] certain part of contemporary humanity does not have sufficient capacity for adaptability to tolerate the increase and expansion of our life work without injury' (Rabinbach 1990, 151). If we continue to label this condition a disorder, we lose all hope and possibility of reclaiming affordance in living and therefore our chance to potentially continue to adapt and thrive, within the environments we simultaneously build and live in.

Drawing from Sendra and Sennett, we must now start making a methodological shift in design-as-being-human. We must start considering disorders as possibilities for change and avenues for reclaiming our affordance and chances for viable eco-systemic living. As such, we must act against what Mark Fisher calls "reflexive impotence," (1992, 21) referring to Deleuze and his essay "Postscript on Societies of Control" and which we are called to de-privatize and make-urban. Urban disorder as affordance is not a radical measure for us to experiment with in hope of salvation against capitalism. On the contrary, embracing disorders of affordance will mean reclaiming the commons and returning to the omnipresent rituals of human life once manifested around Hestia. It will be a fundamentally post-capitalistic move when "all that is left is the consumer-spectator trudging through the ruins and the relics" (Fisher 2009, 4) of an exhausted future. It is a matter of inverting the conditions of the Hegelian "lordship and bondage" (Hegel 1977, 111–38) that our species has become so accustomed to and calling for a re-assembly of humanity by means of a return to the accomplishment and self-consciousness of an independent recognition via labour instead of achievement. We are proposing a flexible and open ecosystemic turn that admits "conflict and dissonance" (Sendra and Sennett 2020, 34) via feast (food) and flame (fire) to address the ever-changing (and therefore meta-constant) needs of urban life and habitation.

4.1 Feasting – Food as Urban Tactics

The fundamental methodological shift is one of ideology connected to disturbance, and risk, which we must start considering as ordinary (think floods and fires). Anthropologist Anna Lowenhaupt Tsing in her book *The Mushroom at the End of the World* follows an ecologists' rather than a humanists' perspective, defining disturbance in its environmental role to pronounce ecosystemic change (Tsing 2015, 160). Tsing uses the matsutake mushroom as edible constant between gift and commodity and paradigm of culture, politics, and life, making some extraordinary and pertinent notes regarding interrogating the role and value of food as token of exchange and consumption in our capitalistic world networks and relationships. She explains that "disturbance opens the terrain for transformative encounters, making new landscape assemblages possible [via] liveability – the possibility of common life on a human-disturbed earth" (Tsing 2015, 160–63), advocating for ecosystemic intelligence. To acquire this kind of intelligence we must reposition the value of food, upsetting its "economic common sense" (2015, 122) towards the formation of relationships, networks, and extensions of the person/human. This implies a decisive shift from food to feast and from convenience to conviviality, reclaiming a strong notion of embodiment and sensory immediacy – what the rigid city is lacking, according to Sendra and Sennett.

The contemporary citizen is Sayaka Murata's *Convenience Store Woman* (2019). This default citizenship presents an established modality of being with powerful rhetoric of food as commodity against its other two dimensions of nature and culture that Eivind Jacobsen, director of the Institute of Consumption Research Norway (SIFO), identifies in *The Politics of Food* (2004, 59–78). Jacobsen highlights the role of food as terminal (instead of constitutional infrastructure in society) and its inscription as such in capitalist societies where it acquires hardly any value beyond its capacity as fungible entity with a presentation for sale (2004, 63). Holding together a rigid hierarchy of taste, commodified and capitalist food serves as social distinction, (Bourdieu 1984) potentially marginalizing citizens exerting dominance according to Jacobsen (2004, 70) who sees food commoning a family affair and not an urban event. This is exactly the space where the proposed disruption can fit, reclaiming the trope of food as commons with a political potential and an almost asynchronous (and therefore constant) chemical sensitivity to which we individually and collectively respond by nature. We propose an urbanized return to the hearth by design of life and by disorder in commodified cultural regimes.

Designing and living the urban must organically or forcefully (and ecosystemically rather than institutionally or formally) be made to re-acquire rituals of feasting. Sharing food around an open fire turns a rhetoric of conflict, danger, and violence into a social account of contemplation and immersion. This is one of the distinctive features of our human selves – "the use of fire, the roots of cooking and sharing food with strangers," archaeologist Martin Jones notes in his book *Feast: Why Humans Share Food,* (2007, 1) and reminds us of Claude Lévi-Strauss and his statement that cooking is what separates us from the rest of the living world in *The Raw and the Cooked: Introduction to a Science of Mythology* (1969) and of Irenäus Eibl-Eibesfeldt's notion of danger signal turned into essence of conviviality that defines humanity as outlined in *Ethology, the Biology of Behaviour* (1970). We must now move beyond ethological treatise and the idealized romanticism of food in relation to nature and tradition, (Jacobsen 2004, 71) no matter how we crave particular cuisines and culinary delights. We must act aiming towards an environmental future anthropology of urbanism and urbanity. To do so, we must – following Rabinbach – deal with the excessive demands of the civilization we created by means of educating our mentally taxed will and affective life. Let us now awaken the will (and desire that in *aboulia* remains impotent) and retell the story of Prometheus for the sake of our future.

4.2 Flaming – Fire as Urban Tactics

Fire is relevant to design both as a human act and as commons. Vitruvius's historical account in *De Architectura* (book II, chapter 1) defines fire as a catalyst agent for the creation of architecture and for socially inscribed human activities. The accidental discovery of fire is what "originally gave rise to the coming together of men, to the deliberative assembly, and to social intercourse" (2003, 38). Clearly rooted in Epicurean evolutionism and Lucretius's philosophy, the possession of fire signifies the transition of the human; a wild caveman turning progressively into a civilized, social being. This widely familiar story is vividly illustrated in ancient Greek mythology with the violent transfer of fire from the divine hands of Zeus and Hephaestus to the humans, through the cheating of the titan Prometheus. The Promethean fire, Jean-Pierre Vernant argues, "represents culture as opposed to wildness" (1989, 38) and its possession presupposes a *technics*, "a technique of transporting, conserving, and lighting the fire, part of the know-how inseparable from human life" (1989, 39). For Gottfried Semper the hearth appears in the centre of the primitive hut. Among the three other fundamental elements of architecture, namely the roof, the enclosure and the mound or foundation, the hearth is the "first and most important, the moral element of architecture" (1989, 102) and "throughout all phases of society [it] formed the sacred focus around which the whole took order and shape" (1989, 102). In *The Spatiality of the City* Schröder says this is an implicitly communal act and notes that the inner-outer space between home and city developed from the single room – what he poetically calls "a blueprint for communal existence" (2018, 15).

The link between fire and the commons was equally strong in the classical world, emerging in rites that dealt with the city or the house. In ancient Greece, the flames of Hestia, the goddess of the hearth and the home, burnt in the *prytaneion* (town hall), an edifice that powerfully symbolizes the epicentre of city life. In ancient Rome, the goddess Vesta ruled both the fire of the household and the civic hearth of the city. Fire in rites is not limited to Greco-Roman culture but as exhaustively demonstrated in James Frazer's *The Golden Bough,* (1993) fire is, and has always been, a key protagonist in festivities and rites around the globe. Shifting our focus to the present day, we produce and inhabit environments that demonize fire. For clear reasons, there is a widespread consensus that fire's catastrophic and obliterating action must be contained, if not entirely suppressed. Buildings are accordingly designed to respond actively and passively to flames and smoke. Cities are equipped to accommodate processes that sustain and promote a fireproof world. Our contact with fire today is compromised, impersonalized, commodified, and plainly calculated in risk assessments and escape routes. We predominantly experience its fierceness in uncomfortable scenes of accidents and violence, usually streamed (though heavily censored) on media, when we witness uncontrolled flames consuming lives, hopes and buildings. The last memory kept from these encounters is what fire by right leaves behind: ashes, a hybrid scene made up of Tarkovsky's *Stalker* (1979) and *The Sacrifice* (1986).

There is one thing that humanity can unanimously and collectively accept with certainty. Our trajectory on this planet became cursed by the discovery of fire. Like capitalism, it is "a malediction which no penitence can ameliorate" (Fisher 2009, 2). In light of this acceptance, we propose a reassessment of our attitude towards fire by appropriating a *technics* that rekindles the flames of its Promethean self. This reclaimed fire no longer burns in a cold, amnesiac and lifeless ground, as precisely experienced in the late-capitalist habitation. Fully determined by design and affordance as conflict and disorder, the city, buildings, and people can now not only promote but also practice controlled but spontaneous processes of combustion. As such, the flames in the reclaimed fireplace will captivate thoughts and dreams and re-establish a long-lost sense of community. We can then potentially experience a sense of conviviality that is abundant in rites and festivities, just like in Aztecs' human sacrifices where "surplus taken from the mass of *useful* wealth … the *accursed share*, destined for violent consumption" (Bataille 2002, 9) and gods received the morbid offering "in the immense confusion of the festival" (Bataille 2002, 60).

5 Digestions and Incinerations – Return to Inauguration

Drawing from work by philosophers, urbanists, archaeologists and (trans)humanists, we – authors – have attempted to spread a whisper and a kind of humanitarian gossip about the archaeology of our futures and of future anthropology. We have hopefully claimed convincingly that fire and food are and should be considered part of the (designed) commons and that the latter are potent when they entail change by disorder and conflict. We also interrogated the possibility and purpose of reclaiming those in the context of urban design practices, inspired by the current flux in how people live the with and around each other in urban environments. We return therefore here to the practical matters of sharing within commoning and urban practices. Food and fire can instigate a shared domain in codesigning, coproducing and cohabiting the commons, through approaching the social change as practice of conflict – stressful affordance. Canonising and vilifying practices and tactics in designing life and living are the two ends of the same rope – that of false normalisation and avoiding critical contemplation as part of praxis.

Reflecting on what was so far discussed, we must also address the "tragedy of commons" (Hardin 1968 and Daniels 2007, 518) and the interchangeable use of the term with that of resources, where particularly fire (and food as related to it) are coincidentally missing from most studies, including the essay "Mapping the New Commons" by Charlotte Hess. In Hess' mapping of the new commons, one can identify the so-called traditional commons such as water, streets, and the atmosphere as well as younger ones such as the internet and antimicrobial resistance, and of course sever markets and welfare infrastructure domains (Hess 2008). The majority of those new commons are examined from the point of view of their mass impact or necessity of the community or the city, leaving out humanity and the human. A little is also said about the design and manipulation of such infrastructural domain beyond the legal aspect of provision and access and whether a reclaim of commons can have a positive impact on the collective and individual thinking (and acting) of humanity and humans respectively. Food (apart from the expected reference to global access to welfare) and fire are commonly(!). An act of return is perhaps too simple and therefore not afforded by the complex systems of super-commons under the advanced capitalism ordnance.

Let us now therefore contemplate – a lost art in the age of capitalism – in ecstasy or in indolence the much-anticipated and ever-promised future of humanity as a radical act of return. Let us immerse in our constitutional substance which is none other than energy: food and fire, granular matter and field, apple, and snake. Let us be human without the imperative to achieve, and which "makes one sick" (Han 2015, 10). When doing so we are recognizing that such a condition is distinctive to humans, because the animal is "incapable of contemplative immersion" (Han 2015, 12–13) and recognising that the myth of Prometheus implies that we can be both liver and eagle. We are also realising that both food and fire as commons can by design and by performance offer a transindividual aspect: "the commonality of a shared capacity assumed to be antecedent," as art theorist Gerald Raunig asserts in *A Thousand Machines* (2010, 116). A shared infrastructure that will allow food and fire to be seen more as tactics that can emancipate a new kind of life. In this mode of living, the expenditure of energy through commons will be regulated by return to ritual and rhythmic labour (Rabinbach 1990, 175); what food and fire stood for at the beginning of (primitive) human time. This will make us again a historical species and will provide us with a collective and individual neurological and immunological defence against the maladies of the turbulent and spiral continuum of what we call civilisation.

References

Alexandre, Laurent, and Jean-Michel M. Besnier. 2018. *Do Robots Make Love? From AI to Immortality*. London: Octopus Books.

Bachelard, Gaston. 1968. *The Psychoanalysis of Fire*. Trans. A. C. M. Ross with a Preface by Northrop Frye. Boston: Beacon Press.

Bataille, Georges. 2002. *The Accursed Share*. Vol. 1. Trans. R. Hurley. New York: Zone.

Bourdieu, Pierre. 1984. *Distinction: A Social Critique of the Judgement of Taste*. Trans. R. Nice. Cambridge, MA: Harvard University Press.

Butler, Judith, and Sunaura Taylor. 2008. *Examined life*. YouTube. https://www.youtube.com/watch?v=k0HZaPkF6qE&t=1s. Accessed 17 Sept 2021.

Colomina, Beatriz. 2015. Privacy and Publicity: The Age of Social Media. In *2000+: The Urgencies of Architectural Theory*, ed. James Graham, 118–131. New York: Columbia Books on Architecture and the City, University of Columbia Press.

Colomina, Beatriz, and Mark Wigley. 2019. *Are We Human? Notes on an Archaeology of Design*. Zürich: Lars Müller.

Daniels, Brigham. 2007. Emerging Commons and Tragic Institutions. *Environmental Law* 37 (3): 515–571. http://www.jstor.org/stable/43267404. Accessed 17 Sept 2021.

Deleuze, Gilles, and Félix Guattari. 2000. Anti-Oedipus *Capitalism and Schizophrenia*. Trans. R. Hurley, M. Seem and H.R. Lane. London: Continuum.

Deleuze, Gilles. 1992. Postscript on the Societies of Control. *October* 59: 3–7.

Dellenbaugh-Losse, Mary, Zimmermann, Nils-Eyk, and de Vries, Nicole. 2020. *The Urban Commons Cookbook: Strategies and Insights for Creating and Maintaining Urban Commons*. http://urbancommonscookbook.com/. Accessed 17 Sept 2021.

Doehlemann, Martin. 1991. *Langeweile? Deutung eines verbreiteten Phänomens*. Suhrkamp: Frankfurt am Main.

Doyle, Hunter, and Sofia P. Belenky. 2019. Space Caviar. In *ED#3 Normal*, 4–5. Pasadena, CA: Archinect.

Eibl-Eibesfeldt, Irenaus. 1970. *Ethology: The Biology of Behaviour*. New York: Holt, Rinehart and Winston.

Fernandez-Galiano, Luis. 2000. *Fire and Memory: On Architecture and Energy*. Trans. G. Cariño. Cambridge, MA: MIT Press.

Fisher, Mark. 2009. *Capitalist Realism*. Alresford: Zer0 Books, John Hunt Publishing.

Frazer, James G. 1993 [1890]. *The Golden Bough: A Study of Magic and Religion*. Ware: Wordsworth Editions.

Han, Byung-Chul. 2015. *The Burnout Society*. Trans. E. Butler. Stanford CA: University Press.

Hegel, Georg W.F. 1977. *The Phenomenology of Spirit*. Trans. A.V. Miller. Oxford: University Press.

Hardin, Garrett. 1968. The Tragedy of the Commons. *Science* 162: 1243–1248. http://www.garretthardinsociety.org/articles/art_tragedy_of_the_commons.html. Accessed 17 Sept 2021.

Hess, Charlotte. 2008. *Mapping the New Commons*. SSRN. https://doi.org/10.2139/ssrn.1356835. Accessed 17 Sept 2021.

Ingold, Tim. 1995. Building, Dwelling, Living: How People and Animals Make Themselves at Home in the World. In *Shifting Contexts: Transformations in Anthropological Knowledge*, ed. Marilyn Strathern, 57–80. London: Routledge.

Institute for the Future (IFTF) for DELL Technologies. 2017. *The Next Era of Human-machine Partnerships.* https://www.delltechnologies.com/en-gb/perspectives/the-next-era-of-human-machine-partnerships/. Accessed 17 Sept 2021.
Jacobsen, Eivind. 2004. The Rhetoric of Food. In *The Politics of Food*, ed. Marianne E. Lien and Brigitte Nerlich, 59–78. Oxford: Berg.
Jones, Martin. 2007. *Feast: Why People Share Food.* Oxford: University Press.
Klein, Naomi. 2019. *On Fire: The Burning Case for a Green New Deal.* London: Allen Lane.
Lawer, Eric. 2019. How Would You Like Your Utopia? In *ED#3 Normal*, 48–52. Pasadena: Archinect.
Guin, Le, and K. Ursula. 2019. *The Carrier Bag Theory of Fiction.* London: Ignota Books.
Lévi-Strauss, Claude. 1969. *The Raw and the Cooked: Introduction to a Science of Mythology.* Trans. J. Weightman and D. Weightman. New York: Harper and Row.
Murata, Sayaka. 2019. *Convenience Store Woman.* London: Granta Books.
Parham, Susan. 2015. *Food and Urbanism: The Convivial City and a Sustainable Future.* London: Bloomsbury.
Plato. 2018. In *Plato: Protagoras*, ed. Cambridge Greek and Latin Classics. Nicholas: Denyer. https://doi.org/10.1017/CBO9780511813023.
Putnam, Robert. 2000. *Bowling Alone: The Collapse and Revival of American Community.* New York: Simon & Schuster.
Rabinbach, Anson. 1990. *The Human Motor: Energy, Fatigue, and the Origins of Modernity.* Berkeley and Los Angeles: University of California Press.
Raunig, Gerald. 2010. *A Thousand Machines.* Los Angeles: Semiotext(e).
Reed, Christopher, ed. 1996. *Not A Home: The Suppression of Domesticity in Modern Art and Architecture.* London: Thames and Hudson.
Roochnik, David. 2009. What is Theoria? Nicomachean Ethics Book 10.7-8. *Classical Philology* 104 (1): 69–82. https://doi.org/10.1086/603572.
Schröder, Uwe. 2018. *The Spatiality of the City.* Rome: DIVISARE.
Semper, Gottfried. 1989. *The Four Elements of Architecture and Other Writings.* Trans. H. F. Mallgrave and W. Herrmann. Cambridge: University Press.
Sendra, Pablo, and Richard Sennett. 2020. *Designing Disorder: Experiments in the City.* London: Verso.
Shotter, John. 1995. *Cultural Politics of Everyday Life.* Milton Keynes: Open University Press.
Spinosa, Charles, Fernando Flores, and Hubert L. Dreyfus. 1997. *Disclosing New Worlds: Entrepreneurship, Democratic Action, and the Cultivation of Solidarity.* Cambridge MA: MIT Press.
Stavrides, Stavros. 2016. *Common Space: The City as Commons.* London: ZEN Books.
Stiegler, Bernard. 1998. *Technics and Time, 1: The Fault of Epimetheus.* Redwood City: Stanford University Press.
Svendsen, Lars. 2005. *A Philosophy of Boredom.* London: Reaktion Books.
Tarkovsky, Andrei (director), and Demidova, Aleksandra. 1979. *Stalker.* Soviet Union: Mosfilm.
Tarkovsky, Andrei (director), and Wiborn, Anna-Lena (producer). 1986. *The Sacrifice.* Sweden: Svenska Filminstitutet (SFI) and Argos Films.
Thrift, Nigel. 1999. Steps to an Ecology of Place. In *Human Geography Today*, ed. Doreen B. Massey, John Allen, and Philip Sarre, 295–322. Cambridge: Polity.
Tsing, Anna Lowenhaupt. 2015. *The Mushroom at the End of the World.* Princeton, NJ: Princeton University Press.
Vernant, Jean-Pierre. 1989. At Man's Table: Hesiod's Foundation Myth of Sacrifice. In *The Cuisine of Sacrifice Among the Greeks*, ed. M. Detienne and J.P. Vernant. (trans. P. Vissing), 23–88. Chicago: University of Chicago Press.
Verzone, Craig, and Christina Woods. 2021. In *Food Urbanism: Typologies, Case Studies, Strategies*, ed. D. Alexander. Basel: Birkhäuser.
Vitruvius. 2003. *Vitruvius, on Architecture.* Trans. T. G. Smith. New York: Monacelli Press.
Zografos, Stamatis. 2019. *Architecture and Fire.* London: UCL Press.

Part IV
Design, the Commons and Transdisciplinarity

Chapter 12
Design and Commons: A Lacanian Approach

Dora Karadima

Abstract This article presents a literature review of the field of social sciences in connection with design theory, more specifically on the issue of collaboration. What we perceive as design today has undergone major changes, mostly in the ways it is expressed by visual and material means. But in the definition of a design ontology, one can trace some common characteristics. Transformation, among the most important of those characteristics, has always acted as core signifier within the various definitions of design practice and theory. With design today being more ubiquitous than ever, we see its contemporary expression acting in tangible ways, traced prominently in the service design field. This has the capacity for various levels of collaboration, which makes the notion central within design studies. Collaboration is also a near tautology to the activity of sharing, a core process of the commons. The commons has been proposed by many scholars as an alternative to capitalism in order to sustain our planet and the human race in general. Can design be linked with the commons? On what grounds? Lacanian psychoanalysis is proposed as an answer to the human tragedy, a tragedy that characterizes human beings and will probably continue to do so. It offers an interpretive mechanism and (complex) ways of operation of our psychic reality within humanity. This article ultimately is a proposal to connect these three seemingly unconnected fields, in that the core of their activity comprises human beings and their distinct existence in the world.

Keywords Collaboration · Design theory · Otherness · Commons · Autonomy · Lacan

In this article I aim to investigate design theory as well as the commons to illuminate the path of their (mutual) theorization under the prism of certain psychoanalytic notions. I embrace Lacanian notions as an interpretive tool, beginning with an

D. Karadima (✉)
School of Architecture, National Technical University of Athens, Athens, Greece
e-mail: dorakaradima@mail.ntua.gr

G. Bruyns, S. Kousoulas (eds.), *Design Commons*, Design Research Foundations, https://doi.org/10.1007/978-3-030-95057-6_12

introduction to Lacan's three orders. What follows is an analysis of the notions of desire, alienation and separation, respectively, which will be presented in separate parts of this article. A univocal definition of design does not exist; thus, it is necessary to dismantle any such definition. An ontological approach will be examined with the aim of finding the common traits and characteristics of design regardless of the way these are presented, and thus locating the framework of conceptual signifiers and understanding how they operate in connection with Lacanian terms.

In order to focus on the common ground between design and the commons, a route to a contemporary expression of design will be presented: namely, the service design direction. Non-material traits, user involvement and constellations of value co-creation show comprehensively why a central signifier of these processes is collaboration through metonymic processes. Linking with the theory of commons through sharing, initially I will investigate what commons and commoning are, resulting in common ground on the issue of autonomy that is proposed as an emancipatory direction. Following the thought of Castoriadis on the imaginary institution of society, it is the project of achieving the autonomy of the subject. He states, regarding the issue of regulation of the subjects and society, that one should aspire to the establishment of rules by themselves, resulting in autonomy versus heteronomy. The linguistic origin derives from the "nomos," which in Greek means rule/law, while the prefixes of auto and hetero reflect oneself and the other respectively. In a nutshell, in order for commons and design to exist in an emancipatory way, I consider autonomy the righteous path. Hence the issue of self-regulation is an important aspect of design and commons; as I will demonstrate, it establishes a dialectic with the Other, not through negation, but rather through making amends with the Other to enable oneself to emerge. Desire will be presented in this article as something that surpasses the Ego; hence I significate this sentence as asking how one can surpass the regulatory drives, so that the subject of desire can emerge: meaning, how could I become a subject where I am not completely driven by my unconscious but take responsibility for my actions. The issue of autonomy will then be analyzed through a political/psychoanalytical lens, resulting in the matter of Otherness. Otherness as a notion is in close connection with subjectivation and its formation; one can simply not exist without the discourse of the Other. The ultimate value of this investigation is identified as Otherness in design practice, any design practice; and a possible direction for future research is proposed at the end of this article.

As I have presented the logical structure of this article, here I would like to provide some explanatory thoughts on the importance of interpreting the mutual theorization of commons and design from a Lacanian perspective. In a general sense I would like to answer whether alternatives to these two practices could exist, other than what has been done so far. In that sense, I mean a practice that manages better the issues of power dynamics and control, both usual problems, as the literature states. Dismantling the three orders could shed some light, beyond conscious discourse, to reveal the unconscious part of design and commons. On the other hand, as I will explain in the article, I deal with the issue of desire, a notion that goes against capitalist contemporary discourse, as it surpasses the narcissistic modes of being. To see through Lacan's lens in this day and age could reveal the synchronicity of the contemporary modes of being, meaning the elements that constitute today's design and commons language.

1 Lacanian Concepts as an Interpretive Tool

Lacan structured our psychic reality into three orders: the Real, the Symbolic and the Imaginary, operating as interwoven and interconnected. If shown in topological terms, they are depicted as three Borromean knots; when one collapses the others collapse as well. One cannot refer to a single order in a self-referential mode, but rather in light of its participation in the whole. In the following, I will use the three orders as an interpretive mechanism to further deepen design theorization, realizing that when one is designing, a psychic mechanism is set in motion to function on an unconscious level. The orders structure human experience within a triadic conception which distinguishes society, spiritual culture, and representations.

The *Symbolic* refers to the symbolic interaction between people, their ultimate points of reference (morals, God, and so on) expressed as internalized codes of communication and behavior (Evans 2003). Lacan, drawing on anthropological studies by Lévi-Strauss, articulated his meaning of the Symbolic. Lévi-Strauss's initial work revealed the basic structural undercurrent of society itself; namely, the elemental structure from which all subsequent social relations originate. This is unconscious, defining people's social status and regulating their relationships without their knowledge. We can infer that the symbolic order intervenes in all aspects of life, as a blank space where it acts on its own terms, so that when we experience reality, we experience it above all as "symbolic reality," which I will refer to from now on as "social reality." Every social activity constitutes a kind of language, leading us to interpret individual acts in connection with others, based on a frame from which meaning is derived. Does design constitute a particular symbolic language? Like any other language or structural system, it refers to the field of the symbolic order (Deleuze 2006). Ernst Cassirer claims that the fundamental notions of science are based on the use of invented symbols, and at a parallel level where even in natural sciences reality is not interpreted as an objective convention, but rather in the light of symbolic mediation (Cassirer 1995). It is precisely the multiplicity of interpretation of meanings that places design in a symbolic universe, the assemblage of which refers to a constructed reality different from the experiential one that we interact with on a daily basis.

To continue with the orders, the *Imaginary* is the relationship of the human psychosoma with identities of images, of persons as well as objects, from the point of view of Ego (Evans 2003). Lacan showed a particular interest in setting apart the Ego from the subject itself and thus developed a conception of subjectivity as alienated. In phenomenological terms, the existence of objects in the world are not independent of our consciousness; instead, our thinking around a subject and the subject itself are codependent. Žižek (2005), in his attempt to explain the imaginary and its linkage with fantasy, offers this analysis: "We must not think of the role of fantasy merely as a realization of a desire in a hallucinatory way: rather, its function is similar to that of Kantian 'transcendental schematism' – a fantasy constitutes our desire, provides its coordinates, it literally 'teaches us how to desire'" (p. 60). He continues with the use of an example which articulates precisely the core of the Imaginary, saying: "To put it in somewhat simplified terms: fantasy does not mean, when I desire a strawberry cake and cannot get it in reality, I fantasize about eating it; the problem is rather, how do I know I desire that strawberry cake in the first place?" (Žižek 2005, 61).

The Imaginary should not be narrowed down to something that is simply illusory and has no consequences. It is far from inconsequential, but has powerful effects in the real, unlike something you can just overcome (Homer 2005), such as the non-realization of a wish. The fundamental fantasy is exactly the point, namely that each subject possesses a unique element that regulates their desire, expressed mostly in sexual terms but nevertheless contributing to our idea of attraction beyond these. For design to be appealing, it is considered vital that it should contribute to this element as well.

Finally, the *Real*, not to be mistaken with reality, is the very body of pleasure/suffering in the world: it can never be fully symbolized or fused into the Ego (Evans 2003). Lacan described the Real as an invisible intense materiality that existed before being symbolized. As such, the Real always returns to its original place in the form of a need – any need – and can be associated with the body prior to its symbolization. In connection with being the need and not the means of satisfying it, it comes into reality as a sign entering the "social reality" previously described. Lacan connects the Real with the concept of trauma. Being familiar with it, approaching it literally as a cut, Lacanian psychoanalysis talks about a psychic trauma, not linked with a "reality." It appears to exist between an external stimulus and the subject: the subject is unable to comprehend it and overcome the excitement accompanying it, leaving a psychological scar which reappears later in life. This scar is suppressed and moves into our unconscious. It stands to reason to ask, "What is suppressed?" To put it simply, the part that is always missing; it is what Lacan calls *objet petit-a*, which literally means no-thing becoming something through the desire of the subject; it refers to the desire to fill the void in a manner that characterizes us as subjects.

Fantasy plays a vital role here, setting the stage for the desire of the subject to be expressed; in this stage we learn how to desire and we are constituted as desired subjects. This exact setting and the pleasure we derive from it are what provides us with *jouissance*, best described as pleasure in pain. *Objet petit-a* represents lack itself, and since desire is always for something that is missing and involves a constant search, a rupture between the subject and the Other occurs, which inaugurates the movement of desire and the advent of *objet petit-a*. In an effort to think we are unified with the Other, to ignore our indivinity, even though the desire for the Other always exceeds it, *objet petit-a* sustains us. Hence, there is a constant search for that which is missing, while *objet petit-a* masks the lack as a residue of the Real. In response, the subject takes responsibility, traversing the fantasy by subjectifying the trauma and assuming its own *jouissance*. Contrary to desire, which moves from one signifier to another, *jouissance* is absolute and certain, like a constant drive that each of us experiences in a unique way, while being unable to say exactly what it is. It causes us to believe that there is always something better, which we generally attribute to the Other, in the belief that we can experience a level of enjoyment beyond our own, with fantasy feeding it (Homer 2005). As design is a symbolic language, the design process, which has material and immaterial derivatives, can be perceived as a negotiation with the Real: an intervention in the domain of symbols and the signifying processes. Within this negotiation, I depict the failures of the Real, as in every design project there could always be something missing, which nevertheless at times drives us to further investigate our stance as designers and how we practice design.

An interesting aspect of Lacanian theory is found in the notions of alienation and separation as human conditions through the construction of subjectivation. Alienation is connected with the mirror stage, following Rimbaud's "I is another" which belongs to the sphere of the Imaginary, perceiving the subject beyond our ego, in the field of the Other. Lacan states: "alienation is destiny," for one cannot avoid it; it constitutes our ego as we identify ourselves with its counterpart. The alienated subject is the subject of the signifier, ruled by the Symbolic and language, and existing in a statutory division. In connection with alienation comes the concept of separation, related to the notion of desire where one is separated from the Other, and leading the subject to have a desire for existence other than the signifying chain. It requires a desire for knowledge beyond the structure of language and the Other, leading us to recognize lack itself and the Other of lack (Homer 2005). For someone to exist they must be recognized by someone else; in that sense our image is mediated through the gaze of the Other in such a way that they become a guarantor of ourselves. We are dependent on the Other as an assurance of our own existence. Through the Other, we recognize our-self (Christodoulídi-Mazaráki 1997).

Through this introduction to the indivisible folds of our psychic reality, as well as to the "stages" where subjectivation is inaugurated, one can spot the connection of design theory and action within them: specifically regarding the issue of identity. Historically, design has acted as a field for demonstrating identity through visual and material means (Sparke 2013), mostly directed towards serving the commodities culture. Extending design's connection to culture beyond its role as a mere partner of industry, it "manifests the progress of human civilization in a material way" (Xianyi and Lingling 2009). Acting in synergy "to the background of the era" (Xianyi and Lingling 2009) among the "cultural background and input of the people, and the system itself," the understanding of design ultimately conveys "a cultural connotation which makes the design be some metaphysics or symbol of a feature in a cultural system" (Xianyi and Lingling 2009). The abovementioned idea could be interpreted in various ways; hence it would be interesting to investigate the matter closely.

Thus emerges a need to understand design ontology, as well as its accompanying unique characteristics and actions, pointing to research on what is, what might be, and how the latter can happen. This process could establish the boundaries of design, holding in synergy the Lacanian theory for interpreting design practice in close connection with the unconscious so as to understand how our unknown truth could be articulated in design creation.

2 Design Ontology and Contemporary Design

Traditionally, design has been linked with commodities culture; simply put: its (stereotypical) main purpose has been to serve the interests of the people who reach out to design practitioners usually to "solve" an everyday-life issue. This direction reinforces the view of design as a mere byproduct of designers' conscious decisions, regulated by a design brief in order to meet human needs. Therefore, design has been perceived solely as means of providing (some) answers to (some) problems. That alone hinders its full potential, narrowing it down to a mainstream narrative of accomplishments and preventing us from conceiving design's full capacity. After this brief presentation of an instance of the design problematic, the direction I choose to follow will be towards discovering the ontology of design going beyond a definition. I am interested in discovering the types of being of design; hence the ontological probe, as a way of discovering being as a condition of presence; what design is constituted of, in a sort of structural approach, but not narrowed down to it, and how I can connect this philosophical point of view with contemporary nuances of service design. A certain common trait seems to emerge when we attempt to define design ontology, as a transformational attribute traversing design creation captured as change in various degrees.

Žižek (2005) detects design as an oscillation between "pure" design in the form of aesthetics such as, for example, the aesthetic form of a product, and design in the sense of construction of the inner core or generic formula. In recognition of the inherent link between them lies the great power of design in relation to the three orders. At first glance, it exists at a Symbolic level as a carrier of meanings that designers contextualize in their designs. At an Imaginary level it offers the ground on which fantasy can be articulated. Finally, at a Real level, it offers the place where needs are formulated, always leaving an undesigned part behind. Creation cannot exist without personal involvement. The way this personal element is formulated in a common, identifiable language defines the type of each design and creation (Sideris 2013), which in turn makes it unique to the designer by depicting his/her desire as a subjective design enunciation. This direction coincides with the psychoanalytic view that understands design as a practice driven from the unconscious, regardless of the design brief in the sense of a "regulator" of creation driving the decisions of the respective designers.

The transformative nature of design acts as a master signifier of its ontology, so that it cannot be fully framed in a unified expression but is linked with what we deposit on it besides the tangible elements accompanying the design creation. In order for this transformation to take place, and in light of the aforementioned multiplicity, a material formation needs to occur. With the acknowledgement of the existence of materiality, we elevate design to its full potential and demonstrate its concrete representation as an expression of the Real.

The notion of material formation plays a vital role in the design scenery in this mode of self-perception: that is, its ontology. A level of autonomy is spotted when materiality is articulated, as the user (of the materiality, such as an object) tends to convey meaning and function to it in his/her own terms. An example would be a stack of books used as a bedside table. Acceptance of the idea that "a form is never a 'mere' form but involves a dynamic of its own which leave traces in the materiality of social life..." (Žižek 2011) epitomizes the complexity regardless of the signifiers, even though they are contained in it to a great extent. They are transcribed, offering clarity as well as the aforementioned synthesis. The result of fantasy is a formatted desire (Sideris 2013) depicted in design creation, with the user having the last word.

The design creation constitutes a privileged process of expression, one of the fundamental fantasies in a transmuted "language" – the design language – which is allowed to express anything. In that vein, due to design's relationship with fantasy, meaning and the impossible, we can understand the capability it holds. It creates alluring images in addition to the symbolism they contain, which has profound effects on a daily basis, along with the inability of the undesigned to be articulated into design language, leaving us startled. In addition, extending the problematic of materiality, drawing from ontological designing, Willis (2006) recognizes a "double movement" within design. Considering design fundamental to being human, she expands this to consider design as something "we deliberate, plan, and scheme in ways which prefigure our actions and makings – in turn we are designed by our designing and by that which we have designed – we design our world while our world acts back on us and designs us". Hence the double movement implies "a circularity in which knowledge comes to be inscribed by being with the 'designing-being' of a tool [my comment: an object as well], this in turn modifying [designing] being of the tool-user' [similarly object-user]" (Willis 2006). This extends the understanding of design beyond that which it would normally be considered "i.e., a mental prefiguration of what is to be made and the pattern or template that guides making" (Willis 2006). This could easily be transferred "to non-material design as well of systems, of organizations, of methods of thinking concluding on most (design) situations both are present and inseparable" (Willis 2006). Tony Fry (Fry 1998, as cited in Willis 2006, 85) makes the point that design is "a meta-category comprised of three elements, each of which gets called design, often to the exclusion of the other two, but all which are connected." These are the design object, meaning the material or immaterial outcome of designing; the design process, significating the system, organization, conduct and activity of designing; and the design agency such as the designer, design instruction in any medium or mode of expression, as well as the designed object itself as it acts on its world.

Most theorizations of design take one of these as their exclusive focus, either paying no attention to the others or viewing them through the filter of "design" that concerns them. There needs to be something fundamental to all three. Thinking of design as unconscious-driven, where the three orders operate, provides this because it implies a condition which is always deployed. Thus, through its acknowledgement a starting point could be found for understanding modes of human being such as purposeful activity and intention.

The interest can now shift to the contemporary practice of design, particularly the service design direction, as an instance of design where collaboration is of great gravity.

What characterizes contemporary society is "the loss of the illusion of control, or the discovery of complexity," (Manzini 2011) depicted in the shift from mass-production to mass-customization. Foglieni et al. (2018) discuss the matter of service design presenting a brief history of it up to the present day. One possible reason why we have arrived at a service-oriented society is that "service satisfies higher needs than goods" (Foglieni et al. 2018, 8) combined with the fact that "manufacturing industries started to increasingly outsource their service activities to firms specialized in the provision of such services" (Foglieni et al. 2018, 8). In the course of the evolution of service design it has been analyzed via different disciplinary perspectives, resulting in two core directions. The first one focuses on "the surface, concentrating on the user experience and on the interface between service providers and customers" (Morelli et al. 2020, 10). The second one sees "service as processes with organizational aspects on the back stage" (Morelli et al. 2020,10). To continue, Kimbell (2011) highlights two tensions in service design; the first one refers to the comprehension of design as problem solving activity. The second one, which Morelli et al. (2020) also embrace, views service design as "enquiry, meaning an exploration of an open problem space involving different actors, including users." Extending this rationale, they see it as activity of value creation from a design-for-services perspective (Morelli et al. 2020). This aligns with Sangiorgi's and Prendiville's view (2017) that "service design... has been revisited with a growing interest in the notion of service as a way to conceive value creation."

The notion of value itself as an evolving and important signifier in the service design scenery has been treated in various ways. The traditional view of it sees value as aligned in a chain to which each actor provides an input, and which stops at the stage of interaction with the final user (Morelli et al. 2020). Later scholars lean towards "the creation of value as the outcome of a constellation of actors rather than a linear production chain" (Morelli et al. 2020). In the latter development it became evident that an inversion had occurred, such that service was defined "as application of resources for the benefit of another actor ... where goods became a medium of service provision" (Foglieni et al. 2018, 12). Foglieni et al. continue: "service value is always co-created by a service provider and a beneficiary, and no more by the sole provider during the production and distribution processes" (2018, 12). This conceptual pivot changes the stance of the user from passive to active, from consumer to co-creator, "from passive receiver of value to an active co-producer that interacts with other actors and also produces and aggregates resources" (Morelli et al. 2020). Corubolo et al. (2018) underpin this with the idea that "the shift… lies at the basis of a widely shared definition of co-production," which points back to Ostrom's original definition: "co-production is a process through which inputs used to produce a good or a service are contributed by individuals who are not in the same organization." Therefore, the process of value is intrinsically based on the participation of various subjects and this implies negotiation (Morelli et al. 2020) and collaboration. Moreover, if accountability in terms of achievements is added, the focus leans towards facilitating the boundaries between subjects (Kimbell and Blomberg 2017). Such engagement will lead either to direct involvement in a co-design process for the definition of new paths or to the creation of facilitation paths (Morelli et al. 2020).

Corubolo et al. (2018) pinpoint that "codesign may be viewed as the first step in a more extensive collaboration process, or as Selloni (2017) states, as an essential pre-condition to co-production, co-management, and co-governance in general." If we take into account that in services "…it is necessary to involve a variety of players who are largely interdependent and therefore who must collaborate in order to achieve any goal" (Meroni et al. 2018) then "the emergence of a co-production, by implication co-designing, economy which enables the sharing of resources reveals both a vertical (providers-users) and a horizontal (among users themselves) collaboration trajectory as well as bottom-up and top-down organizational arrangements" (Corubolo et al. 2018).

There is a growing number of services based on sharing and collaboration, with a huge variation in what is shared as well as how it is shared (Seravalli and Eriksen 2017), which, as expected, have their downsides as well. They are presented as "only constituting an issue of resource management activities that capitalize on spare capacities, whereby the collaborative features turn into marketing strategies or, even worse, involve the gradual commodification of human relationships and trust" (Corubolo et al. 2018).

With this introduction of the shift to the prevalence of services and its distinct characteristics in providing a contemporary design direction, we can understand the gravity of collaboration in this context. As the boundaries of design thinking and action become ever blurrier, the core of this collaboration, to various extents, is the activity of sharing, which is a (near) tautology to commoning. At this point it is necessary to pivot conceptually towards commons and commoning to establish their mutual theorization and possible future direction.

3 Commons and Commoning

Commons are usually perceived in two senses: firstly, as an example of governing and management of resources; and secondly as a set of social practices in virtually all fields of human activity (Bollier and Silke 2015). By social practices I mean the norms, rules and regulations that make this common governance effective. They are mostly perceived and understood in the sense of commoning, the activity of the commons, where commoning is not only concerned with tangible or intangible collective resources but rather the processes of common stewardship of resources held and managed by communities. The nature of these resources varies from natural materials like water, to knowledge, to other forms of sharing. It is exactly when a resource is being taken care of by a community or a network that it becomes common. The community itself, the rules within it, and the resource constitute a comprehensive body, the most important aspects of which are the social commitment, the accompanying knowledge, and the internal practices of management. Commoning could also be interpreted as a design activity, with the goal of creating an ambience of commons. This could only be achieved in partnership with other actors, following a similar direction to that of the contemporary design practice presented earlier. The vast difference is that these two fields operate, usually, on opposite sides of the market. In order to find mutual ground between designing and commoning, by implication between design and commons, one can try to understand the value of commoning as well as the way design can rely on this activity as a social system of collective action operated by the community of people involved in the process (Bollier and Silke 2015). Ultimately it is the design stance within design practice that I consider necessary to change through sharing and collaborating. Stavros Stavrides, discussing ways to expand the activity of sharing, draws on Jacques Rancière's attempt to theorize afresh the notion of community, beginning with the following description of the "common world":

> [The common world] is always a polemical distribution of modes of being and "occupations" in a space of possibilities … For Rancière, what is at stake is a constant redefinition of what is considered as common. This is what creates a common world and this is what, consequently, is at the basis of understanding and symbolizing community. (Rancière 2006, as cited in Stavrides 2018, 58)

"Constant redefinition" does not mean a ceaseless battle around what is common; for the common world to exist, it need not be the result of the homogenization of individuals. Rather, what is suggested is to conceive of the common world as a result of human relationships open to any level of transformation (Stavrides 2018). This requires a level of respect towards Otherness in the process of collaborating and sharing, and by implication in the process of designing as well.

Due to the complementary nature of sharing, collaboration and design practice, I advise that, as a direction within, they be practiced in convivial terms, as defined by Deriu in the context of degrowth. "Conviviality refers to a society in which contemporary tools are used by all in a comprehensive and common way, without being dependent on a body of experts who control them" (Deriu 2015, 126). Such tools are considered symbiotic if they can be used and adapted easily for the purpose chosen by the individual and if they result in the extension of freedom, autonomy and human creation. The symbiotic reconstruction demands limits on the pace of change (Deriu 2015), meaning that we cannot expect the complete abolition of production, any kind of production, but rather a change of heart within it. The alteration could be translated in psychoanalytic terms as a shift from a place of *jouissance* to a place of desire. Recalcati, in his *Ritratti del desiderio*, distinguishes *jouissance* as a quest for forced and perpetually unsatisfactory consumption, serving in that way the paradox of eudemonism of our time (2018). The impulse seems to be provided by an endless dynamic, consolidated at the end as free, unfettered by the limits of the law. However, this freedom cannot create any satisfaction. It is an empty freedom, full of sadness, misery and apathetic superficiality (Recalcati 2018). Desire, as Lacan put it, is that which resists any authoritarian dream, any attempt at homogenization. In this sense, "the desire to have your own desire" remains the coefficient of resistance to all the evocative sirens that offer the promise of an assimilation of human nature into a community of units free to enjoy without limits, in a hereditary community (and therefore without a subject), based on mass enjoyment of the One, such as this madness of the capitalist discourse has tried to accomplish (Recalcati 2018).

In this setting of change, room for personal initiative should be created and would be considered vital for survival. Allowing the co-existence of differences would actually secure the common ground, considered as something that is constantly called into question (Stavrides 2018). Upon that shift, the matter of behavior, by implication the same issue as that of how to apply our design practice, could give us a clue to the kind of transition that should be made. The critical issue here is: how could this be accomplished? The matter of autonomy with regard to the change of stance is important. Autonomy means an internal aptitude to give oneself rules and laws consciously at an individual and collective level, which will be thoroughly explained later (Castoriadis 2005). It aims at a formulation of a discourse within an intersubjective relationship with the Other (Castoriadis 2005). Designers at this point should be treated as political entities in connection with their psychic reality. The core of political entities is the political (*politiko*) and politics (*politiki*), as Cornelius Castoriadis put it in a conversation with the MAUSS team on questions of democracy and relativism (Castoriadis 2015). The political refers to authority within a society, meaning whoever holds the power to make decisions about collective schemes, the violation of which usually means punishment (Castoriadis 2015). Politics refers to collective activity that aspires to project mental clarity and consciousness, as well as to the ability to question the society's prevailing ordinances. The "ideal" society for Castoriadis would be built in ways that allow such questioning as a given (Castoriadis 2015). It seems appropriate that there should be no classification of the designer as persona and the designer as professional. After all, from the reading of our psychic reality, the projection of our unconscious is made clear in various ways, regardless of our consciousness. To see how to achieve this autonomy requires further explanation of this political context as presented by Castoriadis.

4 Is (the Quest for) Autonomy the Answer?

So far I have analyzed the matters of design and commons by proposing the notion of autonomy as the direction in which a mentality change should occur. I refer to the notion of autonomy as elaborated by Castoriadis from the perspective of the subject at an individual and collective level. In addition, investigating the notion of desire, I notice that both theorizations evolve from the same psychoanalytical maxim: "where It was, shall I come to be." In Castoriadis's words, "If to autonomy, that is to self-legislation or self-regulation, one opposes heteronomy, that is legislation or regulation by another, then autonomy is my law opposed to the regulation by the unconscious, which is another law, the law of another, other than myself" (Castoriadis 2005, 102). The aforementioned maxim then, in the author's terms, views autonomy as the construction of the conscious Ego (I) and heteronomy as an alienated discourse of myself, an unconscious one (It). The important here is to clarify that Castoriadis with his magnum opus projected, through a vigorous encounter of the autonomous and the heteronomous, the need for the autonomous to emerge through the encounter with the heteronomous, the Other, the different. In this sense, Others are not a barrier to but rather a part of our co-existence, since, as he states, "human existence is an existence with others" (Castoriadis 2005, 108). In this way, the subject does not eliminate the discourse of the Other, since this would mean an elimination of society: something neither desirable nor possible; on the contrary, the subject needs to find its position and be able to establish a discourse of and with the Other. Hence, as Castoriadis states, "one cannot want autonomy without wanting it for everyone and … its realization cannot be conceived of in its full scope except as a collective enterprise" (Castoriadis 2005, 107). For autonomy to exist it is critical that the participating group of people be regarded as members of the community, as citizens involved alongside functionaries in the deliberations and decisions that concern them (Dardot and Laval 2019), relating in this way to the commons. Only through the autonomous action of the society itself can the autonomous society exist. For the people, this entails investment in something other than ownership of commodities; in a deeper sense it requires a pathos for democracy and freedom, a pathos for the commons that would replace entertainment, cynicism, conformism and the constant struggle to consume (Castoriadis 2015). This is directly linked with the notion of Lacanian desire and its connection with Otherness as an intrinsic condition running our psychic world.

The notion of Otherness, even though it can be binding if expressed in terms of heteronomy, is also important for our existence. It is essential to explain the inauguration of desire to further understand its importance both as a human condition and how it is experienced. The subject is born in a state of helplessness, unable to satisfy its own needs, and hence dependent on the Other to help it satisfy them. In order to get the Other's help, the infant must express its needs vocally; need must be articulated in demand. The primitive demands of the infant may only be articulated as screams, but they serve to cause the Other to minister to the infant's needs. The presence of the Other soon acquires an importance in itself, an importance that goes beyond the satisfaction of needs, since this presence symbolizes the Other's love. Hence demand takes on a double function, serving as both an articulation of need and a demand for love. However, whereas the Other can provide the objects which the subject requires to satisfy its needs, the Other cannot provide that unconditional love which the subject craves. Even after the needs which were articulated in demand have been satisfied, the craving for love which is the other aspect of demand remains unsatisfied, and this residue is desire (Homer 2005). When we say desire, what kind of experience are we referring to?

Simply put, it is the feeling of an experience that surpasses us, hence the reference to the maxim that desire comes into play as deregulating. Whenever there is the experience of desire, "I" feel deprived of the safe governance of myself; I feel as if I am carried away by a force that surpasses me, that exceeds the possibility of governing and controlling the Ego. Desire should not be confused with motives or the driving force of intention; it is not experienced as a result of the will, nor is it determined by the Ego: it is unconscious, that is, it is not a property of the subject. As an experience of loss of dominance, desire is mainly the experience of a slip, an obstacle, a derailment, a fall of the Ego. For this reason, the element that unites the various manifestations of desire is its exaggeration towards the Ego. It transcends the Ego that "I" believe I am; it is a force that transcends and decentralizes the Ego. Lacan connects the term desire with the image of the Other. It is never an experience of the identical, of what I think I am; it is not a self-referential and narcissistic experience. It is always an experience of otherness, an unidentifiable phenomenon, a non-coincidence; as Lacan would say, a division of the subject. This factor of identity disruption reveals that something desires and therefore exists, pushes, tends to emerge beyond the Ego. The point is that the emergence of desire can push the Ego to give in to its faith, its arrogant acceptance of that which transcends it: that is, the unconscious truth of its desire (Recalcati 2018). Therefore, two different perceptions of the subject are contrasted. On the one hand we have the cult of the strong, despotic personality, focused on the principle of performance; it is authoritarian, without uncertainties, as it is based on the crazy faith of his/her Ego – while on the other hand we have another, divided subjectivity, where the emergence of the Otherness of desire forces the Ego to verify all its limits, its narcissistic arrogance, and the ideal of false domination, and to accept the otherness of the Other. This means accepting the psychic institution of desire as stronger; it means taking responsibility for the fact of this psychic institution that transcends us and is at the same time ours.

This reading goes beyond the individual, respecting at a parallel level the danger of losing oneself in the collective. It offers the appropriate balance in collaboration; it is a form of respect towards the otherness of others, together constituting the archipelagos of othernesses. The ways this respect could occur provides a potential research direction at a later stage. In developing the concept of autonomy as a possible response, a mode of self-reflection and self-realization is recommended. Drawing on Aristotelian ethics, we have the notions of poiesis and praxis, poiesis relating to an activity aimed towards an end, and praxis to an activity which is an end in itself. Design as praxis, as identified by Findeli and Bousbaci, is a path worth exploring for future theorization of design and commons (Bousbaci and Findeli 2005) in order to further investigate the ways desire could be applied in practical terms. To conclude, both for design and commons we understand that Otherness contains and is contained in each of us. Hence each of us carries a radical difference, something un-inscribable, that can simultaneously provide us with meaning. Acknowledging this internal aptitude could offer a direction for the issues of power in both fields, ultimately acting as a way to attain harmonious co-existence and equilibrium in collaboration.

5 Conclusion

We need to pay special attention to Otherness, as it represents an evolving signifier in Lacan's work as well as a common signifier in discourses of social and humanitarian disciplines. Otherness demonstrates an intrapsychic as well as an interpersonal condition of the subjects. It expresses "a fundamental paradox of the human condition; for example: "… (it) is structured as the other in the same"", with Lacan continuing "… in true speech the Other is that before which you make yourself recognized. But you can make yourself recognized by it only because it is recognized first" (Lacan 2001). Those who speak share a symbolic responsibility for speech of which neither is the origin, as the Other is the ultimate alteration which we cannot accumulate. Thus, the approach to otherness is a constitutive act of every social encounter, as it serves a double function which is of great importance. Simply put, through the notion of the Other, comes the demand that we articulate to each other, which simultaneously could differentiate us by indicating our own desires. As the aim of this article is to study the notion of collaboration as a "shared" signifier (and process) in both commons and design, it has been made clear that the issue of Otherness has the capacity to be central. On an interpersonal level, Otherness constitutes the dialectical part of our subjectivity, as it poses the question of our identity and identity's relationship with Otherness (Kyriakakēs and Michaēlidou 2006). Its very existence composes and reveals the relationship with ourselves, especially in contexts of an emancipatory nature. The interest of the social sciences now lies in an examination of the social processes and power relations which have the capacity to articulate and promote specific kinds of Otherness in ways that reinforce inequality in contemporary societies (Kyriakakēs and Michaēlidou 2006). Even though the focus is particularly on service design as the non-material element where collaboration could occur more prominently, it remains true that the aforementioned can act as recognition of processes that appear in design practice and thinking in general. Above I have substantiated the connection of Otherness with heteronomy in such a way as to enable the aforementioned issue of autonomy to act as a (research) direction in any design field. Probably the ways collaboration would occur would by implication differ from the ways to achieve autonomy, but the intrinsic heteronomy lies in every design process shown by respective designers. At this point the interpretive tools of Lacanian psychoanalysis will help and hopefully inspire more scholars to find likely connections and ways that unconscious desires are depicted in the design field. I hope it will shed some light on matters of politics in design, by illuminating the nature of practiced power, and so providing the possibility of rendering it fundamentally humane.

The investigation proposed rests mainly on this segregation, which can be found mostly in matters of collaboration as a driving force of the intersection of commons and design. As a possible answer to the differences existing and appearing in all human conditions, the issue of autonomy is proposed in ways imagined by Castoriadis. It constitutes the liberation of the internal laws and regulations that have been imposed on us and that we have chosen unconsciously to obey. In order to move from obedience to choice – which is a matter of ethics in the context of "what should I do?" – first we need to understand how we operate internally, and by implication in our design practice as well, to avoid being in a position where we are ruled by allure, with whatever consequences. The power of design in this context is great, as it has the ability to intrude on our daily lives and transform them in ways not yet imagined. As a byproduct of culture, it would be interesting to see, given its omnipresence, what ground it shares with the commons, as a way of acting for and by the collective. Without its being a new concept, as it is considered a fundamental element of democracy, I believe this investigation can shed some light on its potential formalization and desired continuance.

References

Bollier, David, and Helfrich Silke. 2015. Commons. In *Degrowth: A Vocabulary for a New Era*, ed. Giacomo D'Alisa, Federico Demaria, and Giorgos Kallis, 120–126. New York: Routledge.

Bousbaci, Rabah, and Alain Findeli. 2005. More Acting and Less Making: A Place for Ethics in Architecture's Epistemology. *Design Philosophy Papers*. 3 (4): 245–264. https://doi.org/10.2752/144871305x13966254124996.

Cassirer, Ernst. 1995. *The Philosophy of the Symbolic Forms, Volume 1: Language*. Trans. Ralph Manheim. New Haven and London: Yale University Press.

Castoriadis, Cornelius. 2005. *The Imaginary Institution of Society*. Trans. Kathleen Blamey. Cambridge: Polity Press.

———. 2015. *Dimokratia kai schetikismos, syzitisi me tin omada MAUSS*. Athens: Stasei ekpiptontes.

Christodoulídi-Mazaráki, Aggeliki. 1997. I synkrótisi tou ypokeiménou méso ton domón tis glóssas kai tou politismoú. *Epitheórisi Koinonikon Erevnón* 92 (92–93): 109–125.

Corubolo, Marta, Daniela Selloni, and Anna Seravalli. 2018. *Sharing and Collaborating in Service Design*. Linköping Electronic Conference Proceedings 150:21, s. 237–251.

Dardot, Pierre, and Cristian Laval. 2019. *Common: On Revolution in the 21st Century*. Trans. M. MacLellan. London: Bloomsbury Academic.

Deleuze, Gilles. 2006. "Pōs Anagnorizoume ton strouctouralismo?" Essay. In *Ē Philosophia*, ed. François Châtelet, Trans. K. Papagiōrgēs. Athens: Gnōsē.

Deriu, Marco. 2015. Conviviality. In *Degrowth: A Vocabulary for a New Era*, ed. Giacomo D'Alisa, Federico Demaria, and Giorgos Kallis, 126–131. New York: Routledge.

Evans, Dylan. 2003. *An Introductory Dictionary of Lacanian Psychoanalysis*. Hove: Brunner–Routledge.

Foglieni, Francesca, Beatrice Villari, and Stefano Maffei. 2018. *Designing Better Services: A Strategic Approach from Design to Evaluation*. Cham: Springer.

Homer, Sean. 2005. *Jacques Lacan*. Abingdon: Routledge.

Kimbell, Lucy. 2011. Designing for Service as One Way of Designing Services. *International Journal of Design* 5 (2): 41–52.

Kimbell, Lucy, and Jeanette Blomberg. 2017. The Object of Service Design. In *Designing for Service, Key Issues and New Directions*, ed. Daniela Sangiorgi and Alison Prendiville, 81–94. London: Bloomsbury Academic.

Kyriakakēs, Giannis, and Martha Michaēlidou. 2006. *Hē prosengisē tou allou: Ideologia, methodologia kai ereunētikē praktikē*. Athens: Metaichmio.

Lacan, Jacques. 2001. *Écrits: A Selection*. Trans. Alan Sheridan, ed. Jacques-Alain Miller. New York: Norton, 1977.

Manzini, Ezio. 2011. Introduction. In *Design for Services*, ed. Anna Meroni and Daniela Sangiorgi, 1–6. Surrey: Gower Publishing.
Meroni, Anna, Selloni Daniela, and Martina Rossi. 2018. *Massive Codesign: A Proposal for a Collaborative Design Framework*. FrancoAngeli International.
Morelli, Nicola, Amalia de Götzen, and Luca Simeone. 2020. Core Service Design Capabilities. In *Springer Series in Design and Innovation*, 27–30.
Recalcati, Massimo. 2018. *Portéta tis epithymías*. Trans. Poniros Christos. Athens: Kelefthos.
Sangiorgi, Daniela, and Alison Prendiville. 2017. Introduction. In *Designing for Service: Key Issues and New Directions*. Bloomsbury: Academic.
Selloni, Daniela. 2017. *CoDesign for Public Interest Services*. Springer International Publishing.
Seravalli, Anna, and Mette Agger Eriksen. 2017. Beyond Collaborative Services: Service Design for Sharing and Collaboration as a Matter of Commons and Infrastructuring. In *Designing for Service: Key Issues and New Directions*, ed. D. Sangiorgi and A. Prendiville, 237–250. London: Bloomsbury Academic.
Sideris, Nikos. 2013. *Architecture and Psychoanalysis: Fantasy and Construction*. Athens: Futura.
Sparke, Penny. 2013. *An Introduction to Design and Culture: 1900 to the Present*. London: Routledge.
Stavrides, Stavros. 2018. *Koinós chóros: i póli os tópos koinón*. Athens: Angelus Novus.
Willis, Anne-Marie. 2006. Ontological Designing. *Design Philosophy Papers* 4 (2): 69–92.
Xianyi, Yang, and Gong Lingling. 2009. Design and Culture Innovation. In *IEEE 10th International Conference on Computer-Aided Industrial Design & Conceptual Design, Wenzhou, 2009*, 1696–1699. https://doi.org/10.1109/CAIDCD.2009.5375421.
Žižek, Slavoj. 2005. Design as an Ideological State Apparatus. In Designing Everyday Life, ed. Jan Boelen and Vera Sacchetti, 54–70. Zurich: Park Books.
———. 2011. The Architectural Parallax. In *The Political Unconscious of Architecture: Re-opening Jameson's Narrative*, 253–295. Farnham: Ashgate.

Chapter 13
"Matters of Care" in Spaces of Commoning: Designing In, Against and Beyond Capitalism

Katharina Moebus

Abstract This chapter explores the potentials and possibilities which a commons perspective offers to design practices by drawing on various threads of feminist and Marxist scholarship on commoning, care, and new materialism. As a basis for discussion, the first part of this chapter outlines an expanded definition of design as a transformative and research-based practice, which has emerged over the past years in response to ongoing ecological, economic, and societal upheavals. The second part examines the relationship between designing and commoning as an activity of care and creation of common space, followed by an engagement with, amongst others, María Puig de la Bellacasa's notion of "matters of care" (*Matters of Care: Speculative Ethics in More Than Human Worlds*. Minneapolis: University of Minnesota Press, 2017) as a speculative ethics to nourish new narratives which help reconfigure our relationships to the planet and all of its companions. These theoretical frameworks are then reflected through a "situated" practice research, the neighborhood project Common(s)Lab in Berlin-Neukölln, and three of its ongoing formats – a series of do-it-together (DIT) furniture building workshops, a seasonal gift market, and regular reading groups – in order to investigate the emancipatory potential of such spaces of commons as an "infrastructure for agency" (Petrescu. Being in Relation and Re-inventing the Commons. In *Feminist Futures of Spatial Practice: Materialisms, Activisms, Dialogues, Pedagogies, Projections*, ed. Meike Schalk, Thérèse Kristiansson and Ramia Mazé, 101–9. Baunach: AADR Art Architecture Design Research, 2017; Moebus and Harrison. Caring For the Common and Caring In *Common: Towards an Expanded Architecture/Design Practice*. 8th Biannual Nordic Design Research Society (Nordes) Conference, 2–4 June 2019, Aalto University, Helsinki, 2019). The chapter concludes with a critical reflection on design's own political economy to emancipate itself from coercions set by the market – and liberate its potentials for transformative commoning practices – using J.K. Gibson-Graham's diverse economies framework.

K. Moebus (✉)
School of Architecture, Sheffield University, Sheffield, UK
e-mail: katharina.moebus@gmail.com

G. Bruyns, S. Kousoulas (eds.), *Design Commons*, Design Research Foundations, https://doi.org/10.1007/978-3-030-95057-6_13

Keywords Common space · Commoning · Care · Feminist new materialism · Transformative design · Situated practice

> People are suffering. People are dying. Entire ecosystems are collapsing. We are at the beginning of a mass extinction and all you can talk about is money and fairytales of eternal economic growth. How dare you!
>
> Greta Thunberg, UN Climate Summit 2019

Neoliberal capitalism has taken over our language, social relations, material interactions and democracies. While it had already become evident in the 1970s, with recent reminders from a 16-year-old climate activist, that endless growth is not possible on a finite planet, the "capitalocentric" (Gibson-Graham 1996) fairytale of neoliberal capitalism as the only viable model of the economy still prevails in our minds and actions (Fisher 2009; Springer 2016).[1] The commons concept provides a promising counter-narrative. However, resource-focused approaches to the commons quickly run the risk of being appropriated by capital through a process of "enclosures 3.0" (Haiven 2016). As design and architecture researchers Kim Trogal, Valeria Graziano and Bianca Elzenbaumer point out: "while it is easy to see why the approaches that understand the common as 'commons' or 'common goods' would be especially relevant for design, other, more neglected approaches to this notion open up new possibilities," (2016, 2) such as feminist and Marxist approaches to the commons as an activity and ongoing process – Silvia Federici's commoning with a small "c." Here, the focus on material or immaterial goods is shifted towards the social relations framing and sustaining their (re)production, inseparable from a feminist ethics of care (Tronto and Fisher 1990; Habermann 2016). In line with autonomist thinker and commons author Massimo de Angelis, "the overcoming of capitalism is ultimately the overcoming of a mode of co-producing our livelihoods" (2007, 7), and problematizing what this mode is. In response to the more general question posed, "is design returning to new forms of foundational principles in terms of its material and social practices, and if so, through what means?" I am arguing for feminist commoning as a transformative framework for designing in, against, and beyond capitalism by investigating the former in relation to material, spatial and economic counter-practices in spaces of dissensus in the urban context. This will be reflected through my own collective design (research) practice embodied through a neighborhood project in Berlin-Neukölln.

[1] The Club of Rome commissioned a report on the limits of our world system and the constraints it puts on human numbers and activity from researchers at MIT. The report, called *The Limits to Growth*, published in 1972, became the first significant study to model the consequences of economic growth (Wikipedia contributors 2020).

1 Design as a Transformative Practice

In order to be able to discuss the relationship and potentialities of design and the commons, we first need to establish a common ground for

how design is understood. Taking inspiration from recent framings of design as transformative, politically conceived, and research-based "for the deconstruction and redesign of social relations" (Jonas 2007, 2016), with a critical view on the power relationships and agencies in and of design (Fuad-Luke et al. 2015; Manzini 2015; Keshavarz and Mazé 2013; Latour 2004, 2005), design acts as mediating infrastructure rather than problem-solver, where professional designers and "citizen designers" work together across disciplines and sectors in order to tackle complex and translocal issues. Here, infrastructure serves as a metaphor to describe "the creation of possibilities, in and through which stakeholders can create their own solutions": a process of "infrastructuring" "anything from tools and physical spaces to shared language (and) protocols," (Joost and Unteidig 2016, 140; Star and Griesemer 1989) which allows people to face complex and evolving problems themselves. Thus, rather than building objects, design serves primarily to build democratic design agency to "consciously intervene on the world" – which requires, according to Manzini, diffuse design skills and the "enabling ecosystem" of infrastructure as described above (2019, 38 and 121). This also means to break with prevailing hegemonies of our everyday lives and socio-cultural contexts, in order to "re-design our realities" by imagining, experimenting with, and enacting speculative futures which are based on the commons – not as islands of alternatives, but as embryonic prefigurations of other ways of being and becoming-with to make tangible what a different world could look and feel like (Moebus and Harrison 2019). As design theorist Wolfgang Jonas put it: "design is about what is NOT (yet), which expresses the main epistemological problem the discipline has to face" (2007, 11). Jonas advocates, among others, a transformative design practice that is reflective and grounded in theory: research through design (RTD), or research-based practice, as a particular mode of "designerly enquiry," where, following John Dewey, the reflection of facts and values takes place embedded in situations, rather than from the outside. Methodologically, this implies, according to Jonas, a move from "professional problem-solving expertise to participative projects" (2016, 127) – an emergent paradigm among the creative disciplines which still needs to be firmly established.

Following Pelle Ehn's elaboration of design as entangled cultural-material processes or "things," this chapter explores how such "things" serve as "socio-material frames for controversies, ready for unexpected use, opening up new ways of thinking and behaving" (Ehn 2008, 1). Designer and author Peter Friedrich Stephan similarly sees a need for designers to rediscover objects as "things" through the work of Bruno Latour (Stephan 2014, 118), who claims: "a thing is, in one sense, an object out there and, in another sense, an issue very much in there, at any rate, a gathering ... the same word thing designates matters of fact and matters of concern" (Latour 2005, 233). These matters of concern are, other than matters of fact, a "new conceptual tool to restage things as lively," highlighting "an engaged ethico-political responsiveness ... within the very life of things rather than through normative added values" (Puig de la Bellacasa 2017, 32). In this interpretation, agency and materiality of more-than-human worlds play an important role and go far beyond the idea of only humans being in charge of things. Latour calls for visualizations of these gatherings in order to make them visible – Stephan goes beyond this communicative design task to expand the idea of what design for the 'next society' could and should do: to imagine, disturb, and surprise, thus actively participating in a bottom-up process of collectively generating public discourses and their platforms around concerns, rather than doing that from above (Stephan 2014, 122).

María Puig della Bellacasa extends the notion of matters of concern to "matters of care" as an intervention in order to address a feminist history of unvalued and invisible doings and actors (2017, 52), as I will explore more later in this chapter. Her call for a speculative ethics that is inclusive and dissensual rather than exclusive and moralistic resonates with Jonas's and others' prompt for design "to become less moralistic and ideological … [and] maybe more theoretical, keeping ethics implicit in the theories and methodologies that we are using" (Jonas et al. 2016, 11). In a paper exploring the relationship between design and care, Moebus and Harrison contend care as a design methodology: "as an everyday practice and embedded methodology – or instituting practice – that is inseparable from commoning" (2019, 2). Joan Tronto's generic definition of care is helpful here, as "everything we do to maintain, continue, and repair 'our world' so that we can live in it as well as possible. That world includes our bodies, ourselves, and our environment, all that we seek to interweave in a complex, life sustaining web" (1990, 103) – of which design is a crucial part, actively contributing to what this web looks like. Thus, care can inform both design as research, as a knowledge-producing practice, and as a material practice of engaging-with and making-of our world – as a process of entanglement. This requires a practice which is engaged, situated, and embedded, which relates directly to feminist approaches to knowledge production (Haraway 1988) – moreover, a practice which is reflective and grounded in research, thus engaging in a never-ending loop of practice and theory (Fig. 13.1). How this could look in real life I will explore later by looking more closely at the activities taking place at the neighborhood project Common(s)Lab.

2 Design(ing) and Commoning – Towards New Value Practices

The interest in the commons has been around in academia and activism – in the creative disciplines, particularly in urban studies, architecture, and art, where a concern about space, socialities, and economic justice has a long history. Design has been left relatively untouched by the discourse around the commons, thus there still is ample space to explore the potential the theory and practices offer in relation to

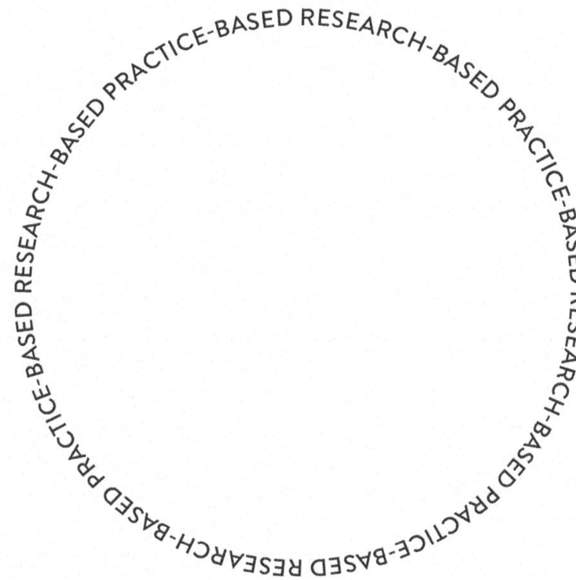

Fig. 13.1 Methodological loop of research-based practice. (Licensed under CC BA-SA by Katharina Moebus, 2020)

design(ing). This part of my text aims to contribute to filling this gap, following Peter Linebaugh's conceptualization of the commons as a process rather than a natural resource:

> to speak of the commons as if it were a natural resource is misleading at best and dangerous at worst – the commons is an activity and, if anything, it expresses relationships in society that are inseparable from relations to nature. It might be better to keep the word as a verb, an activity, rather than as a noun, a substantive (2008, 279).

In this understanding, the commons are always dependent on their own reproduction: on a practice of continuous care – of commoning – by a group of people – the commoners. As Silvia Federici points out:

> we cannot build an alternative society and a strong self-reproducing movement unless we redefine our reproduction in a more cooperative way and put an end to the separation between the personal and the political, and between political activism and the reproduction of everyday life (2010, 147).

Similar to Latour's actor-network theory, which describes relations as unstable connections that need constant reconfiguration, any kind of assemblage is a process that stops as soon as the relations cease to be remade (Latour 2005, 35). Linebaugh and Federici, among many others, belong to a tradition of more radical thought regarding the commons, which regards the process of commoning as an ongoing emancipatory struggle against the continuous enclosure and commodification of our material, immaterial, and social world. A feminist perspective to the commons acknowledges the fact that "capitalist social relations are deeply intertwined with patriarchal ones, thus every struggle from the perspective of politics of commons has to be a feminist one" (Sielert 2013, 4). Thus, as Federici demands in the above quote, we need to put at center stage the question of how we reproduce ourselves in common in order to overcome capitalism. In *Social Reproduction Theory*, editor Tithi Bhattacharya introduces social reproduction as all "the activities and attitudes, behaviors and emotions, and responsibilities and relationships directly involved in maintaining life, on a daily basis and intergenerationally (Brenner and Laslett 1991, 314, in Bhattacharya 2017, 6) – which includes access to food, shelter, and clothing, and also raising children, cleaning, friendships, education, culture, health care, civic engagement, and so on. Feminist-Marxist scholars like Federici criticize the fact that capitalism constantly preys on everything held in common, as it is based on a continuous process of primitive accumulation – the expropriation of our own means of reproduction. Unlike Marx, she and others see this as an ongoing process of enclosure rather than the precondition for capitalism to emerge (Midnight Notes Collective 1990, 1), which is not simply a top-down process, but sometimes even driven by "dispossessed commoners searching for means to survive and thrive in the absence of forms of collective care and support" (Haiven 2016, 279). This means that the latter is what needs to be reconfigured, in order to allow the commons to thrive.

According to De Angelis, the commons are composed of three interdependent elements which constitute a social system: the common good, which can be both material and immaterial; the process of commoning; and the commoner communities taking part in this process (2017, 18). These three elements offer, as Italian design duo Brave New Alps described in 2014, potential entry points for an emancipatory design practice: on the one hand, in order to uncouple the latter from coercions set by the market, and on the other hand, because commons offer exactly those elements whereby other value systems and economies can be actively co-designed – with the good life for all always in sight. Designers could, according to them, intervene experimentally and supportively in these areas in a locally-specific, but translocal manner to strengthen the re-designing of other economic cultures: for example by turning private property common, by fostering heterogeneous communities, and by nourishing practices of commoning in a strategic way (Brave New Alps 2014, 161) – possibilities of intervention are endless. However, they point out the fact that designers also need to emancipate themselves and their own imagination after 250 years of living under capitalism. Thus, bringing designing and commoning to a fruitful coalition also means to unlearn and reshape our own subjectivities and values as practitioners struggling to survive in the market, in order to actively shape our economy differently. As eco-feminist scholars Maria Mies and Veronika Bennholdt-Thomsen write:

> in our view, we cannot simply say 'no commons without community.' We must also say 'no commons without economy,' in the sense of *oikonomia*, i.e. the reproduction of human beings within the social and natural household. Hence, reinventing the commons is linked to the reinvention of the communal and a commons-based economy. (2001, sect. 15)

And hence, bringing commons together with design always entails a rethinking of how we reproduce ourselves, as I will explore further in the concluding part.

Another important aspect is the spatial dimension of commoning, and the physical infrastructures supporting such processes. Architect, design theorist and educator Jesko Fezer sketches a design which could possibly act as a tool to create "common space" as well as room for alternatives in the "post-neoliberal city," "a kind of proto-design, producing fewer solutions (and new problems), but also social situations and processes enabling social imagination, debate, and conflict" (Fezer 2010, 5). Architect and activist Stavros Stavrides talks about space as "not only a product and therefore a stake for commoning but a means of establishing and expanding commoning practices" (2016, 4), and furthermore, describes "space-commoning" as not the simple sharing of space as a resource, but rather, as "a set of practices and inventive imaginaries which explore the emancipating potentialities of sharing" (2016, 7). This emancipatory potential is inscribed in the practice of commoning, as it shapes, according to Stavrides, both the subjects and their means (2016, 35). He points out that commoning should not be reduced to processes of sustaining livelihoods, but rather as a value practice based on the "common" – singular – which already motivates the many political movements, practices, struggles, and discourses committed to building a non-capitalist future (Dardot and Laval 2019, 6; Hardt and Negri 2009). Thus, in line with Federici and others, commoning is not only a means to produce commons, but rather, a way of fostering and sustaining relationships, of doing and becoming different. Consequently, he sees both community and space as processes always in-the-making, "porous" for newcomers across difference, in order to resist enclosure. These assumptions suggest that the commons are not (only) a utopian horizon, but already existent in the here and now – in commons scholar Max Haiven's words: as actuality and spirit (2016, 280).

What does that mean for designing the infrastructures, protocols, and spaces of becoming-in-common though? Perhaps, similar to the emancipatory and relational processes of commoning, these outcomes, too, become more of a verb or process, subject to constant calibration and negotiation, in order to make sure that the values of commoning are being translated into practice (Moebus and Harrison 2019, 3). As such, again, the means also become the end: setting up and sustaining "infrastructures for agency" becomes a collective act of co-creation, a performance and prefiguration of the values that are enacted in the space of commoning itself – as an "outside" that exists "inside" the dominant relations and structures it seeks to challenge (Moebus and Harrison 2019, 1; Van de Sande 2017).

3 Matters of Care as Relational Entanglements

In her book *Matters of Care,* María Puig de la Bellacasa points out a fundamental difference between concern and care:

> As affective states, concern and care are related. But care has stronger affective and ethical connotations. We can think on the difference between affirming 'I am concerned' and 'I care'. The first denotes worry and thoughtfulness about an issue as well as, though not necessarily, the fact of belonging to the collective of those concerned, 'affected' by it; the

> second adds a strong sense of attachment and commitment to something ... One can make oneself concerned, but 'to care' contains a notion of doing that concern lacks (2017, 42).

She continues in a feminist tone, charging care ethically and politically as a practice with a history of being "at the forefront of feminist concern with devalued agencies and exclusions," joining together "an affective state, a material vital doing, and an ethico-political obligation" (2017, 42). Articulating matters of concern as matters of care is meant as an intervention with three dimensions: (1) to problematize the neglect of caring relationalities in an assemblage; (2) to generate care for undervalued and neglected issues; and (3) to generate an attitude of care generally (2017, 56). As Puig de la Bellacasa writes, caring involves a notion of direct action, of becoming actively involved, which the notion of concern does not necessarily generate. Architecture curators Elke Krasny and Angelika Fitz even speak of "critical care," with reference to the medical term to designate the diagnosis and treatment of life-threatening conditions, in order to stress the urgency of the devastating situation of our planet (Krasny and Fitz 2019, 10), furthermore quoting Naomi Klein to explain what is causing its destruction:

> our economy is at war with many forms of life on earth, including human life. What the climate needs to avoid collapse is a contraction in humanity's use of resources; what our economic model demands to avoid collapse is unfettered expansion. Only one of these sets of rules can be changed, and it is not the laws of nature (Klein 2014, in Krasny and Fitz 2019, 11).

This requires a more positive and inclusive narrative, where the focus shifts from economic growth at the expense of those many forms of life as resources towards a positive acknowledgement of the interdependency of all human and other-than-human beings on earth: "interdependency is not a contract, nor a moral ideal – it is a condition. Care is therefore concomitant to the continuation of life for many living beings in more than human entanglements" (Puig de la Bellacasa 2017, 70, in Krasny and Fitz 2019, 13). This interdependence should be, according to Fitz and Krasny, a starting point for architecture and urbanism – as it should be for design and other form-giving and transformative disciplines. Opposed to the positive notion of *independence* promoted by the capitalist paradigm – every individual needs to be self-reliant in order to be a 'free' and 'self-made' person – *interdependence* acknowledges the more-than-human entanglements and relational dependencies that constitute our world. In fact, as economic geographers Ethan Miller and J.K. Gibson-Graham write, the conventional distinction between economy and ecology "severs us from transformative, ethically-infused encounters with our constitutive interdependencies", and promote, instead, a notion of "ecological livelihoods" to bring back together the bifurcation of our *oikos*, or habitat, into two separate domains: one of capitalist market activity and one of nonhuman resources (Miller and Gibson-Graham 2019, 2). They put forth three dimensions of livelihood as a general term and practice indicating the work of sustaining ourselves: being-made by others (*allopoiesis*), making others (*alterpoiesis*), and finally, making ourselves (*autopoiesis*) (Miller and Gibson-Graham 2019, 11), as a counter-narrative to the hegemony of the self-contained individual. According to them, making an "honest living" would not mean to conclude or reduce our dependencies on others, but rather, to take active responsibility for interdependence, for what it means ethically to be reliant on both human and non-human others, and for our partaking in self-making and making others (or not). Commoning is putting these principles into practice by opening "a space of collective negotiation" around matters of care, as a form of engagement generating more caring kinships. As Puig de la Bellacasa writes, "it is not so much a notion that explains the construction of things than it addresses how we participate in their possible becomings. Caring here is a speculative affective mode that encourages intervention in what things could be" (2017, 66). Speculating and experimenting around matters of care can thus act as prefigurative openings (for design) to make tangible values based on the common – and by telling stories otherwise, a favorite method of feminist scholar Donna Haraway. She introduces the age of the "Chthulucene" – not as a posthumanist intervention, but to propose an ontological shift away from the human-centered Anthropocene or Capitalocene. The Chthulucene, instead, describes the interconnectedness between all life forms on the planet in order to reconfigure our multi-species relations: "becoming with, not becoming, is the name of the game … Ontologically heterogeneous partners become who and what they are in relational material-semiotic worlding. Natures, cultures, subjects, and objects do not preexist their intertwined worldings" (Haraway 2016, 13). Like Miller and Gibson-Graham, she advocates a move away from "autopoiesis" towards "sym-poiesis," making-with rather than self-making, and making kin with all "critters" or "multispecies companions" rather than focusing on human relations only, in order to develop a way of thinking which is adaptive to the damaged state of our world and helpful for building more desirable futures. Political theorist Jane Bennett adds another dimension to the multispecies entanglements – like Latour, she addresses the agentic capacities of things in her theory of distributed agency, in order to stretch our thinking beyond the so-called "life-matter binary." She puts the intentionality behind human agency to question by pointing out the agency of assemblages, where, rather than a single subject causing an effect, "there are always a swarm of vitalities at play" (Bennett 2009, 32). Experiencing and interacting with this material vitality in a more horizontal manner might enable us, as Bennett notes, to become more ecologically sensible and to acknowledge the fact that matter is just as much a powerful actant as we might be. It shifts the emphasis from humans being in decisive power towards a collective, "an ecology of human and nonhuman elements" (Bennett 2009, 103; Latour 2005), which I will explore in more detail through concrete examples in the following section.

4 Becoming with Wood, Words, and Garments

At the end of 2017, I co-initiated the neighborhood project Common(s)Lab, nested within the >top project space and community, together with architect Melissa Harrison in Berlin-Neukölln, in order to establish a space where practices of commoning could be explored through creative means and new ways of designing together.[2] My involvement is many-fold as a neighbor living nearby, as a design practitioner, community organizer and design researcher embedded in the research context. Over the course of the past two-and-a-half years, a diverse range of event formats emerged in collaboration with multiple actors, three of which I will analyze in relation to designing and commoning: firstly, a series of do-it-together (DIT)-furniture workshops; secondly, a seasonal gift market; and thirdly, regular academic reading groups about the commons.

The wood workshops took place in collaboration with carpenter and designer Veiko Liis, who is also part of the >top community. Two workshops took place during outdoor festivals, working mostly with children; one ran over a whole weekend to build a collective structure for the neighborhood, and four others took place at our space.[3] For all of the workshops, we had two basic principles: to use only simple hand tools that everybody would be able to use quickly in a safe manner, and to employ only reclaimed wood from the streets of Neukölln, supplemented by what people could bring from home or their basements (Figs. 13.2, 13.3, and 13.4). The limited choice of tools made the available joints limited but also simpler, thus increasing access to building furniture at home without the need for a professional workshop. Consequently, we mostly used screw connections, where the only tools needed were hand drills and screwdrivers, which were co-financed through donations or borrowed from participants and neighbors.[4] The fact that the material was salvaged from the streets fostered a different kind of material-semiotic interaction and relationality, performing its non-human agency: firstly, it took away fears of working with expensive virgin materials, allowing a greater degree of experimentation, imprecision and joy, and secondly, the material always already embodied a rich assemblage of histories, relationships, and uses in itself: whom it might have belonged to, how it ended up on the streets as "trash," how it was retrieved and given a new life elsewhere – not to forget its initial production and distribution processes. Saving the wood from its destiny as junk constituted an act of repair and maintenance, of caring for the unwanted, the broken, the imperfect. Upcycling "trash" also brings a new kind of aesthetics to life, with agentic powers to tell a different story which is empowering in itself – one of replacing consumerism and a *throw-away* mentality with one of re-valuing and re-purposing, becoming less dependent on the market. As Graziano and Trogal describe it, "collective repair can ... be seen as a site

[2] More info at: www.top-ev.de and www.commonslab.de

[3] One of them was part of the Neukölln Nachhaltigkeitsfest in September 2018 on Alfred-Scholz-Platz, organized by and in collaboration with the non-profit organization genug e.V.; the other was part of the Nette Ecke programme, financed by the Quartiersmanagement Flughafenstraße and commissioned by the art collective Artistania e.V. and took place in April 2019.

[4] Similarly, Enzo Mari used nails as the connection method for his open-source furniture to make it as accessible as possible, requiring only a hammer and nails rather than high-end equipment. We experimented once with glue connections, which proved difficult because this method would require a large amount of additional equipment such as clamps, so we stuck to the screw connections.

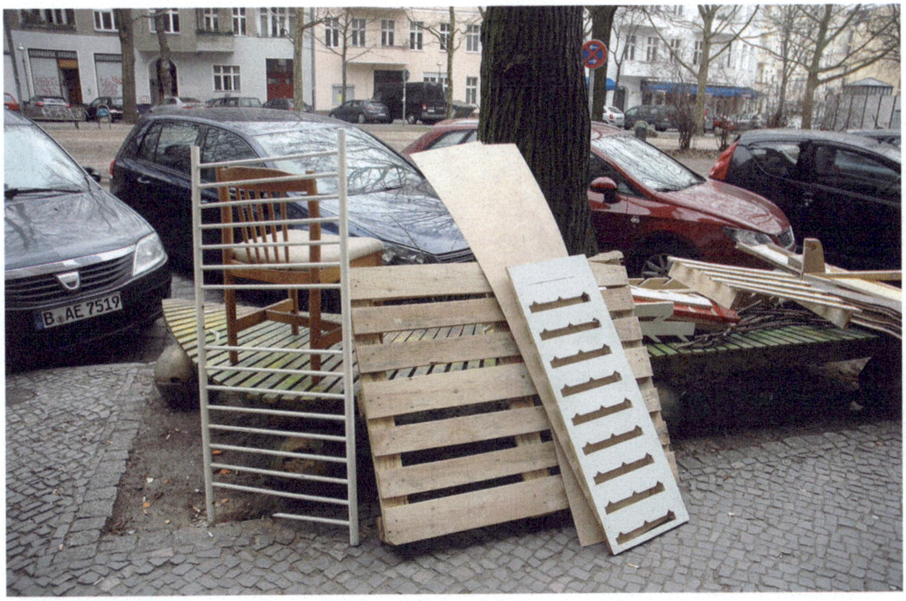

Fig. 13.2 Salvaged wood from the surrounding streets of Berlin-Neukölln. (Licensed under CC BA-SA by Common(s)Lab, 2018)

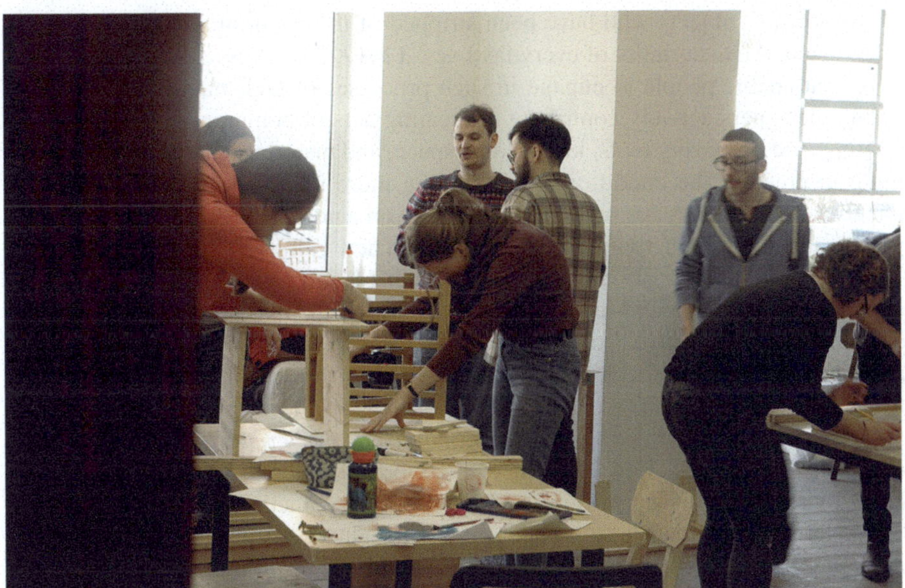

Fig. 13.3 Participants constructing furniture during one of the workshops. (Licensed under CC BA-SA by Common(s)Lab, 2018)

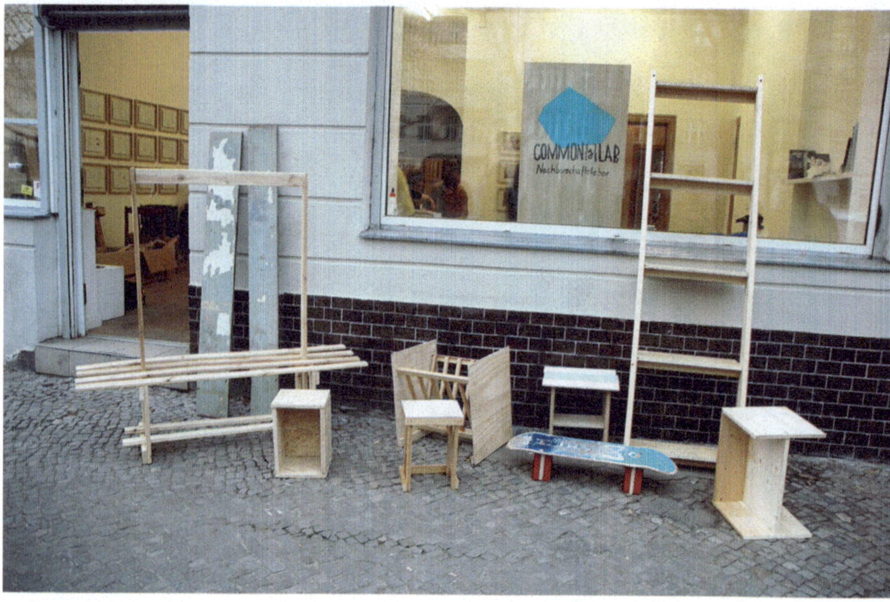

Fig. 13.4 The results from one of the workshops. (Licensed under CC BA-SA by Common(s) Lab, 2018)

for re-skilling subjects who have been stripped of the opportunity to learn how to intervene in the materiality of everyday lives" (2017, 24). Thus, by setting up situations that enable people to engage in such processes of DIT-making and material caring, designers actively contribute to creating acts of commoning – of learning, relating, and of sharing tools, knowledge, space, and materials; and also, of contributing to creating the conditions of becoming more autonomous. While the DIY phenomenon of the 1950s predominantly drew on a nostalgia for manual labor as a private and domestic activity, the shift towards a new culture of DIT or DIO (Do-It-With-Others) (Gelber 1997; Ratto and Boler 2014, in Graziano and Trogal 2017) adds additional value and layers of meaning to the manual process, whereby the sociality of the collective processes becomes paramount – not to be misunderstood as a reduction to the social qualities only, but always with an inherent political motivation.

The book *Die Welt Reparieren: Open Source und Selbermachen als postkapitalistische Praxis* (2016; "Repairing the world: Open Source and DIY as postcapitalist praxis") describes these emerging practices as value practices that build upon the "ruins left by capitalism." "It is the used and devalued that instigates imaginativeness, in order to invent new forms of doing and using, resulting in a new aesthetics comprising the past," two of the co-editors, Karin Werner and Christa Müller, assert in an interview (Moebus and AoA 2020, 45, my translation). Furthermore, they point out the following with reference to new forms of civic engagement: "people want to free themselves of all the things that are impossible to change, in order to experiment with and sustain relationships and forms that are deemed good. Making becomes more prevalent: the concrete, tangible, and experiential become central in the individual and collective act" (Moebus and AoA 2020, 48).

The individual and collective experience of something very concrete and tangible is also the focus during our gift markets. This format is based on a recurring phenomenon in many places in the world: free boxes or gift boxes (often cardboard boxes, sometimes real physical structures) containing things people do not need any more and leave on the streets for others to take away for free. Taking inspiration from Marx's 1875 principle "From each according to her/his ability, to each according to her/his needs" in his *Critique of the Gotha Program* (1875, Part I), we invite people to bring their unwanted clothes and take along whatever they like – without the use of money, barter, or any other logic of exchange, in order to create a space that would stir thought and discussion (Fig. 13.5). Because the format is familiar yet different, bearing many similarities to a flea market, the absence of money or rules of exchange is deliberately confusing, in order to create an atmosphere of generosity and hospitality, generating an entirely new experience of "shopping." People need to establish their own rules when they are interested in the same piece of clothing, negotiating whom it might fit better or who might need it more. Thus, it is not money that regulates access as usual, but instead, the process of negotiation. This creates a different kind of relational interaction between the object and people around it, whereas money usually terminates a relationship when all debts are paid (Graeber 2011; Habermann 2016). According to anthropologist David Graeber, and

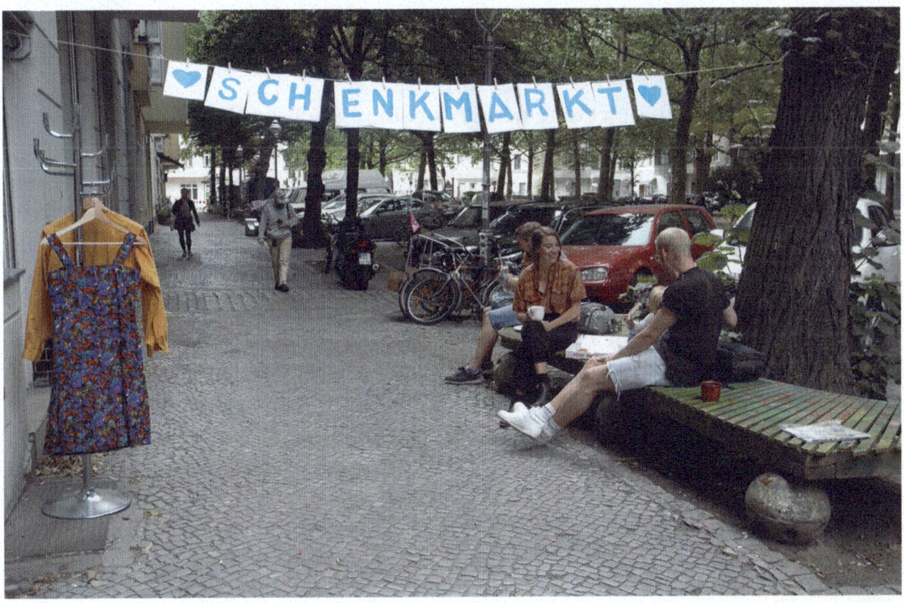

Fig. 13.5 Street view at the beginning of one of the gift markets. (Licensed under CC BA-SA by Common(s)Lab, 2018)

against the commonly accepted myth, barter only evolved later with the introduction of money, not the other way around. Marcel Mauss stated in his seminal book *The Gift* (1925) that earlier societies were mostly based on reciprocity, on "gifting" to each other in order to establish relationships between groups, not necessarily individuals. As artist Jan van Esch, who has worked with the principle of the gift in the realm of charity (Van Esch explores and points towards the hierarchies and power relationships inherent in the gift by employing donated clothes sent to Tanzania in his art) claims "gifts that never allow a return" create a feeling of debt for the receiver that is difficult to be resolved; he continues: "a present given always expects one in return".[5] In relation to swap parties and similar phenomena, this fact made us move away from direct exchange as a principle regulating access towards one based on need and ability to contribute. Possibly, the name "gift market" sometimes created confusion (some visitors would take along as much as they could carry without even trying anything or wanted to take along the plants from the space and toys from the kids' corner!) – but most people were pleasantly surprised by the basic principles of trust, generosity, and openness.

The format attracted a diverse audience: many neighbors and passers-by, but also teenagers and young adults with a critical attitude towards mass production and consumption in times of "Friday for Futures" protest marches. One of the younger participants, who later got involved in organizing a market, indeed said:

> I really like how diverse the audience is that the gift markets attract, as it is accessible to such a wide range of people from different social backgrounds. I am sure there are many people who might first think the basic principle that everything is free doesn't work out in practice, but then they can see that it actually does! Also, the clothes are much more diverse and fun than the stuff which you can get at the stores, and I like the idea that somebody else will continue wearing the things which I brought. (Interviewed by Katharina Moebus in August 2020, Berlin)

Design can consciously intervene here by creating a setting that facilitates discussion and supports the emergence of new protocols and socialities for sharing, and by providing the infrastructure for these commoning practices to take place. Regular formats like the gift market can foster new habits and relationships with our everyday needs such as access to clothing and help us re-examine what we deem valuable in the process.

The reading groups deal with an immaterial commons: knowledge. By sharing and discussing texts in an informal setting, the reading group format creates a counter-proposal to the established institutionalized access to knowledge that starts and ends within a predetermined period of time in our lives – in many countries outside of Europe only available to the privileged through very high fees (as is the case in for example the UK and the US), turning universities, academic staff and students into entities driven by capital and debt.[6] Instead, the reading groups aim to create a transversal space for growing knowledges and translocal communities of

[5] More info on his work can be found here: https://www.janvanesch.com/About

[6] Student debt in the US currently makes up over $1.6 trillion, collectively held by 44 million Americans (Hess 2020).

interest, which come together based on a shared concern and a desire for continued learning beyond institutions. While the first reading groups were organized mainly by us, soon participants would take over the curation, facilitation and organizing. Thus, participating in the readings both encouraged and enabled people to initiate events themselves – eventually, commoning is a process of becoming and learning how to organize collectively.

Most reading groups took place in English rather than German, since one non-German speaker would suffice to switch to English. This, of course, influenced the composition of the reading groups – mostly expats, temporary or long-term migrants would join, or German-speakers and locals who would feel comfortable having an academic discussion in English. To overcome this issue, we tried to make the readings as welcoming and unacademic as possible, and established certain protocols; for example no name-dropping without explaining who that person is; writing down questions and terms people would not be familiar with to learn together using the collective knowledge of the group, if necessary, supported by information found on the internet; taking turns in reading out loud and discussing together, so that people didn't have to come prepared and would each have a chance to speak; and choosing future texts together in a democratic process of decision-making (Fig. 13.6). One of the participants said in an interview in 2020:

> At Common(s)Lab, everyone has an equal chance to read, to comment... and when you have to say something, then everybody listens to you. I think this equal way of talking to each other in a group setting and really listening to each member is something I really learned from the reading groups. (Interviewed by Katharina Moebus in August 2020, Berlin)

To conclude, wood, words and garments, acting as things not objects, invite us to become part of an emancipatory process of becoming-with: making furniture from reclaimed wood, sharing and negotiating about (un)wanted clothes, and engaging with academic texts regarding the commons, are more than simple gestures of everyday life – they are processes of commoning around matters of care, where the ends indeed become the means. They are acts that instigate diverse communities of practice, place, and interest, which are sometimes temporary, sometimes stable, but always open and always evolving, in opposition to the nostalgic idea of what once constituted a community. Rather, we can speak of translocal "communities of urban practice" where social relations take various shapes (Blokland 2017, 11), and collectively performing things differently becomes decisive to the creation of fluid commoner communities: of struggle, of dissensus, and of urban emancipation (Butler 2015; Stavrides 2019).

5 Conclusions

In this chapter, we have looked at design from a transformative perspective, how it may work together with feminist commoning as methodology and value practice, and how an attitude of care towards the current state of our world may help to design

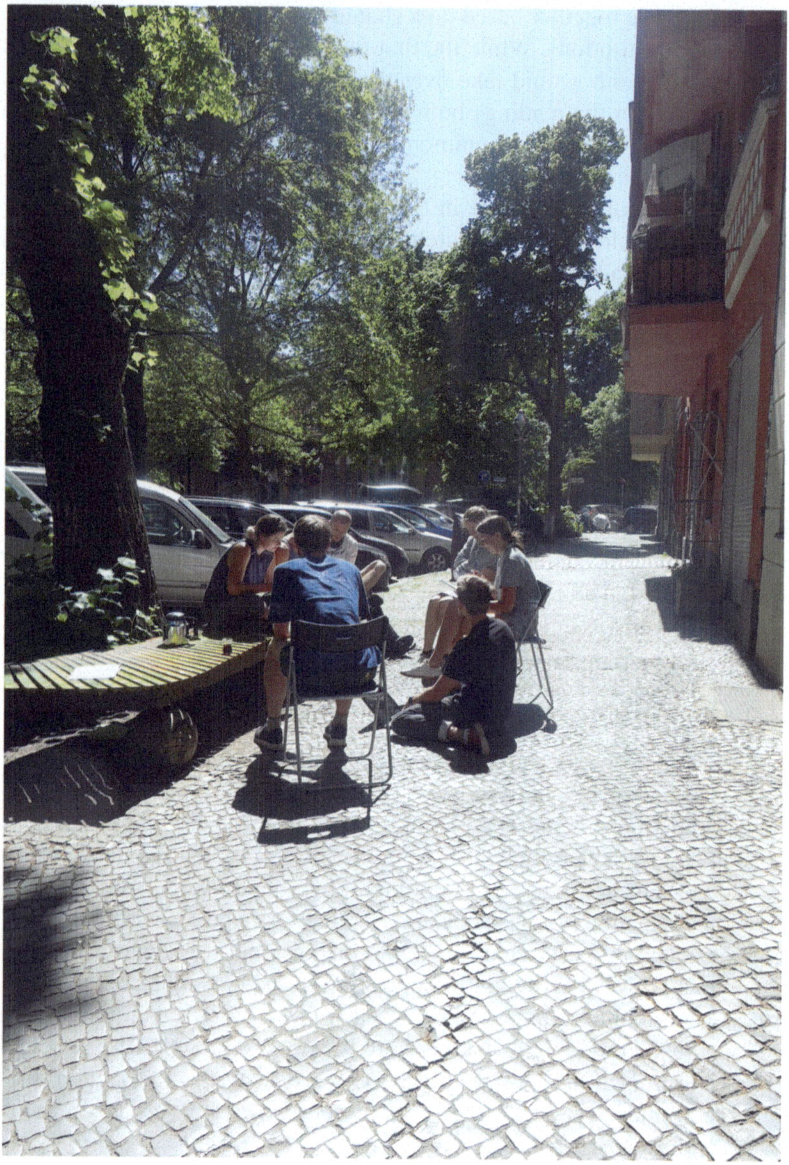

Fig. 13.6 Reading group outside of our project space. (Licensed under CC BA-SA by Common(s) Lab, 2018)

"in, against, and beyond capitalism." The latter slogan originates from John Holloway's 2016 lectures bearing the same title and acknowledges three facts: that we are currently still living under capitalism and have to deal with the conditions it sets; that there are plenty of struggles and opportunities to oppose capitalism in the present; and, last but not least, that prefigurative action of life beyond capitalism can already enact possible futures in the here and now, and help move us to a postcapitalist era (Chatterton and Pusey 2020). As I have argued, design has the potential to contribute to the redesign of how we reproduce ourselves. Following J.K. Gibson-Graham's and others's call to "take back the economy" and start recognizing ourselves as significant actors and shapers of the economy (2013), we all have the potential to change the narrative towards ecological livelihoods based on interdependent connections of care with our human and more-than-human worlds. What does this mean in practice though? We may take inspiration from Gibson-Graham's iceberg of the diverse economy, which exemplifies and makes visible the already existent diversity of economic practices that sustain our human and more-than-human worlds. This is a helpful first step to overcome the market/non-market binary which currently gives value (or not) to all the (re)productive activities taking place in our habitats (Fig. 13.7). Practices of commoning means bringing reproductive

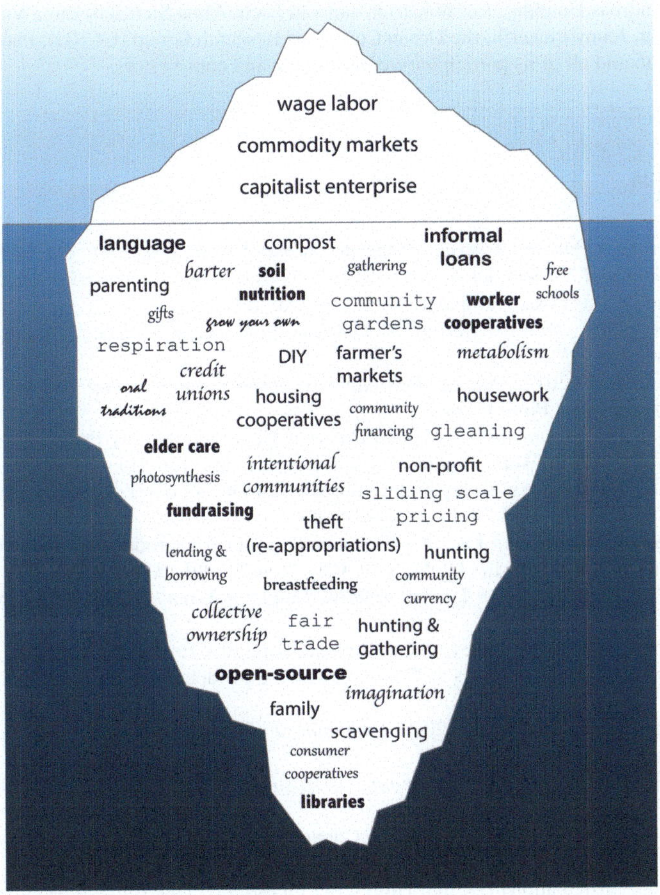

Fig. 13.7 Diverse economies iceberg by Community Economies Collective. (Licensed under CC BA-SA)

activities to the public, thus making visible and giving value to the already existent variety of economic practices that are often taken for granted but are vital for keeping our societies alive. Practices such as life-long learning and educating ourselves in a mutual and horizontal manner, making things ourselves, working with materials from the ruins of capitalism, and establishing new protocols about the access to daily material necessities such as clothing might seem trivial at first sight – but the way we pursue these 'trivial' activities is, in the end, a politically-charged question of how we reproduce ourselves. Designing in, against, and beyond capitalism is always both means and end, and a feminist commoning which puts exactly the latter question at the forefront of this struggle might help us to really become a strong movement.

Acknowledgements This chapter is based on collective research, discussions, and endeavors. I am thankful to the many inspiring discussions and collaborations that contributed to this text, particularly with: Melissa Harrison, Bianca Elzenbaumer, Fabio Franz, Flora Mammana, Alastair Fuad-Luke, Andreas Unteidig, Lucas Kuster, Jan van Esch, Alma Siemsen, Liina Viil, Veiko Liis, Doina Petrescu, Jenny Pickerill, the Urban Commons Research Group (UCRG), Peter Breuer and Common(s)Lab and all of its participants, contributors, and conspirators.

References

Baier, Andrea, Tom Hansing, Christa Müller and Karin Werner, eds. 2016. *Die Welt reparieren: Open Source und Selbermachen als postkapitalistische Praxis*. (Repairing the World: Open Source and DIY as Postcapitalist Praxis.) Bielefeld: transcript.

Bennett, Jane. 2009. *Vibrant Matter: A Political Ecology of Things*. Durham, NC: Duke University Press.

Bhattacharya, Tithi. 2017. *Social Reproduction Theory: Remapping Class, Recentering Oppression*. London: Pluto Press.

Blokland, Talja. 2017. *Community as Urban Practice*. Cambridge: Polity Press.

Brave New Alps. 2014. Werte-Praktiken für DesignerInnen von Heute und Morgen. (Value Practices for Designers of Today and Tomorrow). In *Design der Zukunft*, ed. Cornelia Lund and Holger Lund, 152–164. Stuttgart: avedition.

Brenner, Johanna, and Barbara Laslett. 1991. Gender, Social Reproduction, and Women's Self-Organization: Considering the US Welfare State. *Gender and Society* 5: 314.

Butler, Judith. 2015. *Notes Toward a Performative Theory of Assembly*. Cambridge, MA: Harvard University Press.

Chatterton, Paul, and Andre Pusey. 2020. Beyond Capitalist Enclosure, Commodification and Alienation: Postcapitalist Praxis as Commons, Social Production and Useful Doing. *Progress in Human Geography* 44: 27–48.

Dardot, Pierre, and Christian Laval. 2019. *Common: On Revolution in the 21st Century*. London: Bloomsbury.

De Angelis, Massimo. 2007. *The Beginning of History: Value Struggles and Global Capital*. London: Pluto Press.

———. 2017. *Omnia Sunt Communia: On the Commons and the Transformation to Postcapitalism*. London: Zed Books.

Ehn, Pelle. 2008. Participation in Design Things. In *Proceedings of the 10th Anniversary Conference on Participatory Design, October 2008; Bloomington*, 92–101. New York: ACM Press.

Federici, Silvia. 2010. Feminism and the Politics of the Common in an Era of Primitive Accumulation. In *Revolution at Point Zero: Housework, Reproduction, and Feminist Struggle*, ed. Silvia Federici, 138–148. Oakland: PM Press, 2012.

Fezer, Jesko. 2010. Design for a Post-Neoliberal City. eflux journal. https://www.e-flux.com/journal/17/67367/design-for-apost-neoliberal-city/. Accessed 26 Feb 2020.

Fisher, Mark. 2009. *Capitalist Realism: Is There Really No Alternative?* Hampshire: Zero Books.

Fuad-Luke, Alastair, Anja-Lisa Hirscher, and Katharina Moebus, eds. 2015. *Agents of Alternatives: Re-designing Our Realities*. Berlin: AoA.

Gelber, Steven M. 1997. Do-It-Yourself: Constructing, Repairing and Maintaining Domestic Masculinity. *American Quarterly* 49: 66–112.

Gibson-Graham, J.K. 1996. *The End of Capitalism As We Knew It: A feminist Critique of Political Economy*. Minneapolis: University of Minnesota Press.

Gibson-Graham, J.K., Jenny Cameron, and Stephen Healy. 2013. *Take Back the Economy: An Ethical Guide for Transforming our Communities*. Minneapolis: University of Minnesota Press.

Graeber, David. 2011. *Debt: The First 5 000 Years*. Brooklyn: Melville House.

Graziano, Valeria, and Kim Trogal. 2017. The Politics of Collective Repair: Examining Object-Relations in a Postwork Society. *Cultural Studies* 31: 634–658.

Habermann, Friederike. 2016. *Ecommony: UmCARE zum Miteinander.* (Ecommony: Turn to Togetherness). Sulzbach/Taunus: Ulrike Helmer Verlag.

Haiven, Max. 2016. The Commons Against Neoliberalism, The Commons of Neoliberalism, the Commons Beyond Neoliberalism. In *The Handbook of Neoliberalism*, ed. Simon Springer, Kean Birch, and Julie MacLeavy, 271–283. London: Routledge.

Haraway, Donna. 1988. Situated Knowledges: The Science Question in Feminism and the Privilege of Partial Perspective. *Feminist Studies* 14: 575–599.

———. 2016. *Staying with the Trouble: Making Kin in the Chthulucene*. Durham, NC: Duke University Press.

Hardt, Michael, and Antonio Negri. 2009. *Commonwealth*. London: Belknap Press.

Hess, Abigail. 2020. How Student Debt Became a $1.6 Trillion Crisis. CNBC Make it. https://www.cnbc.com/2020/06/12/how-student-debt-became-a-1point6-trillion-crisis.html. Accessed 18 Aug 2020.

Holloway, John. 2016. *In, Against, and Beyond Capitalism: The San Francisco Lectures*. Oakland: PM Press.

Jonas, Wolfgang. 2007. Design Research and Its Meaning to the Methodological Development of the Discipline. In *Design Research Now*, ed. Ralf Michel, 187–206. Basel: Birkhäuser Verlag.

———. 2016. Social Transformation Design as a Form of Research Through Design (RTD): Some Historical, Theoretical, and Methodological Remarks. In *Transformation Design: Perspectives on a New Design Attitude*, ed. Wolfgang Jonas, Sarah Zerwas, and Kristof von Anselm, 114–133. Basel: Birkhäuser Verlag.

Jonas, Wolfgang, Sarah Zerwas, and Kristof von Anselm. 2016. Introduction. In *Transformation Design: Perspectives on a New Design Attitude*, ed. Wolfgang Jonas, Sarah Zerwas, and Kristof von Anselm, 9–22. Basel: Birkhäuser Verlag.

Joost, Gesche, and Andreas Unteidig. 2016. Design and Social Change: The Changing Environment of a Discipline in Flux. In *Transformation Design: Perspectives on a New Design Attitude*, ed. Wolfgang Jonas, Sarah Zerwas, and Kristof von Anselm, 134–148. Basel: Birkhäuser Verlag.

Keshavarz, Mahmoud, and Ramia Mazé. 2013. Design and Dissensus: Framing and Staging Participation in Design Research. *Design Philosophy Papers* 11: 7–30.

Krasny, Elke, and Angelika Fitz, eds. 2019. *Critical Care: Architecture and Urbanism for a Broken Planet*. Vienna: Architekturzentrum Wien and MIT Press.

Latour, Bruno. 2004. Why Has Critique Run Out Of Steam? From Matters of Fact to Matters of Concern. *Critical Inquiry – Special issue on the Future of Critique* 30: 225–248.

———. 2005. *Reassembling the Social: An Introduction to Actor-Network-Theory*. Oxford: University Press.

Linebaugh, Peter. 2008. *The Magna Carta Manifesto: Liberties and Commons for All*. Berkeley: University of California Press.
Manzini, Ezio. 2015. *Design, When Everybody Designs: An Introduction to Design for Social Innovation*. Cambridge, MA: MIT Press.
———. 2019. *Politics of the Everyday: Designing in Dark Times*. London: Bloomsbury.
Marx, Karl. 1875. Critique of the Gotha Programme. *Marxists Internet Archive*. https://www.marxists.org/archive/marx/works/1875/gotha/index.htm. Accessed 7 Oct 2020.
Midnight Notes Collective. 1990. The New Enclosures. *Midnight Notes* 10.
Mies, Maria, and Veronika Bennholdt-Thomsen. 2001. Defending, Reclaiming and Reinventing the Commons. *Revue canadienne d'études du développement* (Canadian Journal of Development Studies) 22:997–1023.
Miller, Ethan, and J.K. Gibson-Graham. 2019. Thinking with Interdependence: From Economy/Environment to Ecological Livelihoods. In *Thinking in the World Reader*, ed. Jill Bennett and Mari Zournazi. London: Bloomsbury Academic.
Moebus, Katharina, and Melissa Harrison. 2019. Caring For the Common and Caring In Common: Towards an Expanded Architecture/Design Practice. 8th Biannual Nordic Design Research Society (Nordes) Conference, 2–4 June 2019, Aalto University, Helsinki.
Moebus, Katharina, and AoA, eds. 2020. *Economies of Commoning | Ökonomien des Gemeinschaffens: Interview Reader #1*. Berlin: AoA.
Petrescu, Doina. 2017. Being in Relation and Re-Inventing the Commons. In *Feminist Futures of Spatial Practice: Materialisms, Activisms, Dialogues, Pedagogies, Projections*, ed. Meike Schalk, Thérèse Kristiansson, and Ramia Mazé, 101–109. Baunach: AADR Art Architecture Design Research.
Puig de la Bellacasa, María. 2017. *Matters of Care: Speculative Ethics in More Than Human Worlds*. Minneapolis: University of Minnesota Press.
Ratto, Matt, and Megan Boler. 2014. *DIY Citizenship: Critical Making and Social Media*. Cambridge, MA and London: MIT Press.
Sielert, Deborah. 2013. Commons That Care: Feminist Interventions in the Construction of the Commons. (Un)usual Business Reader 2013. https://unusualbusiness.nl/en/theory/commons-care/index.html. Accessed 20 Aug 2020.
Springer, Simon. 2016. *The Discourse of Neoliberalism: An Anatomy of a Powerful Idea*. London: Rowman and Littlefield.
Star, Susan Leigh, and James R. Griesemer. 1989. Institutional Ecology, Translations, and Boundary Objects: Amateurs and Professionals in Berkeley's Museum of Vertebrate Zoology, 1907–39. *Social Studies of Science* 19: 387–420.
Stavrides, Stavros. 2016. *Common Space: The City as Commons*. London: Zed Books.
———. 2019. *Common Spaces of Urban Emancipation*. Manchester: University Press.
Stephan, Peter Friedrich. 2014. Matters of Concern: Designaufgaben der Next Society (Matters of concern: design tasks of the next society). *Revue – Magazine for the Next Society* 16: 118–123.
Trogal, Kim, Valeria Graziano, and Bianca Elzenbaumer. 2016. The Politics of Commoning and Designing. In *Proceedings of DRS2016: Design + Research + Society – Future-Focused Thinking*, ed. Peter Lloyd and Erik Bohemia, vol. 10, 4005–4015.
Tronto, Joan, and Berenice Fisher. 1990. Toward a Feminist Theory of Caring. In *Circles of Care: Work and Identity in Women's Lives*, ed. Emily K. Abel and Margaret K. Nelson, 36–54. Albany: State University of New York Press.
Van de Sande, Mathijs. 2017. The Prefigurative Power of the Common(s). In *Perspectives on Commoning: Autonomist Principles and Practices*, ed. Guido Ruivencamp and Andy Hilton, 25–64. London: Zed Books.
Wikipedia Contributors. 2020, Nov 25. Degrowth. In *Wikipedia, The Free Encyclopedia*. https://en.wikipedia.org/wiki/Degrowth. Accessed 26 Feb 2020.

Chapter 14
Design as Commoning: Drawing Together with Care

Lőrinc Vass, Roy Cloutier, and Nicole Sylvia

Abstract This chapter places the evolving discourse of the commons in architecture and urbanism in relation to broader cosmopolitical questions concerning the agency and response-ability of design. It contends that more than a socially held or produced resource, the commons must be reconceived in a twofold manner: from a discrete locus (the commons) to a process (commoning, or the *politics of connection*), and in turn, from concerning primarily human decisions to explicitly involving a more-than-human ensemble. In this shift to a systems-relational approach, common*ing* becomes a cipher for rethinking our relation to *relation*. The first section of the chapter addresses the polemics of commons discourse and its reflections in architecture and urbanism. It follows several historical threads leading to the more recent re-conceptualization of commoning as the politicization of livelihood relations. The second part situates these trajectories within the feminist material-semiotics of *care*, positing it as a lens to refract current discussions of commoning in design. Incorporating contemporary examples, the chapter concludes with two postulations for drawing as a material-semiotic practice of commoning, building from the Latourian provocation of "drawing together." *Drawing-together those who have nothing in common* involves representing the manifold entities assembled in, and affected by, design. And *drawing, together, common-enough worlds* entangles these heterogeneous agencies in new constellations beyond received roles and hierarchies, recasting design as a collective production. Both examples begin a renegotiation of the ways in which the designer's place in webs of life and matter *matter*, and how we might act *care*-fully with(in) them.

L. Vass (✉)
Department of Architecture, Tokyo Institute of Technology, Tokyo, Japan
e-mail: lorinc.vass@gmail.com

R. Cloutier · N. Sylvia
School of Architecture + Landscape Architecture, University of British Columbia, Vancouver, Canada

G. Bruyns, S. Kousoulas (eds.), *Design Commons*, Design Research Foundations, https://doi.org/10.1007/978-3-030-95057-6_14

Keywords Commons · Commoning · Entanglement · Care · Collective production · Design

> What world is it that you are assembling, with which people do you align yourselves, with what entities are you proposing to live?
> Bruno Latour, (2015) *What is Cosmopolitical Design?*

1 The Polemics of (Design) Commons

"The commons" has begun to reappear in design discourse of late, following a meandering journey across a broad spectrum of theoretical inquiry – from political economy and legal studies to radical political thought and feminist theory. This propagation is evinced by the incorporation of the term – and its next of kin, "collectivity" and "sharing" – in the title of numerous conferences, exhibitions and publications in architecture and urbanism over the past decade.[1] Across these forums, the figure of the commons has been deployed with a broad spectrum of intent and precision: from vague calls for finding a "common ground" in architecture to defining this "common" via typology; from managing "ecology and technology commons" in the posthuman city to a critical apprehension of "commoning" as a relational spatial practice; and from the espousing of "participatory design" to the more somber (and rare) accounts of its limits and exclusions. Indeed, a common definition may be neither possible, given the manifold historical and disciplinary trajectories of the term, nor desirable, insofar as what is needed is a conception "that recognises that there isn't even a common concept of the common" (Amin and Howell 2016, 11).

Underlying these polemics are a number of recurring assumptions in the broader literature on commons, such as its relegation to a resource waiting to be "unlocked" or "activated" by commoners (and designers); ahistorical undercurrents that reflexively valorize commons as virtuous and inclusive by conjuring up fantasies about "reclaiming"; and anthropocentric, objectifying presuppositions that neglect the agency of other-than-human entities involved. This chapter contends that more than a socially held or produced resource, the commons needs to be conceived – following in the footsteps of Haraway, Latour, Stengers and others – to encompass the *politics of connection* across the multiplicity of entities, human or otherwise, that

---

[1]An incomplete list includes the Venice Biennale of Architecture "Common Ground" (2012) and the upcoming "How Will We Live Together?" (2021); the conferences "Designing the Urban Commons" in London (2015) and "Constructing the Commons" at TU Delft (2016); the ongoing travelling exhibition "An Atlas of Commoning: Places of Collective Production" (2018–); two instalments of the Seoul Biennale of Architecture and Urbanism, "Imminent Commons" (2017) and "Collective City" (2019); and the Tbilisi Architecture Biennial "What Do We Have In Common" (2020).

comprise “the social.” Common*ing* thus becomes a cipher for rethinking our relation to *relation*. In other words, understanding commoning as the politicization of livelihood relations offers an opportunity to consider the broader cosmopolitical questions about the agency and response-ability of design(ers) in a more-than-human world. This chapter examines these questions through the most fundamental disciplinary activity of design: *drawing* – understood in an expansive sense as a material-semiotic practice, a means of representing and composing our matters of concern and care.

The chapter comprises two parts. The first section revisits the well-known polemics of the commons and the ways in which its multiple discursive trajectories have been reflected in architecture and urbanism. It seeks to reorient the discussion from the complimentary notions of *management* of common-pool resources and the *reclaiming* of commons from enclosure, via their critique and prolongation in the discourse on *urban commons*, and ultimately, towards a cosmopolitical conceptualization of commoning as the politicization of livelihood relations. The second section situates these trajectories within the feminist material-semiotics of *care*, positing it as a guiding ethos for design in an entangled, more-than-human world. Discussions of care build from Latour and actor-network theorists and the notion of drawing together *matters of concern,* which implies representing – aesthetically, scientifically and politically – the heterogeneous and conflicting entities that are assembled in the process of design. Yet, as María Puig de la Bellacasa and other feminist thinkers argue, also underlying this undertaking are *matters of care* – the empathy required for the making of open-ended, convivial relations with one another and the things with which we share a world. This lens of care is used to refract the Latourian provocation of “drawing together,” in order to explore, through contemporary examples, two interrelated postulations of drawing as a material-semiotic practice of commoning. *Drawing-together those who have nothing in common* encompasses the representation of the manifold entities assembled in, and affected by, design. And *drawing, together, common-enough worlds* involves partaking in the composition of heterogeneous agencies in new constellations beyond received roles and hierarchies, thus recasting the practice of design as a collective production. Both examples begin a renegotiation of the ways in which the designer’s place in webs of life and matter *matters*, and how we might act *care*-fully with(in) them.

2 (Un)common Trajectories

The past decades have seen a robust theorization of the commons across a broad disciplinary spectrum, resulting in a multiplicity of overlapping, and at times contradictory, trajectories. These include “environmental commons” via political economy; “legal commons” via critical legal studies; “(neo-)Marxist commons” via radical political thought and social movements; “urban commons” via critical urban studies; and the more recent turn to “commoning” (Kirwan et al. 2016). While these approaches are marked by significant ideological and methodological divergences, some scholars draw attention to the prevalence of defining commons through variations of the following conceptual triad: “(a) *common* resources, (b) institutions (i.e. *commoning* practices) and (c) the communities (called *commoners*) who are involved in the production and reproduction of commons” (Dellenbaugh et al. 2015, 13). This tripartite structure has also been widely adopted in recent architectural and urbanism discourse (viz. Ngo et al. 2018; Kimmel et al. 2018; Petrescu et al. 2020), with a frequent emphasis on the participatory nature of commoning, and an alternative to both state- and market-led space-making:

> Commoning is the act of sharing and managing resources – cultural and natural – with minimal reliance on the market or state, and where each stakeholder has an equal interest. User-managed governance of the environments we inhabit – from land ownership, to buildings, to domestic spaces – enables residents to be key agents in how resources are distributed, valued, and maintained. (Bhatia 2019, 95)

Notwithstanding the aspirations to render the production of space more equitable and empowering, this conceptual triangulation raises a host of questions concerning the predominantly human-centered perspective behind *resource management*; the potential limits and exclusions underlying *participation* and notions of *community*; the degree of autonomy of commoning practices from the capitalist *state* and *market*; the constitutive and performative role of *spatiality* in this process; and perhaps most importantly, concerning the specific agencies and responsibilities of design(ers) situated amidst more-than-human entanglements. The following section unpacks these questions through three conceptual constellations: the contrasting trajectories of managing and reclaiming commons, the problematization of the (re)production of urban commons, and the cosmopolitical understanding of commoning as the politicization of more-than-human livelihood relations.

2.1 Managing and Reclaiming the Commons

Two narratives have been particularly omnipresent since the mid-1980s: one centered on the collective management of environmental resources, the other on the reclaiming of commons from capitalist enclosure. The notion of environmental commons, implied in Garret Hardin’s (1968) oft-quoted “tragic” narrative on the inevitable depletion of shared resources, has directly shaped the so-called (neo-) institutionalist perspective. Central to this approach is the “common-pool resource” (CPR), a subtractable and nonexcludable reserve whose sustainable utilization necessitates regulatory institutions. Unlike Hardin, who argued that such regulation was only possible through market privatization or state intervention, CPR researchers – most famously the political economist Elinor Ostrom (1990) – have demonstrated the long-term management of local resources according to a third model: collective governance. Based on a large number of empirical studies on forests, fisheries, irrigation systems and other local resources, Ostrom deduced a list of design principles for the successful collective management of CPRs, most notably the necessity of clearly defined commons boundaries, both physically and in terms of membership (Huron 2018, 23).

The institutionalist perspective has been critiqued both for its predominant focus on "natural" resources assumed to be asocial, and its lack of historical positioning in relation to capitalism. On one hand, "thinking about commons as spatial demarcations of natural resources … positions them as outside and prior to social relations, so that the question of their (mis)management and maintenance comes to the fore" (Kirwan et al. 2016, 7) – a move that reinforces long-held problematic assumptions concerning the division of nature and culture, and the subordination of the former as an exploitable reserve. On the other hand, the tendency of the institutionalist approach to focus on the management, but not the formation or reclamation of commons, evinces a lack of theoretical distinction between the commons and capitalist relations of production (Huron 2018, 26). While the CPR perspective seeks an alternative to the (neo)liberal narrative of self-interested economic subjects and the ideology of private property, it "begins with the same 'naturalized' assumption as do the 'tragedyists': without proper rules and norms, individuals will degrade and ultimately destroy common resources" (Bresnihan 2016, 93).

Refractions of the neo-institutionalist approach can be found in architecture and urbanism in various guises. It informs the view that commons are resources to be *managed*, *unlocked*, *activated*, *constructed* or *designed* by architects, with the city and its architecture constituting such a resource (viz. Avermaete 2018). On a more fundamental level, a managerial mentality risks turning the (re)production of space into a technocratic, post-political undertaking – exemplified by the Trojan Horse-like collectivity rhetoric of the "smart city" and "sharing economy" movements of late, as well as by the omnipresent practice of quantifying and "optimizing" the heterogeneity of the lived-in environment in a digital "managerial surface" (May 2012). These tendencies serve as strong reminders that the institutions, rules and protocols connected to commons often reproduce existing power relations, and that transformative spatial practices necessitate "open institutions of expanding commoning" (Stavrides 2016, 39).

The 1980s also saw the emergence of an explicitly anticapitalist, neo-Marxist approach to commons. Under this "alterglobalizationist" perspective, "the commons" is conceived as an essential entity of subsistence, collectively managed and embedded in social relations (Huron 2018, 28). Contrary to the institutionalist focus on management, the alterglobalizationist approach focuses on the *reclamation* and defense of commons against the enclosures of commodification and privatization that deprive populations from their means of subsistence in order to create exploitable wage labor and surplus wealth – a process termed "primitive accumulation" by Marx, and "accumulation by dispossession" by David Harvey (2003). Thus, while the institutionalist and alterglobalizationist approaches concern the same phenomena, the latter focuses on the "dynamic relationship between commons and enclosure [that] points to the social nature of commons … as an activity, not as a 'resource'" (Huron 2018, 31).

Accordingly, Hardt and Negri (2009) distinguish between "the commons," an inert notion that refers to natural resources, and the dynamic "common" which involves "both the product of labor and the means of future production. This common is not only the earth we share but also the languages we create, the social practices we establish, the modes of sociality that define our relationships" (p. 139). While contextualized as a critique of late capitalism and its "new forms of exploitation of biopolitical labor" (p. 139) that thrive on the creative capacity of the common, this capacity also stretches to *before* and *beyond* commodifying enclosures, a "collective production that is not necessarily but is often currently tied to capitalism, is not dependent on a logic of scarcity" (Kirwan et al. 2016, 13). Hardt and Negri's emphasis on new forms of cognitive and "immaterial" labor, however, has been critiqued for obscuring "the question of the reproduction of everyday life ... the material requirements for the construction of a commons-based economy" (Federici 2019, 107).

A further point of criticism, levelled at the alterglobalizationist approach in general, concerns the tendency to romanticize the commons as an inclusive, open-for-all realm, harboring the promise of "some sort of mythical precapitalist society" (Huron 2018, 32). As urban historian Leif Jerram opines, the commons frequently appears in "a *good* mood … something from the past which has the character of a solution to something in the present, rather than a problem; and the solution that it offers is almost invariably desirably consensual, peaceful and socially just" (2015, 48, emphasis in original). This mood also frequently characterizes contemporary architecture and urbanism discourse on commons, positioning it alongside other purportedly anti-capitalist and bottom-up space-making practices such as the various urbanisms of the "everyday," "participatory" or "tactical" variety. Yet others remind us that such initiatives are all-too-often idealized as inclusive, authentic, and free of internal boundaries and contestation (viz. Krivý and Kaminer 2013).

2.2 The Contradictions of Urban Commons

Over the last decade, "urban commons" has emerged as a discrete analytical category, and a lens for revisiting the contradictions pertaining to the commons at large. Much of this scholarship also embraces a shift from an "objectified notion of commons" (Borch and Kornberger 2015, 5) to *commoning* – a term popularized by historian Peter Linebaugh (2008) to imply a durational activity rather than a static entity. Key to the urban commons discourse is the Lefebvrian conceptualization of "the urban" as not a predominantly territorial condition, but a multi-scalar and mediated form of social organization characterized by a series of contradictions between openness and enclosure (Kip 2015, 44).

Following social theorist Amanda Huron (2015), the urban way of life, first of all, is marked by the co-existence of "strangers in saturated space": a densely clustered and heterogeneous population which gives rise to both potentials for conflict and opportunities to learn to live together in diversity, including concerning access to and exclusion from commoning. Contrary to the scarcity-centered discourse on environmental commons, urban commons are often *produced through consumption*: "far from being a resource waiting for the appropriator to deploy it, the city constitutes its subjects … particular urban commons are not simply out there, waiting to be exploited; rather they must first be produced and then constantly reproduced" (Borch and Kornberger 2015, 12). Moreover, given the heterogeneity and contestation that underlies urbanity, the notion of a static commons tied to a clearly defined community also becomes problematic. In this sense, the community involved in (urban) commoning might be better recast "as an obligation (*communis*) rather than a shared identity or interest (*com-unus*)" (Amin and Howell 2016, 11).

Secondly, the urban as a site of capital accumulation – less a particular place than "a concentrated process of surplus and exchange" constituted by social relations (Huron 2018, 49), with most inhabitants engaged in non-subsistence activities – precipitates a tension between commoning and capitalism. As noted above, in the Marxist tradition the commons constitute a counterpoint to the enclosure of the market (Kirwan et al. 2016, 22), whereas it would be more accurate to regard commoning as "a practice that exists in a dialectical relationship with capitalism, often but not always in explicit opposition to it" (Huron 2018, 40). On one hand, it is essential to recognize the plurality of actually existing economic modalities – everyday practices of non-capitalist life that constitute forms of experimentation with other possible worlds, articulated for instance in J. K. Gibson-Graham's (2008) notion of "diverse economies." Conversely, the metaphor of enclosure no longer circumscribes all the possible ways the commons can be made productive and profitable – one only has to think of the ongoing proliferation of "sharing" platforms and other forms of "digital commonism" (Pasquinelli 2008, 208).

Thirdly, urbanity is also a "concentrated process of statecraft" (Huron 2018, 49), a site of regulation and surveillance of the population and space itself by the state – contributing to a further point of tension between the commons and the notion of "the public." Whereas the public, and its corollary, the "individual," are the products of the modern nation-state, the "common" is seen as predating these terms and representing a form of social relation oblique to the public/private divide. It is in this sense that Hardt and Negri, following Spinoza, mobilize the emancipatory potential of the "multitude," a heterogeneous collective of immanent difference; with the common seen "as the production of the multitude, the actualisation of its practices" (Jalón Oyarzun 2016, 51). Commoning practices are thus seen as forms of "differentiated publicness" that challenge existing socio-spatial frameworks and "facilitate the emergence of differentiated forms of social and political subjectivity" (Sohn et al. 2015, 2). The question remains, however, whether commoning risks absolving the retreating welfare state of its responsibilities; and conversely, whether it could exist not in opposition to, but supported by the state (Huron 2018, 60).

Huron calls for adopting a feminist approach that allows one to work through these contradictions by shedding light on the labors required to both create and maintain (urban) commons, including the historically overlooked activity of social reproduction; on the networks of care involved in this process; and on questions of differentiated subjectivity and freedom in urbanity. By dissolving the conceptual binary between reclaiming and management that underlie the alterglobalizationist and institutionalist perspectives, a feminist lens allows for equal consideration to everyday processes and their relationship with larger social and power structures

(Huron 2018, 62). Tracing the link between the enclosure of physical resources and the conceptual separation of reproductive work from economic production, feminist Marxist scholar Silvia Federici (2019) argues that

> if commoning has any meaning, it must be the production of ourselves as a common subject. This is how we must understand the slogan 'no commons without community.' But 'community' has to be intended not as a gated reality, a grouping of people joined by exclusive interests separating them from others, as with communities formed on the basis of religion or ethnicity, but rather as a quality of relations, a principle of cooperation, and of responsibility to each other and to the earth, the forests, the seas, the animals (p. 110).

The performative, subject-producing nature of urban commoning has been most elaborately examined in its spatial dimensions by architect and activist Stavros Stavrides (2014), who remarks that "urban commoning neither simply 'happens' in urban space, nor does it simply produce urban space as a commodity to be distributed. Urban commoning treats and establishes urban space as a medium through which institutions of commoning take shape" (p. 83). Elsewhere, Stavrides (2016) also notes that

> space is an active form of social relations, a constituent aspect of social relations and a set of relations itself … [and] common space cannot be fixed in the form of a product (no matter how collectively it was produced) because it keeps on producing those who produce it. The production and uses of common space cannot be separated (p. 260).

2.3 *Commoning as Politicizing Livelihood Relations*

Recognizing the interrelation of socio-spatial production and reproduction, and the call for responsibility towards not only fellow humans but entanglement at large, sets the stage for a largely unexplored perspective: commoning as a more-than-human activity (viz. Metzger 2015). The widely used conceptual triad of commons, commoners and commoning maintains an implicit anthropocentric bias, being driven and dominated by human agency. The starting point for a more-than-human perspective, for environmental scholar Patrick Bresnihan, "is not an individual subject separated from other people and the world around them, but a *relational subject* who is always already caught up in a world that is intimately shared" (2016, 99, emphasis added). This view is echoed by activist-scholar Ethan Miller, who theorizes the political ecology of livelihood as radically relational, eclipsing the conceptual triad of economy, society and environment. Miller (2019) defines commoning as the "ethico-political explicitation of habitat," in other words, the *politicization* of the relations of sustenance that unfold "as the composition of an innumerable set of variable 'spheres' of concern and negotiation that articulate livelihood triads together in explicitly common habitat assemblages" (p. 189). "Habitat" comprises and organizes the relay of *self-making* (autopoiesis), *being-made by others* (allopoiesis) and *making others* (alterpoiesis) – a relational ensemble that also resonates with Donna Haraway's (2016) notion of *sympoiesis,* making-with, as a radically non-anthropocentric form of production. Habitat is thus not an "environment" in the passive sense of a resource, but a dynamic realm both composed by and composing – or, following Haraway, *composting* – relations. Commoning, then,

> refers to the myriad ways in which livelihood relations are rendered into shared spaces of mutual exposure and negotiation through which living singularities actively respond to the ethical demands posed by specific instances of an ontological in-common. This is to say that commoning constitutes shared 'matters of concern' (Latour 2005), or, better yet, 'matters of care' (Puig de la Bellacasa 2017), where 'matter' should be taken in both senses of the word at once. It is not that all things shared are commoned, but that all shared matters are commoned to the extent that they appear as active questions, concerns, or sites of struggle. (Miller 2019, 189)

Miller makes an important distinction between livelihood and commoning, terms that are often equated in commons scholarship and positioned in opposition to capitalist accumulation (viz. De Angelis 2017). For Miller (2019), the opposite of commoning is not *enclosure*, but *uncommoning*, or depoliticization: "if commoning is a making-explicit of the negotiations of the common (habitat), then uncommoning is an *anesthetization of the common*, its ethico-political closure, a rendering-nonnegotiable of habitat relations" (p. 190, emphasis in original). A more-than-human perspective on commoning, then, not only challenges capitalist exploitation, but also unsettles and eclipses the question of human political participation and agency, to become that of *cosmopolitics*. As succinctly articulated by Isabelle Stengers (2005):

> As for the cosmopolitical perspective, its question is twofold. How to design the political scene in a way that actively protects it from the fiction that 'humans of good will decide in the name of the general interest'? … But also how to design it in such a way that collective thinking has to proceed 'in the presence of' those who would otherwise be likely to be disqualified as having idiotically nothing to propose, hindering the emergent 'common account'? (p. 1002).

The task, then, is to recognize that "whatever 'common ground' there is, it cannot be fixed or settled, but is always-emerging" (Amin and Howell 2016, 11–12), and "to challenge all articulations that appear to determine beforehand what the shape of common habitat is and might become, and to open new spaces for shared recognition and negotiation of our actual, complex livelihood relations" (Miller 2019, 192).

3 Drawing Together Matters of Care

As the material-semiotic politics of composing, and being-composed by, a more-than-human habitat, commoning has profound relevance for design practice. As Bruno Latour (2004a) writes, "the only way to compose a common world … consists precisely in not dividing up at the outset and without due process what is common and what is private, what is objective and what is subjective" (p. 93). It is not sufficient to follow "matters of fact"; rather, *things* need to be pursued in their polemical and lively entanglements, as the "matters of concern" that gather *because* they divide us (Latour 2005, 23). Design plays a key role in this composition:

> The typically modernist divide between materiality on the one hand and design on the other is slowly being dissolved away. The more objects are turned into things – that is, the more matters of facts [sic] are turned into matters of concern – the more they are rendered into objects of design through and through ... artefacts are becoming conceivable as complex assemblies of contradictory issues. (Latour 2008, 2–4)

The question of design, then, can be posed by prolonging the etymological meaning of "drawing": "How can we draw together matters of concern so as to offer to political disputes an overview ... of the difficulties that will entangle us every time we must modify the practical details of our material existence?" (Latour 2008, 12). And further, what are new modes of drawing that allow us to "adequately represent the conflicting natures of all the things that are to be designed?" (p. 13). Central to this undertaking is *representation*, in all three meanings of the term: (1) political, "*Who* is to be concerned?" (2) scientific, "*What* is to be considered?" and (3) aesthetic, "How to represent, and through which medium, the sites where people meet to discuss their matters of concern?" (Latour 2005, 16, emphasis in original).

These questions can be further developed by situating them in the context of the feminist material semiotics of *care*. Feminist philosopher of science María Puig de la Bellacasa (2017) advances the notion of "matters of care" as both a critique and a prolongation of Latour's "matters of concern," arguing that care adds a "critical edge" to the politics of gathering through an imperative to *both* assemble neglected matters *and* generate caring relationalities (p. 66). Indeed, care already figures in Latour's (2004b) original formulation: "Can we devise another powerful descriptive tool that deals this time with matters of concern and whose import then will no longer be to debunk but to protect and to care, as Donna Haraway would put it?" (p. 232). Puig de la Bellacasa conceptualizes *care* as a key ontological foundation for acting in/on the entangled relationships that comprise the world. Extending earlier notions that centered on care work and issues of gender, she adopts the definition articulated by political scientist Joan C. Tronto and feminist educator Berenice Fisher (1990):

> On the most general level, we suggest that caring be viewed as a species activity that includes everything we do to maintain, continue, and repair our 'world' so that we can live in it as well as possible. That world includes our bodies, our selves, and our environment, all of which we seek to interweave in a complex, life-sustaining web (p. 40).

This broader definition allows for the recognition that a politics of care amounts to more than a moral stance and encompasses practical, ethical, and affective dimensions: "care stands for necessary yet mostly dismissed labors of everyday maintenance of life, an ethico-political commitment to neglected things, and the affective remaking of relationships with our objects" (Puig de la Bellacasa 2017, 66). "Matters of care" is thus

> a proposition to think with: rather than indicating a method to 'unveil' what matters of fact are, it suggests that we engage with them so that they generate more caring relationalities.

> It is thus not so much a notion that explains the construction of things than it addresses how we participate in their possible becomings (p. 66).

As noted in a recent volume on urban commoning, "there is a globally emergent culture of commoning based on an ethos of care … based on our shared existential vulnerability" (Özkan and Büyüksaraç 2020, 17). Drawing on Tronto and Puig de la Bellacasa's theorization, cultural theorists Angelika Fitz and Elke Krasny (2019) reflect on the question of care in design, remarking that not only are architecture and urbanism interwoven in the life-sustaining web that comprise our environment,

> they are very much part of weaving this web. In our view, architecture and urbanism are central to caring for the habitat, its inhabitation and continued livability. With habitat we refer to all possible scales of inhabitation, from the living room to the region, from the schoolyard to the city, from the refugee camp to the planet. An ethics of care in architecture and urbanism is based on local-planetary interconnectedness (p. 13).

However, as Tronto (2019) notes, "using care as a critical concept will require a fundamental reorientation of the disciplines of architecture and urban planning" (p. 26). A "caring architecture" would involve a change of focus from buildings as objects in space towards relations within habitats, among more than human inhabitants, and across time. Tronto outlines five aspects of caring practice: *caring about*, an attentiveness to the needs that have to be addressed; *caring for*, accepting and allocating responsibility for addressing a caring need; *care giving*, attention to the actual acts of giving care; *care receiving*, the evaluation of the caregiving process, and the recognition of further needs that may arise; and *caring with*, the development of trust and solidarity upon reliable acts of care over time (pp. 30–31). Each of these aspects have manifold refractions in design theory and practice, from the diverse and often conflicting needs of the more-than-human constituents of the built environment, to the labors involved in the processes of planning, construction and maintenance.

Whereas the influence of actor-network theory, including the call for pursuing matters of concern, has precipitated the so-called ethnographic turn in architecture, thereby expanding the range of actors considered and affirming the socially situated nature of design, matters of care prompt further questions about responsibly acting with(in) more-than-human entanglements. Considering this predicament in the context of urban commons planning, Jonathan Metzger (2016) identifies a triple task: to recognize and *care for* other-than-human needs; to *cosmopoliticize* by raising questions about what is relevant and what is excluded; and to *assume responsibility* for the "necessary exclusions and Otherings" involved in planning (p. 145). Underscoring that commoning unavoidably involves the definition, and patrolling, of boundaries (material, as well as ontological), Metzger urges

> to not only recognize that boundaries sometimes must be drawn, and to assume responsibility for how we decide to draw them, but also to recognize that we must always be conscious of the effects of these practices, and prepared to constantly challenge and reconsider them (p. 145, emphasis added).

A key task for design, then, involves increased attentiveness to drawing as a political act, and reflection on who/what to *care about* in the first place and how to *care-with – become-with, draw-with* – others, including other-than-humans. “Drawing” in this context must be understood in an expansive sense, as a material-semiotic practice involving both matter and meaning, enaction and figuration, actualization and speculation – in short, as a means to represent and (re)compose our world. Reorienting the question of commoning towards the key disciplinary tool and activity of design, then, is not a last resort move to recover its “lost” autonomy, but a proposition to partake in the politics of livelihood relations through drawing.

Seen in this light, the Latourian provocation of “drawing together” comprises a double-entendre for design as commoning. On one hand, it is a matter of *assembly*: drawing-together, in the sense of representing the manifold entities who comprise our habitat. On the other hand, it is a program for *action*: drawing, together with others, to compose new linkages across the “relay of livelihoods” (Miller 2019). These propositions constitute two sides of the same coin: tracing existing entanglements and mapping new relationships. And underlying both are questions concerning the care and response-ability required to “maintain, continue, and repair our ‘world’ so that we can live in it as well as possible” (Tronto and Fisher 1990, 40).

3.1 Drawing-Together Those Who Have Nothing in Common

The first task involves gathering a heterogeneous assemblage across the human/non-human divide: *drawing-together those who have nothing in common.* The phrase is borrowed from philosopher Alphonso Lingis (1994), who distinguishes between the “rational community” produced by the logic of reason and a common language that casts individuals as members of a universal collective; and the “other community” that emerges during an encounter with the other, the “intruder,” who demands an ethical response:

> The other community forms when one recognizes, in the face of the other, an imperative. An imperative that not only contests the common discourse and community from which he or she is excluded, but everything one has or sets out to build in common with him or her (pp. 10–11).

A similar ethos informs Stengers’s (2005) proposal for a cosmopolitical thinking that proceeds *in the presence of* those others who may have “idiotically nothing to propose, hindering the emergent ‘common account’” (p. 1002). Navigating this predicament demands an “ecology of practices”: the acknowledgment of a multiplicity of knowledge-practices and their situatedness in distinct “habitats.” As Stengers (2013) notes, “there is no identity of a practice independent of its environment … the very way we define, or address, a practice is part of the surroundings which produces its ethos” (p. 187). An ecology of practices is thus a tool for thinking, underlied by the recognition that “nobody can speak in the name of [a challenging] situation. Indeed borders are involved and there is no neutral, extra-territorial way of defining what matters in the situation” (p. 193).

Commoning – that is, cosmopoliticizing – a heterogeneous assemblage through drawing-together informs two recent figurations of "architectural ethnography." On one hand, recent *ethnographies on architecture* focus on the "socio-material dimension of architectural practice," foregrounding it "as a collective process of negotiation, and one that is also shared with a variety of nonhumans" (Yaneva 2018, 18). An ecology of practices, according to architectural theorist Albena Yaneva,

> pays specific attention to the texture of the ordinary life of different participants in architecture making without an a priori ontological distribution of entities that matter in design: architects, city planners, digital software, users, clients, models, renderings, sustainable technologies, drawings, design materials, bricks (p. 19).

In this sense, "architectural ethnography" seeks to reorient the commonplace focus on the objects and products of design towards its social and temporal processes and more-than-human protagonists.

An example of this undertaking is architect Pauline Lefebvre's (2018) ethnographic account of "two days in the life of an object in the making." Interspersed with a series of experimental narratives written from the point of view of a wooden turntable stand in the process of being designed, the study foregrounds the variable and *situated* capacity of this agential object to *affect* the design process, as it metamorphoses across media (sketches, drawings, digital models, physical mockups) and physical milieus, forming assemblages with other actors. As Lefebvre remarks,

> turning objects into subjects is not supposed to deprive the designers and users of any active contributions and responsibilities. On the contrary, training our descriptive apparatus increases our chances to 'ecologize' (Latour) the design scene, allowing for richer exchanges, and a deeper sense of accountability. This also concerns the practice of design: designers' responsibility increases as they engage in more intricate transactions with the objects they deal with (p. 114).

Insofar as such ethnographic projects draw together elements of design practice as matters of concern and care *amidst* its broader material-semiotic context, a parallel line of investigation expands these matters *within* architecture's representational repertoire, as *ethnographies in architecture*. This is at work in the exhibition and accompanying book *Architectural Ethnography* (2018), which surveys 42 drawing-based investigations from around the world from the past two decades that exemplify a distinct approach towards representing the social realm. In these projects, drawings figure both "as instruments to document, discuss, and evaluate architecture in a critical feedback-loop" and as ethnographic tools that "allow usages, needs, and aspirations to be investigated through the lens of the various actors – both human and non-human" (Kaijima et al. 2018, 7). Classified as drawings *of, for, among*, and *around* architecture, they include depictions of buildings and their surroundings as rich sites of everyday inhabitation; visualizations of the relations between construction and economics, lifestyle and craft; illustrations of the objects, tools and spaces that comprise local environments; and mappings of larger ecological networks and systems. Such representations expand and redefine the boundaries of concern and care in architecture and can also serve as a basis for response-able practices. As co-curator and architect Momoyo Kaijima writes, drawing "the world from the standpoint of daily life, or an ethnographic point of view, [makes it possible] not only to observe the existing situation but also reconnect pieces of our disconnected world" (p. 11).

3.2 *Drawing, Together, Common-Enough Worlds*

Making connections is at the heart of the second, complementary task, of enfolding manifold actors and agencies in an open-ended collective production: *drawing, together, common-enough worlds.* This proposition is closely informed by the work of Donna Haraway (2016), who, similarly to Stengers, refutes the contemporary overemphasis on connectivity in favor of situated practice, emphatically remarking that "nothing is connected to everything; everything is connected to something" (p. 31). Understanding relations as immanent yet partial and contingent, in turn, precipitates the creation of new material-semiotic entanglements through SF – a term that variously stands for *speculative fabulation, science fact, science fiction, string figures, speculative feminism, so far*. Haraway asserts SF as "a mode of attention, a theory of history, and a practice of worlding" (p. 213, note 8) – in other words, as a simultaneously critical and affirmative undertaking. Speculative fabulation is an inherently political production: not a mere responsiveness but a matter of care and response-ability, involving both connecting and disconnecting, knotting and untying, living and dying. It is a pursuit of the partial connections that *matter*, through "learning to compose possible ongoing-ness inside relentlessly diffracting worlds … focusing on those practices that can build a common-enough world" (Haraway and Wolfe 2016, 288).

The ethos of SF directly informs *Geostories: Another Architecture for the Environment*, a compendium of drawing-based investigations of social-ecological issues related to climate change by the collaborative practice Design Earth. For founding partners Rania Ghosn and El Hadi Jazairy (2018), "climate change demands urgent transformations in the ways we care for and design the Earth, moving away from a visual rhetoric of crisis … and its politics numbed by 'matter of fact' aesthetics" towards forms of representation that incorporate the various "things, spaces, and scales that are erased from the geographic imagination" (pp. 11–12). Accordingly, they adopt the method of visual storytelling, arguing that instead of serving as rallying points around a cause, stories "entreat their readers to envision threats and opportunities together and offer them the means by which to begin to care and respond" (p. 20). Ghosn and Jazairy thus remind us that "it matters what drawings we draw to inscribe the Earth," and that drawing is

> an argument about the world; it is at once descriptive, synthetic, critical, and always speculative. Drawings make worlds; they encompass scales linked to brief organic lifetimes and the immensity of geologic time. They are also important in part because of the other drawings that they could make possible. Much like storytelling, drawing is always re-drawing – borrowing, misusing, appropriating. They matter for our disciplinary and political futures, as educators, designers, and citizens (p. 22).

Serving as graphical repositories of speculation and invention, and as relational catalysts of imagination, communication and debate, such narrative drawings can comprise both sources of agency and vehicles of action, allowing us to cosmopoliticize and reimagine our common habitat.

Others seek to extend the performative and sympoietic potentials of drawing even further, in order to fully subvert the etymological connotation of design (via the Italian *disegno*, "drawing") as the illustration of ideas towards predetermined outcomes. For instance, architect Simone Ferracina (2018) calls for new modes of drawing that serve as a dynamic "conversational platform" between a multitude of agencies. Critiquing the so-called "monologue-drawing," a fundamentally linear and unidirectional undertaking that proceeds from author to audience and from plan to building, he asks whether drawing could "become a medium for vulnerable, open-ended and materially charged conversations across disciplines and beings" (p. 142). Through the case of an interdisciplinary research project aiming to develop "living bricks" – modular construction blocks that incorporate microorganisms and their metabolic processes – Ferracina suggests that an open-ended approach and ethic to drawing constitutes a generative method of working with a diversity of contributors. He makes three propositions towards this end: *medium-drawing*, the "ongoing and precarious interweaving and colliding of agents, directions, times and media" (p. 142); *exaptation-drawing*, the distribution of intentionality and invention across agents; and *seed-drawing*, the releasing of context from authorial control and allowing it to become a generative ground for differentiation. All of these modalities of drawing "demand that the architect relinquish a measure of authorship and control to engage in conversations with the other – large and small, disciplinary and non-disciplinary, human and nonhuman, alive and inert" (p. 137). Drawing, in short, becomes a means and locus of a speculative, open-ended, collective production.

4 Conclusion: Towards Design as Commoning

The notion of drawing together with a more-than-human ensemble brings full circle the discussion on the limits of the predominant figuration of commoning in architecture and urbanism. In spite of its ethico-political objectives to engage in a more equitable and democratic spatial production, the notion of commoning as the collective management of resources privileges human commoners and falls short of engaging in a cosmopolitics of partial, asymmetrical, more-than-human relations. This chapter ventured to offer an alternative narrative, contending that as the cosmopolitics of connection, commoning can serve as a cipher for rethinking our relation to *relation*: shedding light on the agency and response-ability of design(ers) in a more-than-human world.

Through revisiting – within the limited space allotted – some of the key trajectories of commons discourse, the text aimed to prolong and reorient the architectural-urbanistic discussion from metaphors of resource management and reclaiming commons from enclosure, towards contested nature-cultural habitats such as the more-than-human city, and ultimately, towards a conceptualization of commoning as the politicization of livelihood relations. As Bresnihan (2016) notes, despite the history, and even historical continuity, of the commons, "it is not a pre-modern form of life or cosmology that is lost forever. Understood as a more-than-human commons, it not only challenges capitalist modes of production but humanist (dualist) assumptions of agency and social change" (pp. 105–6). Grounded in this understanding and committed to the feminist ethos of caring as a practical, ethical, and affective undertaking, the chapter suggested that a key task of design as a form of commoning is to partake in the material-semiotic politics of habitat with response-ability and care. One of the fundamental tools for this task is drawing – not as a noun but an active verb, a non-innocent and speculative practice. Through a commitment to drawing, performed *amidst* and *together with* heterogeneous and heterodox matters, designers can partake in representing and composing existing and yet-to-come worlds.

References

Amin, Ash, and Philip Howell, eds. 2016. *Releasing the Commons: Rethinking the Futures of the Commons*. Abingdon: Routledge.

Avermaete, Tom. 2018. Constructing the Commons: Architecture as Unlocking Shared Resources. In *Living Lab: Constructing the Commons*, ed. Tom Avermaete, Lisa Schmidt-Colinet, and Daniela Herold, 13–15. Vienna: IKA.

Bhatia, Neeraj, ed. 2019. *New Investigations in Collective Form: The Open Workshop*. New York: Actar.

Borch, Christian, and Martin Kornberger, eds. 2015. *Urban Commons: Rethinking the City*. Abingdon: Routledge.

Bresnihan, Patrick. 2016. The More-Than-Human Commons: From Commons to Commoning. In *Space, Power and the Commons: The Struggle for Alternative Futures*, ed. Samuel Kirwan, Leila Dawney, and Julian Brigstocke, 93–112. London: Routledge.

De Angelis, Massimo. 2017. *Omnia Sunt Communia: On the Commons and the Transformation to Postcapitalism*. London: Zed Books.

Dellenbaugh, Mary, Agnes Katharina Müller, Martin Schwegmann, Markus Kip, and Majken Bieniok, eds. 2015. *Urban Commons: Moving Beyond State and Market*. Basel: Birkhäuser.

Ferracina, Simone. 2018. Designing Living Bricks: The Architectural Drawing as Conversational Platform. *Ardeth* 2: 137–155.

Federici, Silvia. 2019. *Re-Enchanting the World: Feminism and the Politics of the Commons*. Oakland: PM Press.

Fitz, Angelika, and Elke Krasny, eds. 2019. *Critical Care: Architecture and Urbanism for a Broken Planet*. Cambridge, MA: MIT Press.

Ghosn, Rania, and El Hadi Jazairy. 2018. *Geostories: Another Architecture for the Environment*. New York: Actar.

Gibson-Graham, J.K. 2008. Diverse Economies: Performative Practices for 'Other Worlds'. *Progress in Human Geography* 32 (5): 613–632.

Haraway, Donna J. 2016. *Staying with the Trouble: Making Kin in the Chthulucene*. Durham, NC: Duke University Press.

Haraway, Donna J., and Cary Wolfe. 2016. *Manifestly Haraway*. Minneapolis: University of Minnesota Press.

Hardin, Garreth. 1968. The Tragedy of the Commons. *Science* 162 (3859): 1243–1248.
Hardt, Michael, and Antonio Negri. 2009. *Commonwealth*. Cambridge, MA: Belknap Press.
Harvey, David. 2003. *The New Imperialism*. Oxford: Oxford University Press.
Huron, Amanda. 2015. Working with Strangers in Saturated Space: Reclaiming and Maintaining the Urban Commons. *Antipode* 47 (4): 963–979.
———. 2018. *Carving Out the Commons: Tenant Organizing and Housing Cooperatives in Washington, D.C.* Minneapolis: University of Minnesota Press.
Jalón Oyarzun, Lucia. 2016. Common Spatialities: The Production of the Multitude. *Footprint* 16 (1): 51–67.
Jerram, Leif. 2015. The False Promise of the Commons: Historical Fantasies, Sexuality and the 'Really-Existing' Urban Common of Modernity. In *Urban Commons: Rethinking the City*, ed. Christian Borch and Martin Kornberger, 47–67. Abingdon: Routledge.
Kaijima, Momoyo, Laurent Stalder, and Yu Iseki. 2018. *Architectural Ethnography*. Tokyo: TOTO.
Kimmel, Jens, Till Gentzsch, and Sophie Bloemen. 2018. *Urban Commons, Shared Spaces*. https://www.commonsnetwork.org/wp-content/uploads/2018/11/SharedSpacesCommonsNetwork.pdf. Accessed 16 Feb 2022.
Kip, Markus. 2015. Moving Beyond the City: Conceptualizing Urban Commons from a Critical Urban Studies Perspective. In *Urban Commons: Moving Beyond State and Market*, ed. Mary Dellenbaugh, Agnes Katharina Müller, Martin Schwegmann, Markus Kip, and Majken Bieniok, 42–59. Basel: Birkhäuser.
Kirwan, Samuel, Leila Dawney, and Julian Brigstocke, eds. 2016. *Space, Power and the Commons: The Struggle for Alternative Futures*. Abingdon: Routledge.
Krivý, Maroš, and Tahl Kaminer. 2013. Introduction: The Participatory Turn in Urbanism. *Footprint* 7 (2): 1–6.
Latour, Bruno. 2004a. *Politics of Nature: How to Bring the Sciences into Democracy*. Cambridge, MA: Harvard University Press.
———. 2004b. Why has Critique Run Out of Steam? From Matters of Fact to Matters of Concern. *Critical Inquiry* 30: 114–248.
———. 2005. From Realpolitik to Dingpolitik or How To Make Things Public. In *Making Things Public: Atmospheres of Democracy*, ed. Bruno Latour and Peter Weibel, 14–41. Cambridge, MA: MIT Press.
———. 2008. *A Cautious Prometheus? A Few Steps Toward a Philosophy of Design: (With Special Attention to Peter Sloterdijk)*. "Networks of Design", Annual International Conference of the Design History Society, University College Falmouth, Cornwall, UK.
———. 2015. Waiting for Gaia: Composing the Common World Through Arts and Politics. In *What is Cosmopolitical Design?* ed. Albena Yaneva and Alejandro Zaera-Polo, 21–32. Farnham: Ashgate.
Lefebvre, Pauline. 2018. I, T.T. Stands. Two Days in the Life of an Object in the Making. *Ardeth* 2: 97–117.
Linebaugh, Peter. 2008. *The Magna Carta Manifesto: Liberties and Commons for All*. Berkeley: University of California Press.
Lingis, Alphonso. 1994. *The Community of Those Who Have Nothing in Common*. Indianapolis: Indiana University Press.
May, John. 2012. Logic of the Managerial Surface. *Praxis* 13: 116–126.
Metzger, Jonathan. 2015. The City is Not a Menschenpark: Rethinking the Tragedy of the Urban Commons Beyond the Human/Non-Human Divide. In *Urban Commons: Rethinking the City*, ed. Christian Borch and Martin Kornberger, 22–46. Abingdon: Routledge.
———. 2016. Expanding the Subject of Planning: Enacting the Relational Complexities of More-Than-Human Urban Common(er)s. In *Space, Power and the Commons: The Struggle for Alternative Futures*, ed. Samuel Kirwan, Leila Dawney, and Julian Brigstocke, 133–149. London: Routledge.
Miller, Ethan. 2019. *Reimagining Livelihoods: Life Beyond Economy, Society, and Environment*. Minneapolis: University of Minnesota Press.

Ngo, Anh-Linh, Mirko Gatti, Stefan Gruber, Christian Hiller, and Max Kaldenhoff, eds. 2018. *An Atlas of Commoning: Places of Collective Production*. Stuttgart: IFA.
Ostrom, Elinor. 1990. *Governing the Commons: The Evolution of Institutions for Collective Action*. Cambridge: University Press.
Özkan, Derya, and Güldem Baykal Büyüksaraç, eds. 2020. *Commoning the City: Empirical Perspectives on Urban Ecology, Economics and Ethics*. Abingdon: Routledge.
Pasquinelli, Matteo. 2008. *Animal Spirits: A Bestiary of the Commons*. Rotterdam: NAi Publ.
Petrescu, Doina, Constantin Petcou, Maliha Safri, and Katherine Gibson. 2020. Calculating the Value of the Commons: Generating Resilient Urban Futures. *Environmental Policy and Governance* 2020: 1–16.
Puig de la Bellacasa, María. 2017. *Matters of Care: Speculative Ethics in More Than Human Worlds*. Minneapolis: University of Minnesota Press.
Sohn, Heidi, Stavros Kousoulas, and Gerhard Bruyns. 2015. Commoning as Differentiated Publicness. *Footprint* 16 (1): 1–8.
Stavrides, Stvaros. 2014. On Urban Commoning: The City Shapes Institutions of Sharing. In *Make_Shift City: Renegotiating the Urban Commons*, ed. Francesca Ferguson and Urban Drift Projects, 83–85. Berlin: Jovis.
———. 2016. *Common Space: The City as Commons*. London: Zed Books.
Stengers, Isabelle. 2005. The Cosmopolitical Proposal. In *Making Things Public: Atmospheres of Democracy*, ed. Bruno Latour and Peter Weibel, 994–1003. Cambridge, MA: MIT Press.
———. 2013. Introductory Notes on an Ecology of Practices. *Cultural Studies Review* 11 (1): 183–196.
Tronto, Joan C. 2019. Caring Architecture. In *Critical Care: Architecture and Urbanism for a Broken Planet*, ed. Angelika Fitz and Elke Krasny, 26–32. Cambridge, MA: MIT Press.
Tronto, Joan C., and Berenice Fisher. 1990. Toward a Feminist Theory of Caring. In *Circles of Care: Work and Identity in Women's Lives*, ed. Emily K. Abel and Margaret K. Nelson, 35–62. Albany: SUNY Press.
Yaneva, Albena. 2018. New Voices in Architectural Ethnography. *Ardeth* 2: 17–32.

Index

G. Bruyns, S. Kousoulas (eds.), *Design Commons*, Design Research Foundations, https://doi.org/10.1007/978-3-030-95057-6

D

The manufacturer's authorised representative in the EU is Springer Nature Customer Service Centre GmbH, Europaplatz 3, 69115 Heidelberg, Germany. If you have any concerns regarding our products, please contact ProductSafety@springernature.com

Printed and bound by CPI Group (UK) Ltd, Croydon, CR0 4YY
22/09/2026
02220961-0008